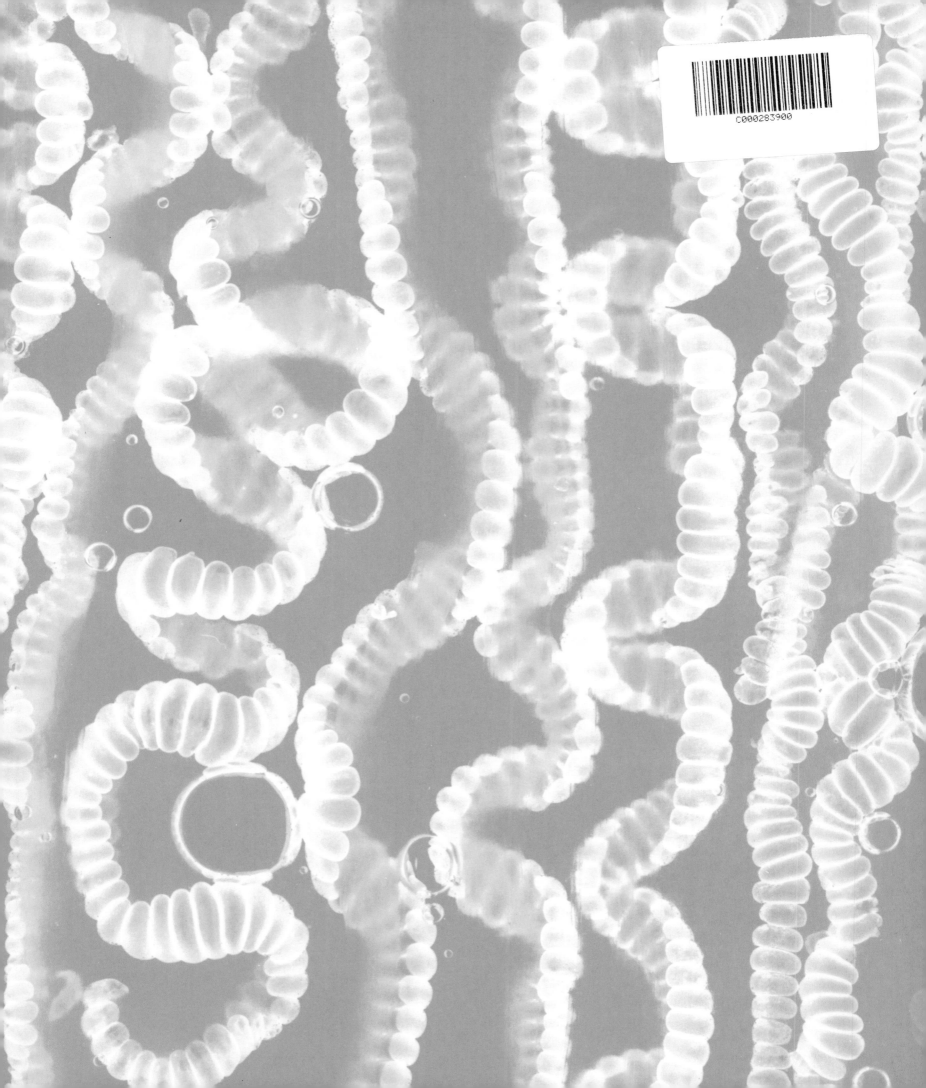

The SCIENCE of ANIMALS

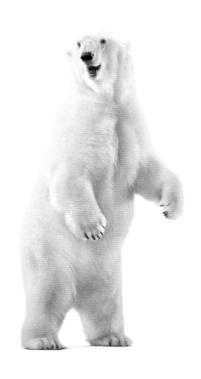

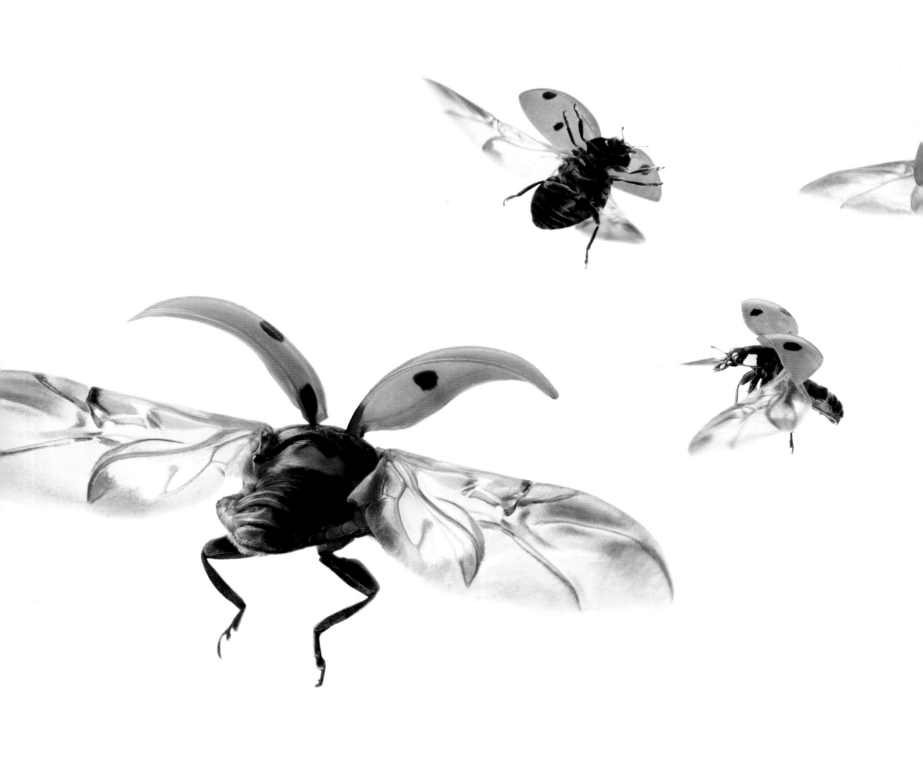

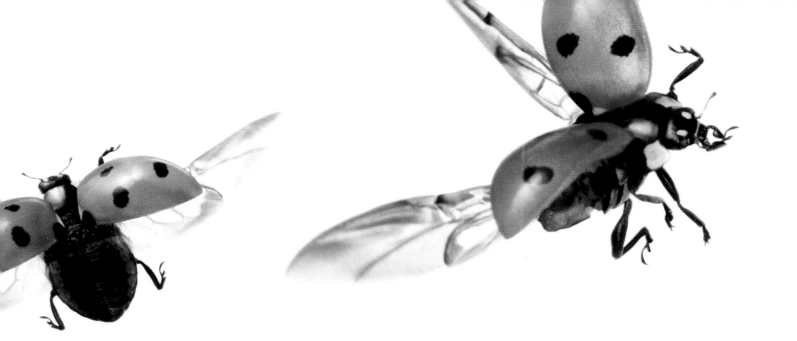

N NATURAL HISTORY MUSEUM

The SCIENCE of ANIMALS

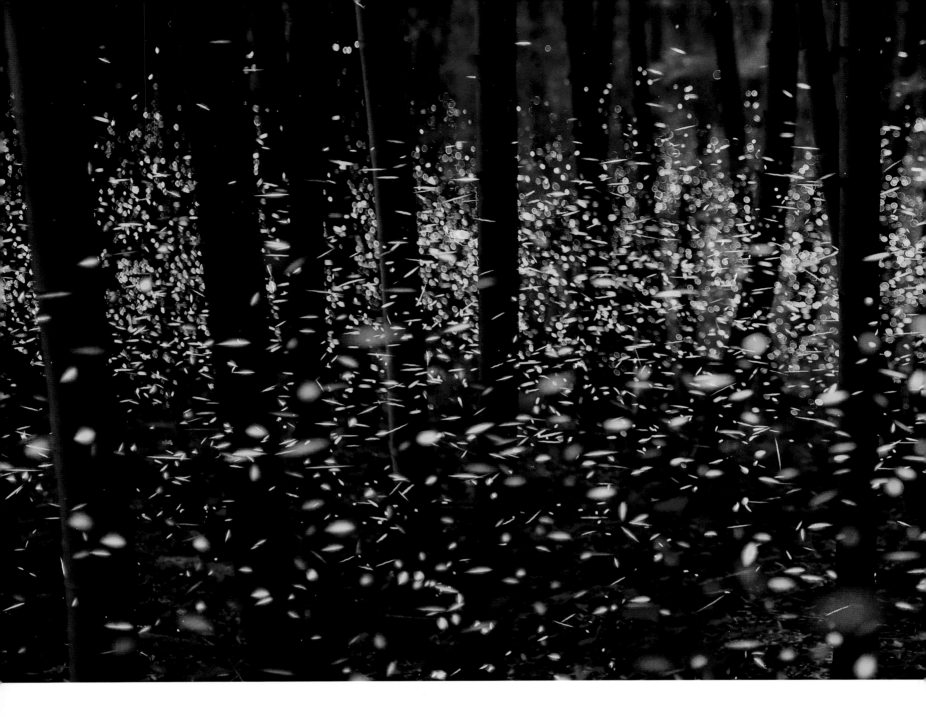

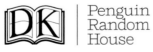 Penguin Random House

DK LONDON

Senior Editor Rob Houston
Editors Jemima Dunne, Steve Setford,
Ruth O'Rourke-Jone, Kate Taylor
Photographer Gary Ombler
Producer, Pre-Production Rob Dunn
Senior Producer Meskerem Berhane
Managing Editor Angeles Gavira Guerrero
Associate Publishing Director Liz Wheeler
Publishing Director Jonathan Metcalf

Senior Art Editor Ina Stradins
Project Art Editors Simon Murrell,
Steve Woosnam-Savage
Design Assistants Briony Corbett, Bianca Zambrea
Jacket Design Development Manager Sophia MTT
Jacket Designer Akiko Kato
Managing Art Editor Michael Duffy
Art Director Karen Self
Design Director Phil Ormerod

DK INDIA

Senior Art Editor Chhaya Sajwan
Art Editors Anukriti Arora, Shipra Jain, Jomin Johny
Senior Managing Art Editor Arunesh Talapatra
Senior Picture Researcher Surya Sarangi
Project Picture Researcher Aditya Katyal

Senior DTP Designer Jagtar Singh
DTP Designer Jaypal Singh Chauhan
Production Manager Pankaj Sharma
Pre-Production Manager Balwant Singh

First published in Great Britain in 2019
by Dorling Kindersley Limited
80 Strand, London, WC2R 0RL

Copyright © 2019 Dorling Kindersley Limited
A Penguin Random House Company
10 9 8 7 6 5 4 3 2 1
001–310754–September/2019

A CIP catalogue record for this book is available from the British Library.
ISBN: 978-0-2413-4678-5

Printed and bound in China

A WORLD OF IDEAS:
SEE ALL THERE IS TO KNOW
www.dk.com

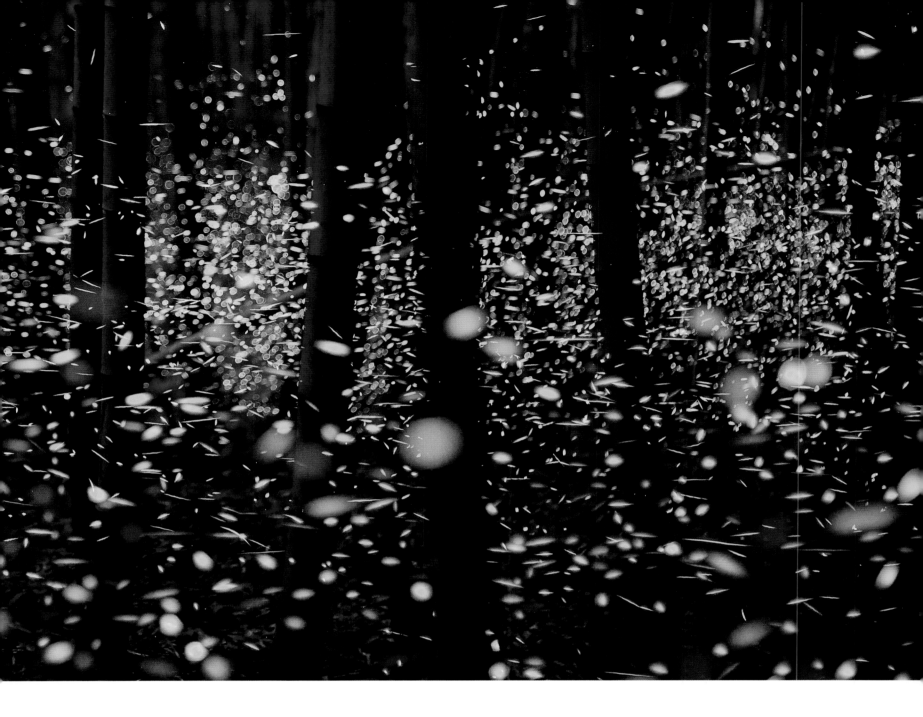

Contributors

Jamie Ambrose is an author, editor, and Fulbright scholar with a special interest in natural history. Her books include DK's *Wildlife of the World*.

Derek Harvey is a naturalist with a particular interest in evolutionary biology, who studied Zoology at the University of Liverpool. He has taught a generation of biologists, and has led student expeditions to Costa Rica, Madagascar, and Australasia. His books include DK's *Science: The Definitive Visual Guide* and *The Natural History Book*.

Esther Ripley is a former managing editor who writes on a range of cultural subjects, including art and literature.

Half-title page Polar bear (*Ursus maritimus*)
Title page Seven-spot ladybird (*Coccinella septempunctata*)
Above Fireflies in a forest on the island of Shikoku, Japan
Contents page Blue iguana (*Cyclura lewisi*)

ℕ **NATURAL HISTORY** MUSEUM

The Natural History Museum, London, looks after a world-class collection of over 80 million specimens spanning 4.6 billion years, from the formation of the Solar System to the present day. It is also a leading scientific research institution, with ground-breaking projects in more than 68 countries. Over 300 scientists work at the Museum, researching the valuable collections to better understand life on Earth. Every year more than five million visitors, of all ages and levels of interest, are welcomed through the Museum's doors.

contents

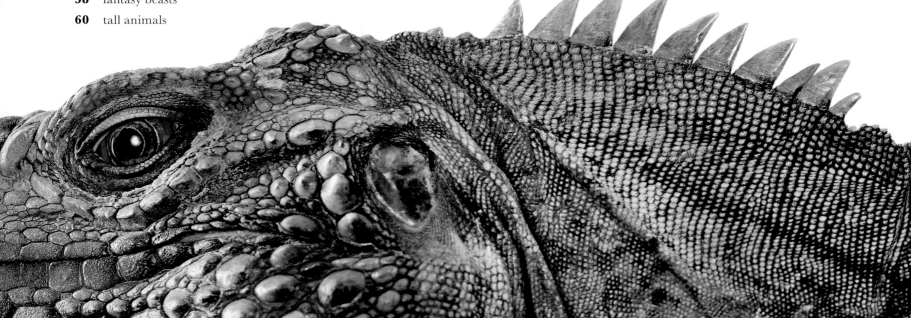

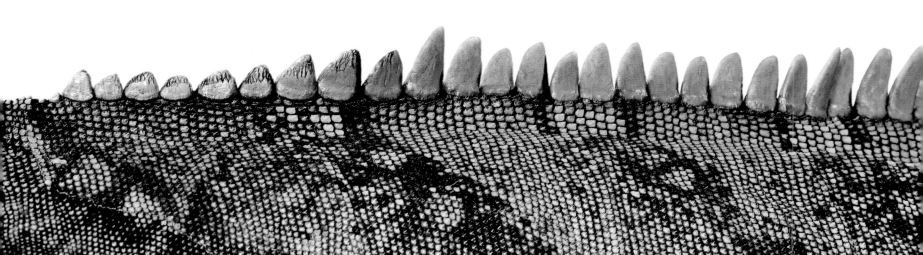

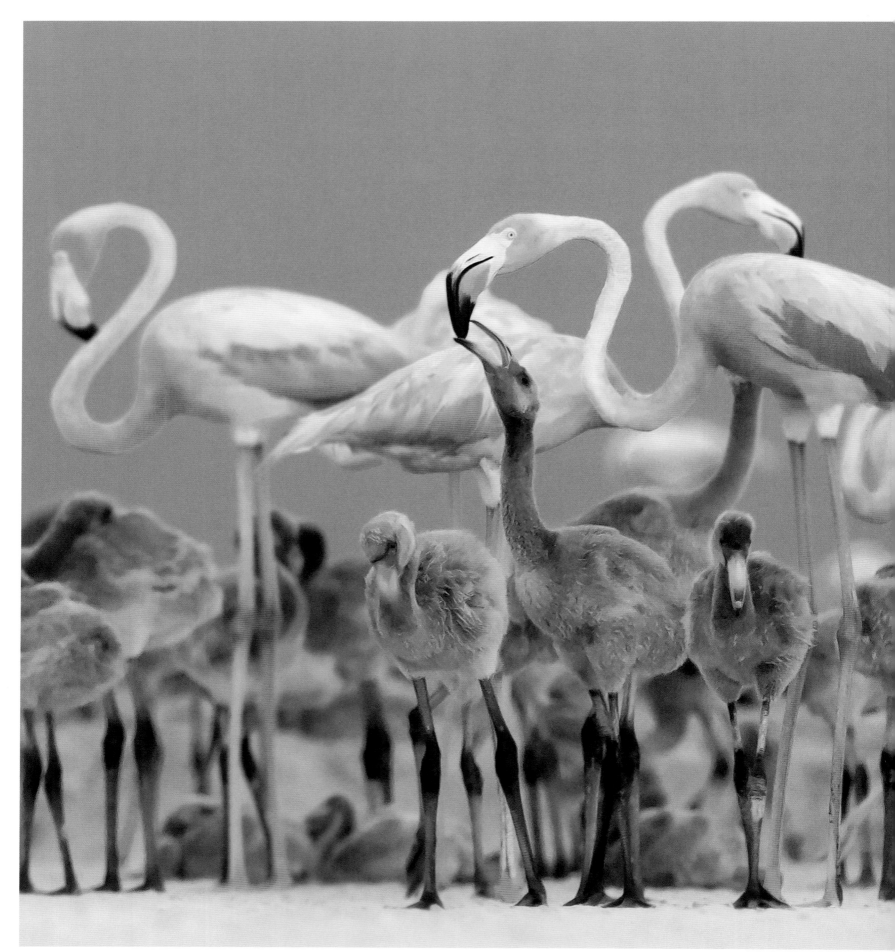

AMERICAN FLAMINGO *Phoenicopterus ruber*

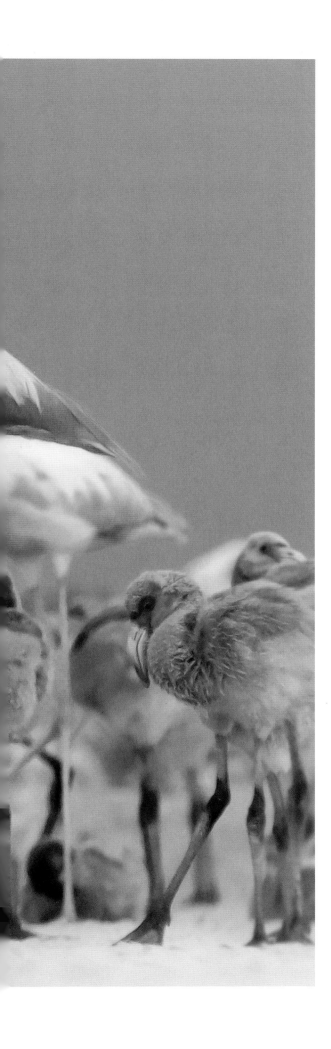

foreword

Beauty is alluring, truth is essential, and art is surely humanity's purest reaction to these ideals. But what of that other irrepressible human trait – curiosity? Well, for me, curiosity is the fuel for science, which in turn is the art of understanding truth and beauty. And this beautiful book presents a perfect fusion of these paragons, celebrating art, revealing remarkable truths, and igniting a curiosity for natural science.

In life, all form must have a function, it can be in transition, but it's never redundant. This means that from childhood onwards, we can study and question the shapes and structures of natural forms and try to determine what they are for and how they work. I remember examining a feather, weighing it, preening it, bending it, twisting it to watch its iridescence flash from green to purple, all the while working through a process of understanding why it was an asset to bird flight and behaviour. Such investigation is perhaps the most fundamental skill of a naturalist and the essential technique of a scientist. And then I tried to paint it, its simple beauty the inspiration for art.

Natural forms also allow us to identify relatedness between species, throwing light on their evolution, which in turn opens our eyes to how we group them together. Of course there are tricksters – mammals with beaks that lay eggs! It's certainly fun to discover how our predecessors were fooled by these strange anomalies, but it's even more satisfying to uncover the truth as to why animals evolved such apparently bizarre forms.

This book reveals that nature is not short of such fascinations and revels in the joy that we can never completely satisfy our curiosities, as there is always more to learn and know about life.

CHRIS PACKHAM
NATURALIST, BROADCASTER, AUTHOR, AND PHOTOGRAPHER

the animal kingdom

animal. a living organism that is made of many cells that usually collaborate to form tissues and organs, and that ingests organic matter – such as plants or other animals – for nutrition and energy.

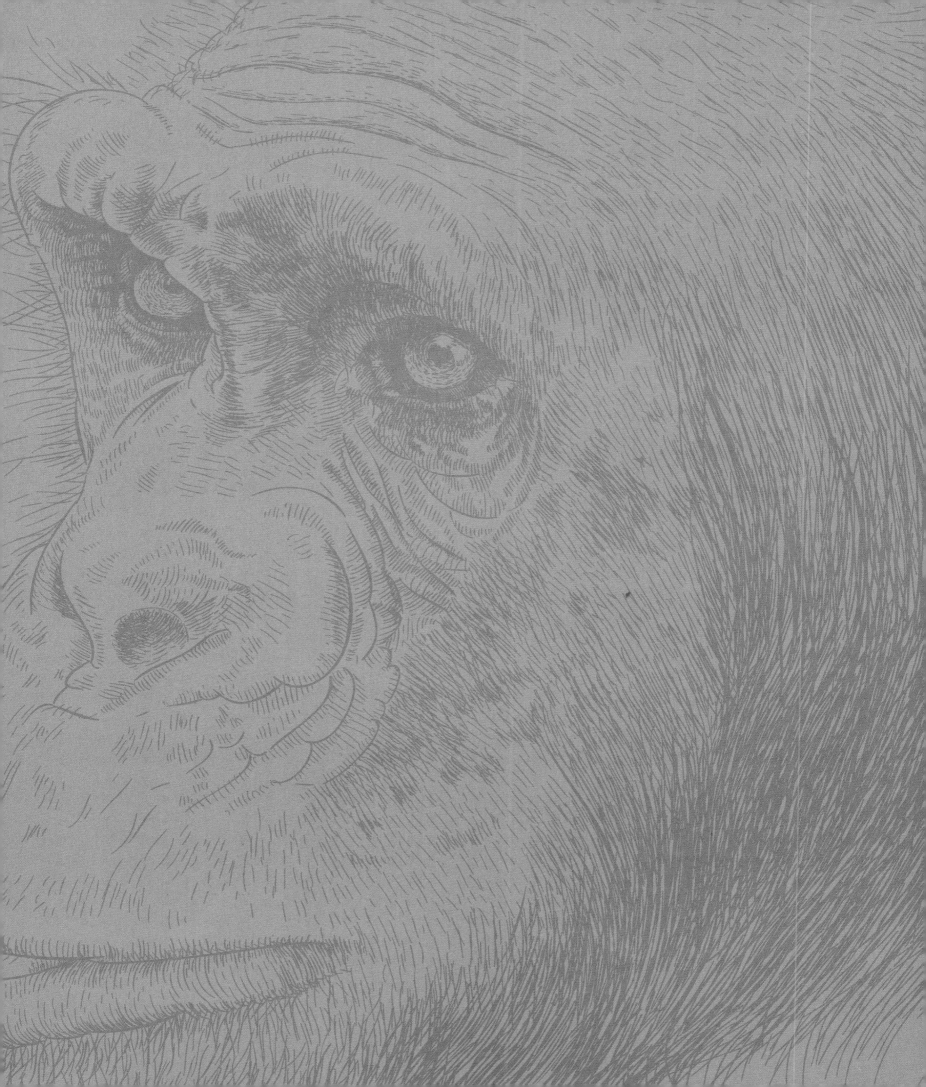

Single-celled relatives
Many complex single-celled organisms, such as this ciliate, *Paramecium bursaria*, were once called "protozoa" and classified as animals. But DNA evidence shows they are only distant relatives of the animal kingdom.

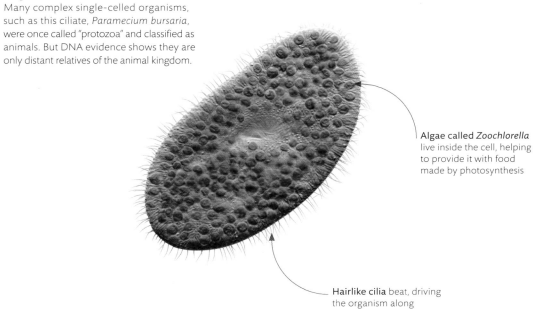

Algae called *Zoochlorella* live inside the cell, helping to provide it with food made by photosynthesis

Hairlike cilia beat, driving the organism along

what is an animal?

Animals differ from the two other major kingdoms of multicellular organisms – fungi and plants – in the fabric of their bodies. Animals have collagen protein holding their cells together into tissues, and all but the simplest have nerves and muscles that help them move. And although some are forever rooted to one spot like plants, most animals forage for food: they derive their nourishment by feeding on other organisms – not absorbing dead matter like a fungus or photosynthesizing like a plant.

THE SIMPLEST ANIMALS

Sponges are the simplest animals alive today. Unlike more complex animals, whose cells become fixed for purpose in the adult body, sponge cells are totipotent – each capable of regenerating the entire body. Some sponge cells develop beating hairlike flagella that create a current for filter-feeding. The hairs are almost identical to those of single-celled organisms called choanoflagellates, suggesting that the first animals evolved from similar organisms.

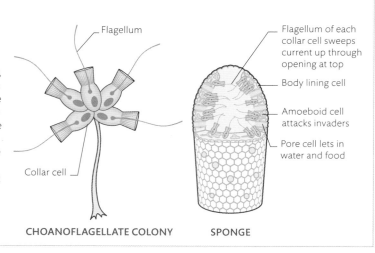

Flagellum

Collar cell

Flagellum of each collar cell sweeps current up through opening at top

Body lining cell

Amoeboid cell attacks invaders

Pore cell lets in water and food

CHOANOFLAGELLATE COLONY   **SPONGE**

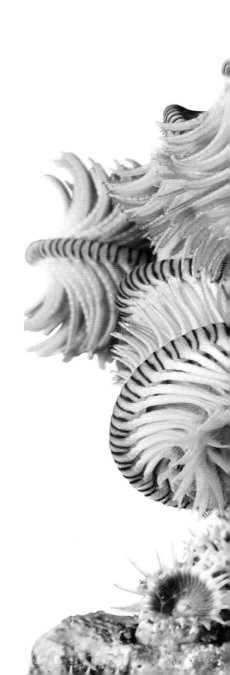

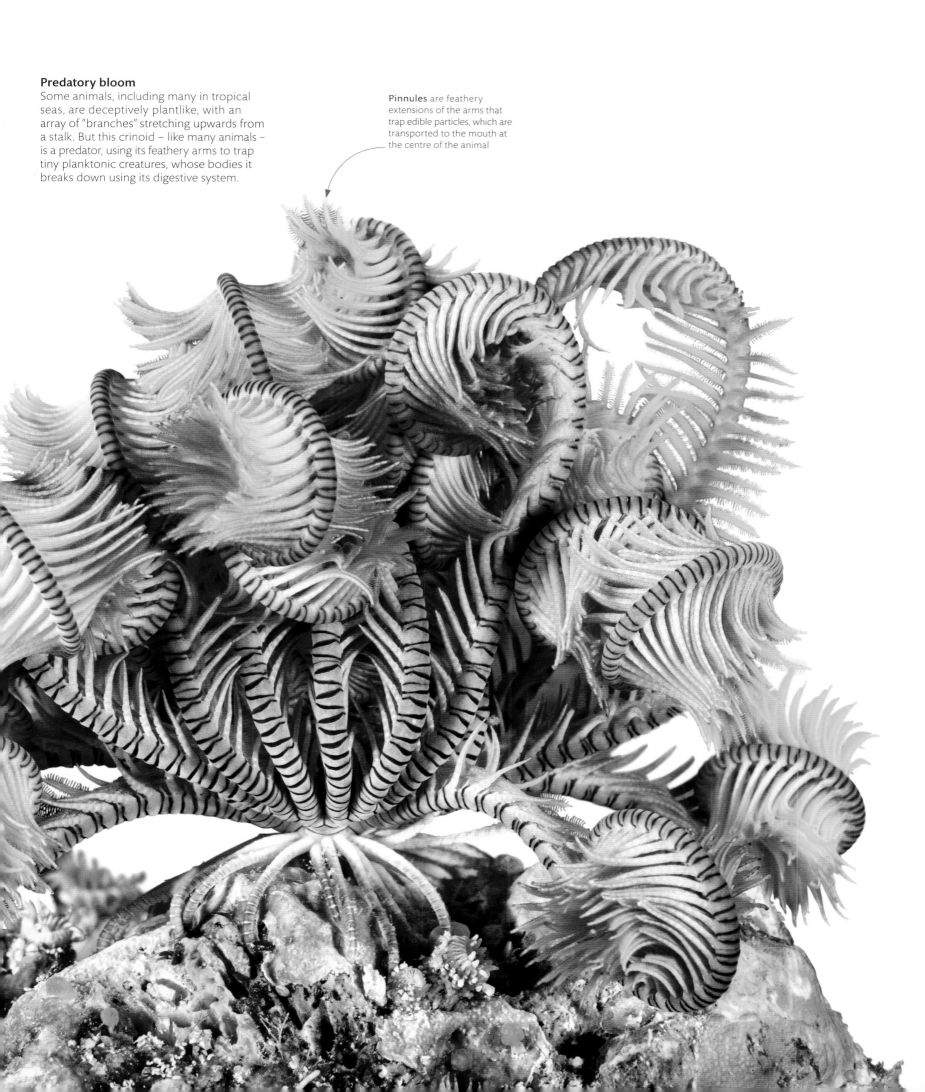

Predatory bloom

Some animals, including many in tropical seas, are deceptively plantlike, with an array of "branches" stretching upwards from a stalk. But this crinoid – like many animals – is a predator, using its feathery arms to trap tiny planktonic creatures, whose bodies it breaks down using its digestive system.

Pinnules are feathery extensions of the arms that trap edible particles, which are transported to the mouth at the centre of the animal

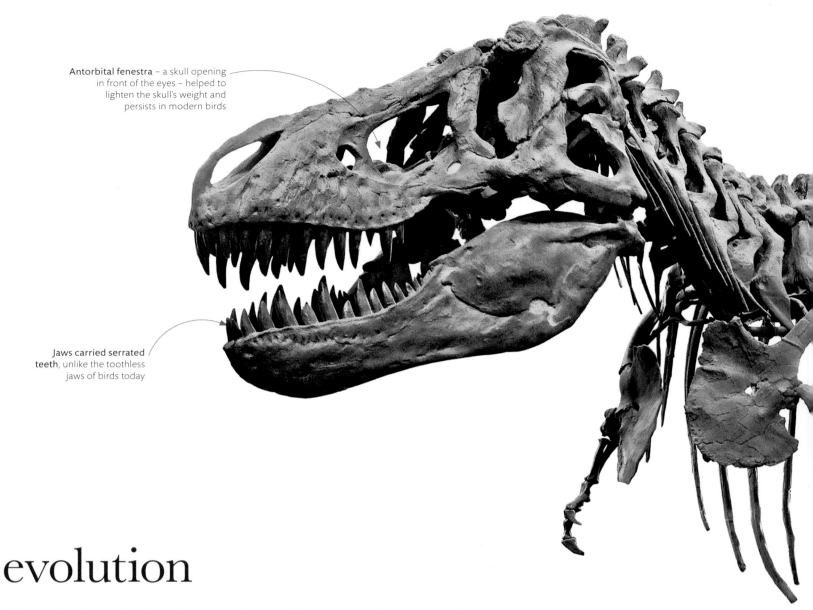

Antorbital fenestra – a skull opening in front of the eyes – helped to lighten the skull's weight and persists in modern birds

Jaws carried serrated teeth, unlike the toothless jaws of birds today

evolution

All animals alive today have descended from different animals in the past by a process of evolution. No single individual evolves, rather entire populations accumulate differences over many generations. Mutation – random replication errors in the genetic material – is the source of inherited variation, while other evolutionary processes, notably natural selection – the driving source of adaptation – determine which variants survive and reproduce. Over millions of years small changes add up to bigger ones, helping to explain the appearance of new species.

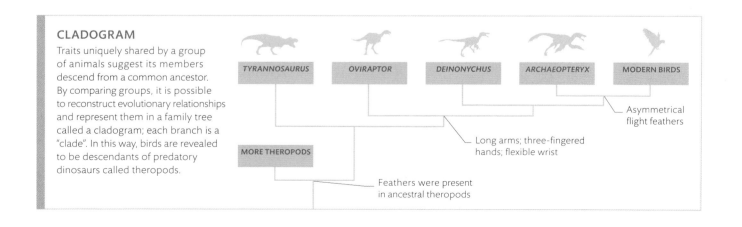

CLADOGRAM

Traits uniquely shared by a group of animals suggest its members descend from a common ancestor. By comparing groups, it is possible to reconstruct evolutionary relationships and represent them in a family tree called a cladogram; each branch is a "clade". In this way, birds are revealed to be descendants of predatory dinosaurs called theropods.

TYRANNOSAURUS   OVIRAPTOR   DEINONYCHUS   ARCHAEOPTERYX   MODERN BIRDS

Asymmetrical flight feathers

Long arms; three-fingered hands; flexible wrist

MORE THEROPODS

Feathers were present in ancestral theropods

Vertebrae and some other bones were air-filled, or pneumatized, which saves weight and in today's birds, also aids breathing efficiency

Animal from the past
The fossils of prehistoric animals can be dated, compared with animals alive today, and used to establish evolutionary relationships. *Tyrannosaurus*, from 66 million years ago, was an upright dinosaur with bladed teeth – indicating a predator. But details of its skeleton point to it being a distant cousin of living birds.

Bipedal, or two-legged, locomotion, was common to all theropods and inherited by birds

Early birds
Fossils of *Archaeopteryx* show the skeleton of a small theropod dinosaur, but other features – including well-developed feathered wings – show it was able to fly.

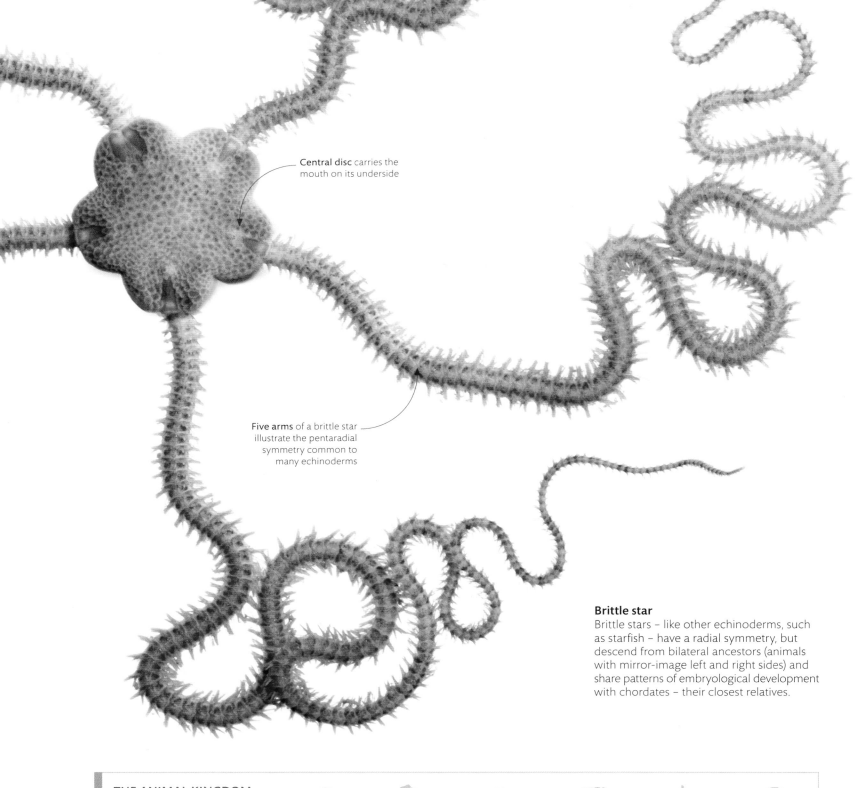

Central disc carries the mouth on its underside

Five arms of a brittle star illustrate the pentaradial symmetry common to many echinoderms

Brittle star
Brittle stars – like other echinoderms, such as starfish – have a radial symmetry, but descend from bilateral ancestors (animals with mirror-image left and right sides) and share patterns of embryological development with chordates – their closest relatives.

THE ANIMAL KINGDOM

Traditionally, animals were divided into invertebrates (without backbones) and vertebrates (with backbones), but this neglects the true pattern of their evolutionary relationships. Most animals lack backbones, and a cladogram (family tree) shows that all vertebrates make up just a subgroup of the chordates – only one of the multiple branches of the family tree. The deepest, oldest splits in the tree occurred with shifts in body symmetry.

| SPONGES | CNIDARIANS include jellyfish | SPIRALIANS include molluscs and many worms | ECDYSOZOANS include insects and other arthropods | ECHINODERMS include starfish | CHORDATES include mammals and other vertebrates |

Return to radial symmetry

Radial symmetry

Bilateral symmetry

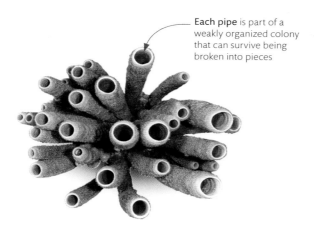

Each pipe is part of a weakly organized colony that can survive being broken into pieces

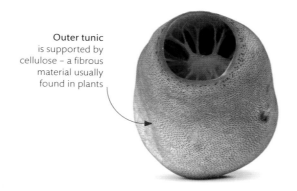

Outer tunic is supported by cellulose – a fibrous material usually found in plants

Pipe sponge
Lacking permanent organization of body tissues, sponges diverged early from the base of the animal kingdom family tree and are regarded as the "sister group" of all other groups of animals with more complex tissues.

Sea squirt
Vertebrates belong to the same group as sea squirts – the chordates. While most adult sea squirts are fixed to the sea bed, they have tadpolelike swimming larvae with bodies supported by a rodlike notochord – a precursor of a vertebrate backbone.

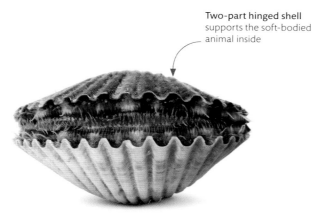

Two-part hinged shell supports the soft-bodied animal inside

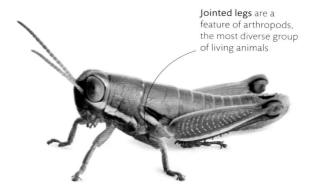

Jointed legs are a feature of arthropods, the most diverse group of living animals

Clam
Molluscs, such as this clam, are united with earthworms and allies in a group called spiralians, according to DNA evidence. Many share a particular form of early development that involves spiral arrangement of cells in the developing embryo.

Grasshopper
The most species-rich group of animals – the ecdysozoans – include jointed-legged arthropods, such as insects and crustaceans, as well as nematode worms. Their growth involves "ecdysis" – the moulting of a tough cuticle or exoskeleton.

types of animal

Scientists have described some 1.5 million species of animals. They organize this diversity into groups of animals whose shared characteristics suggest common ancestry. Some, such as echinoderms, have a starlike radial symmetry; others have bodies arranged like our own – with head and tail ends. Most are popularly called invertebrates, because they lack a backbone. But invertebrates as different as a sponge and an insect otherwise have nothing uniquely in common, so scientists do not recognize them as a natural group.

Family groups

There are around 200 beetle families; the four largest are represented here. The biggest, the rove beetles, has 56,000 described species – nearly as many as all backboned animals combined. The other three groups have 30,000 to 50,000 each. It is likely that 90 per cent of beetle species are still yet to be discovered.

Carnivorous beetle with a powerful jaw

Metallic-coloured wing case common in leaf beetles

Short wing case

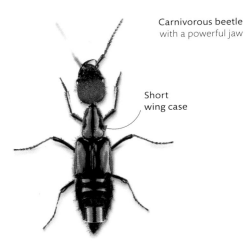 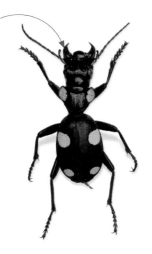 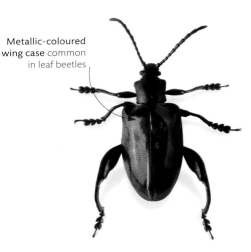

ROVE BEETLES (STAPHYLINIDAE)
Metallic blue rove beetle
Plochionocerus simplicicollis

GROUND BEETLES (CARABIDAE)
Six-spot ground beetle
Anthia sexguttata

LEAF BEETLES (CHRYSOMELIDAE)
Violet frog-legged beetle
Sagra buqueti

Diversity by habitat

Almost every land or freshwater habitat capable of supporting life – with the exception of the coldest polar regions – has its own species of beetles. The rich rainforest habitats could contain hundreds of thousands of undiscovered species. Only oceans, largely unconquered by insects, lack beetles, but some do survive by the shore.

Fog condenses dew on body, which beetle drinks, hence its name

Large eyes used for hunting in daylight

Exoskeleton reflects light making beetle appear gold

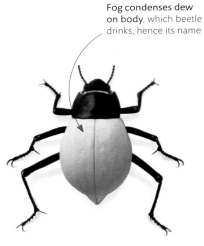 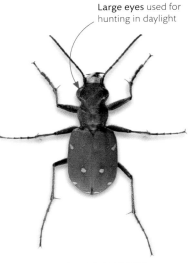 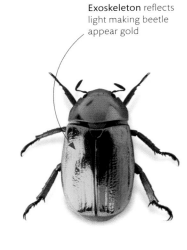

DESERT HABITAT
Black-and-white fog beetle
Onymacris bicolor

MOORLAND HABITAT
Green tiger beetle
Cicindela campetris

RAINFOREST HABITAT
Golden jewel scarab
Chrysina resplendens

Beetle behaviour

Like any successful group of animals, the triumph of the beetles comes down to their ability to make a living in different ways. In each case their versatile mouthparts have the potential to chew through anything. Some species eat leaves and plant matter, others hunt prey, and there are beetles that thrive on foods such as beeswax, fungi, and animal waste.

Shield and horn protects beetle when competing for dung

Fanlike antenna supports sensors that detect food

Head case is irridescent; colour changes according to angle viewed

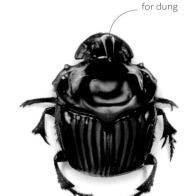 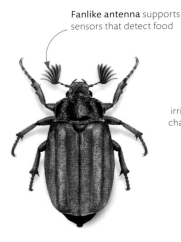 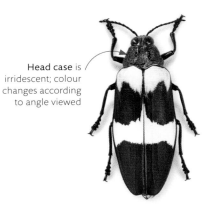

DUNG-ROLLER
Green devil dung beetle
Oxysternon conspicillatum

LEAF AND ROOT EATER
Cockchafer
Melolontha melolontha

WOOD BORER
Banded jewel beetle
Chrysochroa rugicollis

beetle diversity

Despite the extraordinary range of animal life in the sea and on land, one-quarter of all described species belong to a single group of insects – beetles. Like any taxonomic group, they are built on a common theme. They all have hardened wing cases, called elytra, and chewing mouthparts. But in 300 million years of evolution, they have diversified into a multitude of forms.

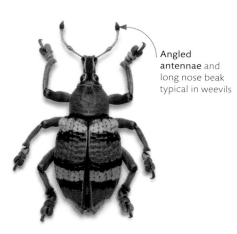

Angled **antennae** and long nose beak typical in weevils

WEEVILS (CURCULIONIDAE)
Schönherr's blue weevil
Eupholus schoenherri

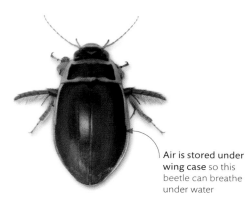

Air is stored under wing case so this beetle can breathe under water

FRESHWATER POND
Great diving beetle
Dytiscus marginalis

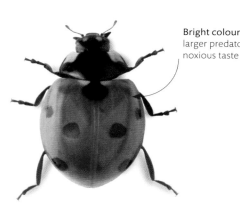

Bright colour warns larger predators of noxious taste

PREDATOR
Seven-spot ladybird
Coccinella septempunctata

Collector's pieces
Inspired by their diversity, many naturalists are passionate collectors of beetles. Austrian specialist, Karl Heller, named this longhorn *Rosenbergia weiskei* in honour of the German explorer Emil Weiske who discovered it in the New Guinea jungle in 1898.

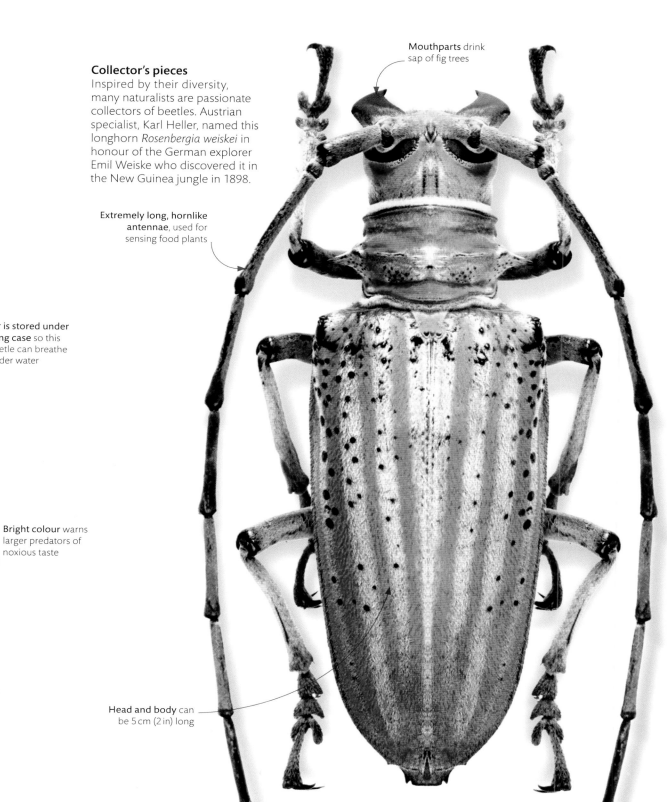

Mouthparts drink sap of fig trees

Extremely long, hornlike **antennae**, used for sensing food plants

Head and body can be 5 cm (2 in) long

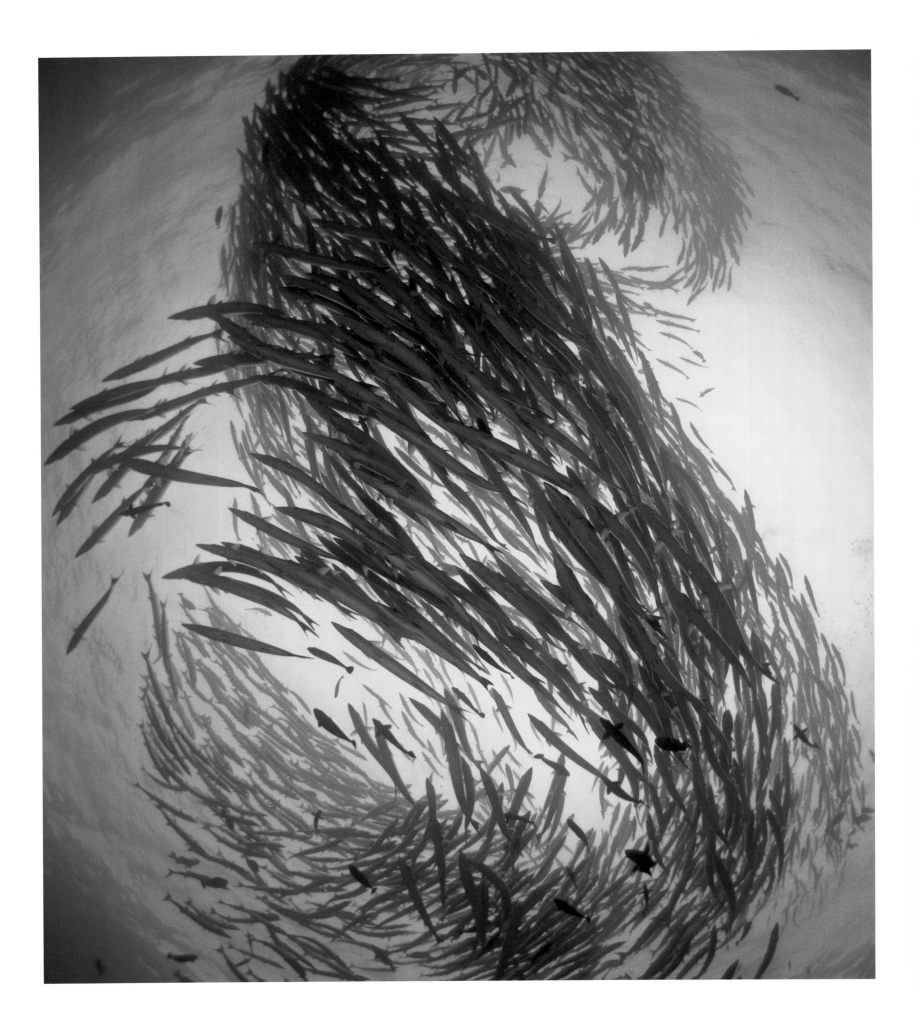

fish and amphibians

The first backboned animals were fish, and around half of today's 69,000 or so vertebrate species have retained the fish body form. The archetypical fish has a hydrodynamic shape that aids movement in water, scaly skin, fins that stabilize and control movement, and gills that absorb oxygen – but many forms have deviated from this plan. Amphibians are descendants of a group of fleshy-finned fish, and were the first vertebrates to walk on land.

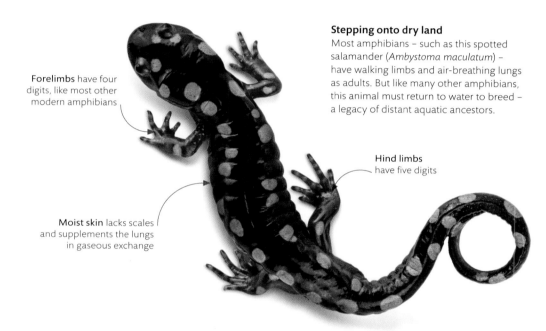

Forelimbs have four digits, like most other modern amphibians

Moist skin lacks scales and supplements the lungs in gaseous exchange

Stepping onto dry land
Most amphibians – such as this spotted salamander (*Ambystoma maculatum*) – have walking limbs and air-breathing lungs as adults. But like many other amphibians, this animal must return to water to breed – a legacy of distant aquatic ancestors.

Hind limbs have five digits

Ocean origins
The first fish swam in the oceans half a billion years ago. The saw-toothed barracuda (*Sphryraena putnamae*) and other modern fish are very different from those early jawless pioneers. Faster muscles, a swim bladder controlling buoyancy, and jaws make barracudas and other predatory shoaling fish masters of their underwater world.

FROM WATER TO LAND

Fish do not form a natural group, or "clade". Clades contain all descendants of a common ancestor, but fish descendants include land-living vertebrates. Fish are an evolutionary "grade" – a stage in an evolutionary trend, in this case a vertebrate body with fins and gills. Amphibians are the sister group of lobe-fins – a clade of fleshy-finned fish that today include only the lungfish and coelacanths.

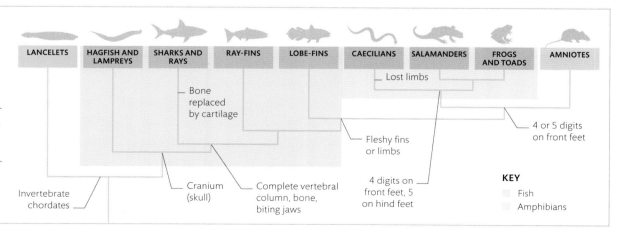

LANCELETS	HAGFISH AND LAMPREYS	SHARKS AND RAYS	RAY-FINS	LOBE-FINS	CAECILIANS	SALAMANDERS	FROGS AND TOADS	AMNIOTES

Lost limbs

Bone replaced by cartilage

4 or 5 digits on front feet

Fleshy fins or limbs

Cranium (skull)

Complete vertebral column, bone, biting jaws

4 digits on front feet, 5 on hind feet

Invertebrate chordates

KEY

Fish

Amphibians

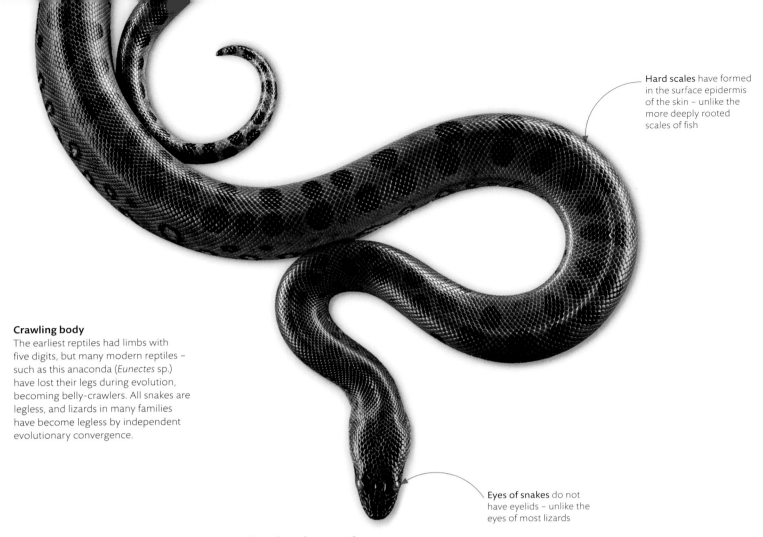

Hard scales have formed in the surface epidermis of the skin – unlike the more deeply rooted scales of fish

Crawling body
The earliest reptiles had limbs with five digits, but many modern reptiles – such as this anaconda (*Eunectes* sp.) have lost their legs during evolution, becoming belly-crawlers. All snakes are legless, and lizards in many families have become legless by independent evolutionary convergence.

Eyes of snakes do not have eyelids – unlike the eyes of most lizards

reptiles and birds

Reptiles represent a major shift in body plan as vertebrates became better adapted for life out of water. Compared with their amphibian ancestors, reptiles acquired a harder, scaly skin that resists drying out, and hard-shelled eggs that can develop on land. Giant reptiles, in the form of dinosaurs, dominated the world for 150 million years, and the descendants of dinosaurs – the birds – today match living reptiles in number of species.

Features of flight
With its coat of feathers, a grey crowned crane (*Balearica regulorum*) is unmistakably a bird. Its aerial lifestyle is made possible by forelimbs modified as wings and lightweight, hollow bones.

REPTILES AND DESCENDANTS

Like fish, reptiles represent a grade of animal life (see p.25), not a clade – because both birds and mammals are among their descendants. Some of the earliest reptiles split into two main branches, one of which gave rise to the mammals. The other branch produced all living reptiles, as well as prehistoric forms, such as aquatic plesiosaurs, flying pterosaurs, and dinosaurs. Birds descend from a group of upright, predatory dinosaurs called theropods (pp.14–15).

AMPHIBIANS	TURTLES AND TORTOISES	LIZARDS, SNAKES, AND TUATARA	CROCODILIANS	BIRDS	MAMMALS

Feathers; hollow bones; warm-blooded

Skull openings lost; shelled body

Scaly skin, waterproof eggs

Teeth set more firmly in sockets

KEY
Reptiles

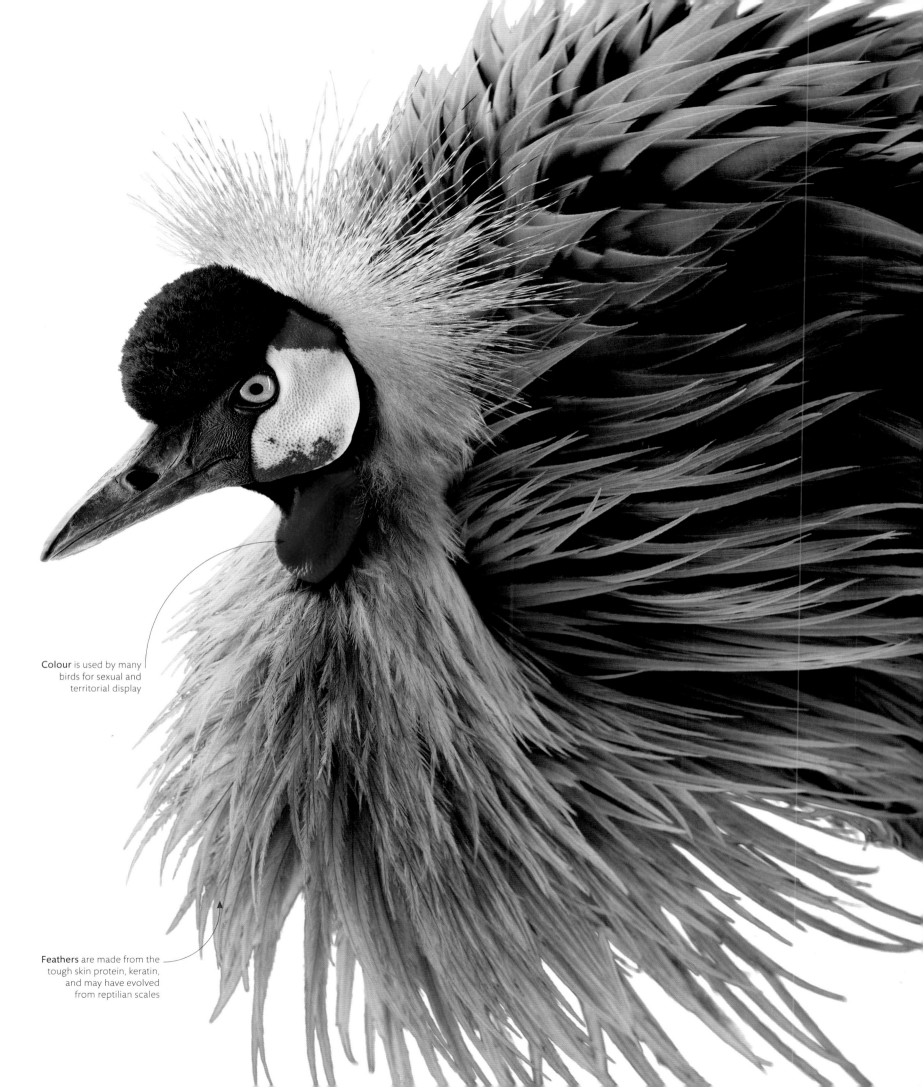

Colour is used by many birds for sexual and territorial display

Feathers are made from the tough skin protein, keratin, and may have evolved from reptilian scales

MAMMAL RELATIONSHIPS

Like birds and living amphibians – but unlike reptiles and fish – mammals constitute a clade: a natural group containing all the descendants of a common ancestor. The most ancient division among living mammals is between egg-laying monotremes (platypus and echidnas) and mammals with live-birth – marsupials and placentals. Today, placentals make up 95 per cent of mammal species.

REPTILES	MONOTREMES	MARSUPIALS	PLACENTAL MAMMALS

Live birth

Milk secretion

KEY

Mammals

Brain power

Warm-bloodedness provides consistently optimal conditions for the functions of the body's tissues – something that undoubtedly helped to increase mammal brain size. A mandrill (*Mandrillus sphinx*) has better problem-solving skills than a lizard and is capable of being a more accomplished parent.

mammals

Mammals descended from a group of reptiles that flourished before the time of the dinosaurs. The first true, furry mammals lived alongside dinosaurs, but during this time remained small and shrewlike. They diversified only after the dinosaurs went extinct. Like birds, mammals became warm-blooded, and regulation of a constant body temperature helped them stay active even in the cold. But, unlike birds, most mammals abandoned egg-laying to give birth to live young, nourished by milk produced from their eponymous mammary glands.

Herds

To a certain degree, mammals filled many of the ecological niches vacated by the dinosaurs. Mammals became the largest land animals, and herds of herbivores, such as this plains zebra (*Equus quagga*), became some of the biggest gatherings of animal biomass on the planet.

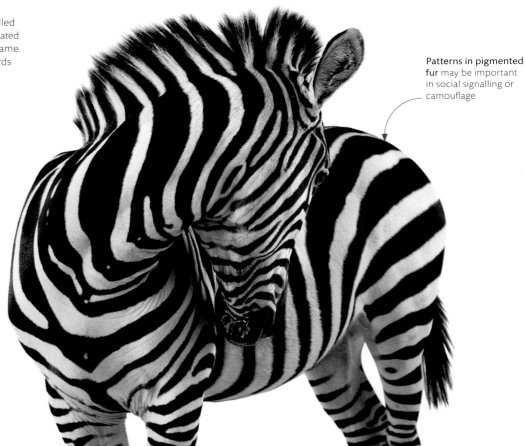

Patterns in pigmented fur may be important in social signalling or camouflage

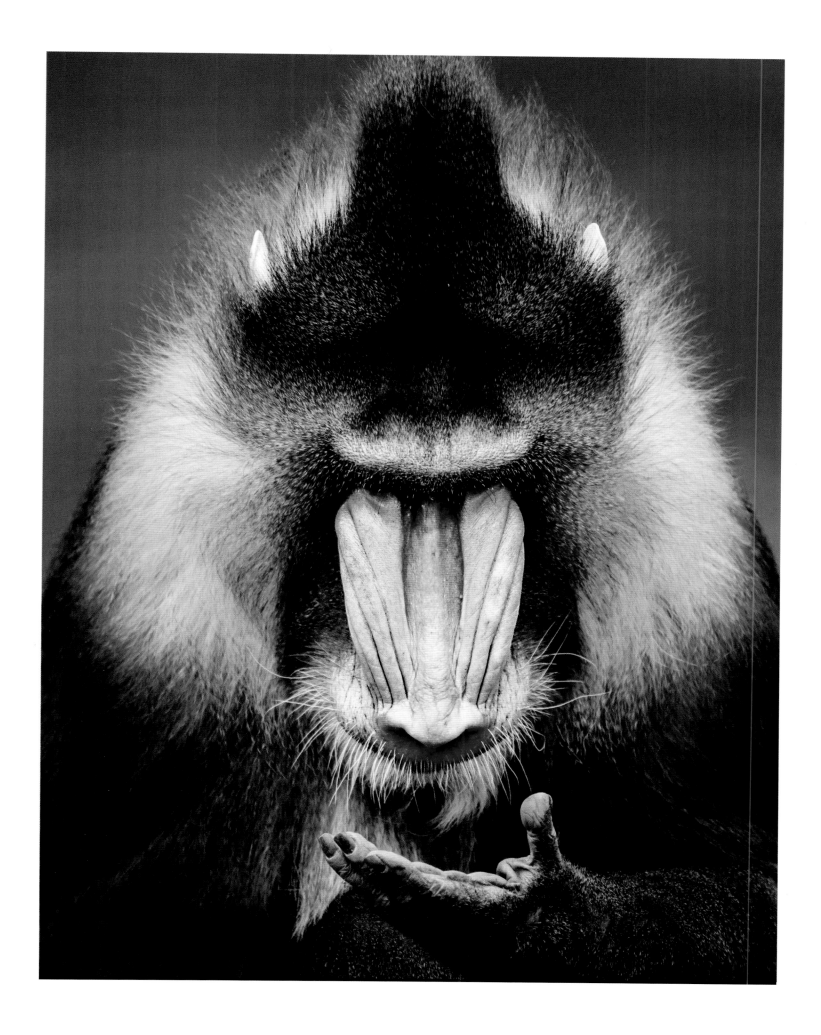

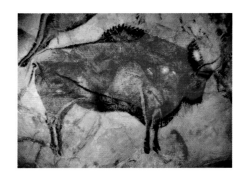

Female bison (c.16,000–14,000 BCE)
Sophisticated techniques in Spain's Altamira cave include linework on the bison's horns and hooves and scraped paint to produce shading.

Horses and bull (c.15,000–13,000 BCE)
In Lascaux's great hall of the bulls, a red and black horse is flanked by the figure of a bull and a frieze of small horses. The world's largest cave art animal, a bull 5.2 m (17 ft) long is in this hall.

animals in art

prehistoric paintings

Discoveries over the past 200 years reveal a sophistication in prehistoric painting that dissolves the difference between Paleolithic and present-day art. The pictures of animals and people that adorn hundreds of caves in western Europe and around the world go far beyond simple representation. They suggest that for more than 30,000 years cave painting has been a human imperative driven by mystic purposes.

It took more than 20 years for experts to confirm that the vibrant visions of animal life cycles on the walls and ceilings of the caves in Altamira were prehistoric works. After their discovery in Cantabria, Spain, in the 1870s, one critic maintained that the paintings were so recent he could wipe the paint away with a finger.

Some 70 years later, four teenagers squeezed through a foxhole into the caves of Lascaux, near Montignac, southern France. They became the first humans in 17,000 years to walk through its 235 m (770 ft) of galleries depicting nearly 2,000 animals, human figures, and abstract signs. At the two sites, sacred works of animal art were created by lamp light using pigments made from ochres and manganese oxide in the soil on surfaces that were almost inaccessible, suggesting a private devotional intent. Unsurprisingly, both caves have earned the epithet "the Sistine Chapel of Prehistoric Art."

> 66 With the cave of Altamira, painting reached a peak of perfection which cannot be bettered. 99

LUIS PERICOT-GARCIA, *PREHISTORIC AND PRIMITIVE ART*, 1967

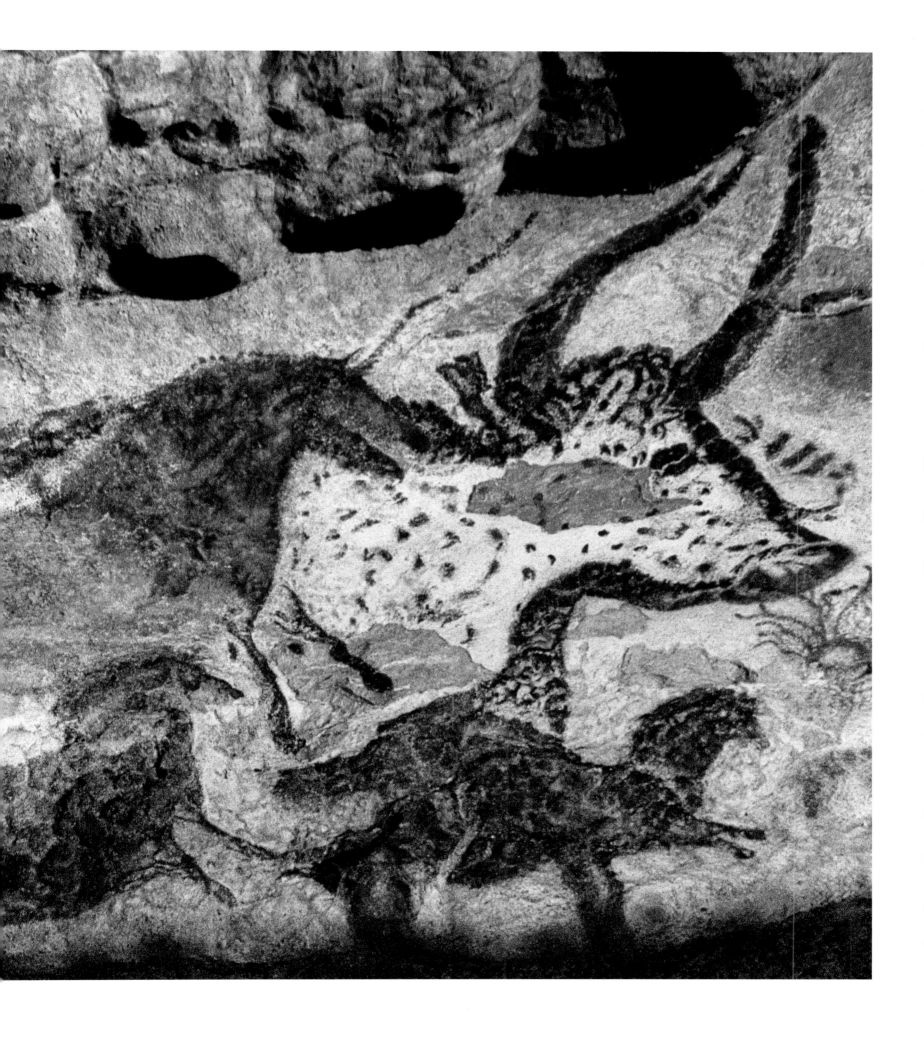

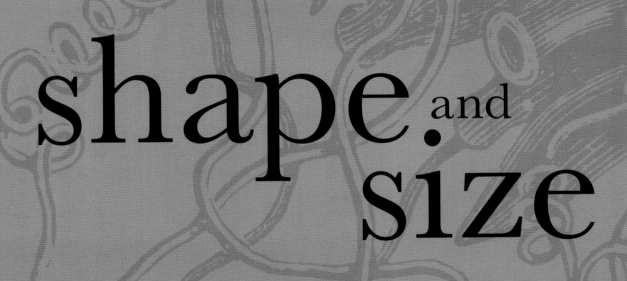

shape.and
size

shape. the external physical form
or outline of an animal.

size. the spatial dimensions, proportions,
or extent of an animal.

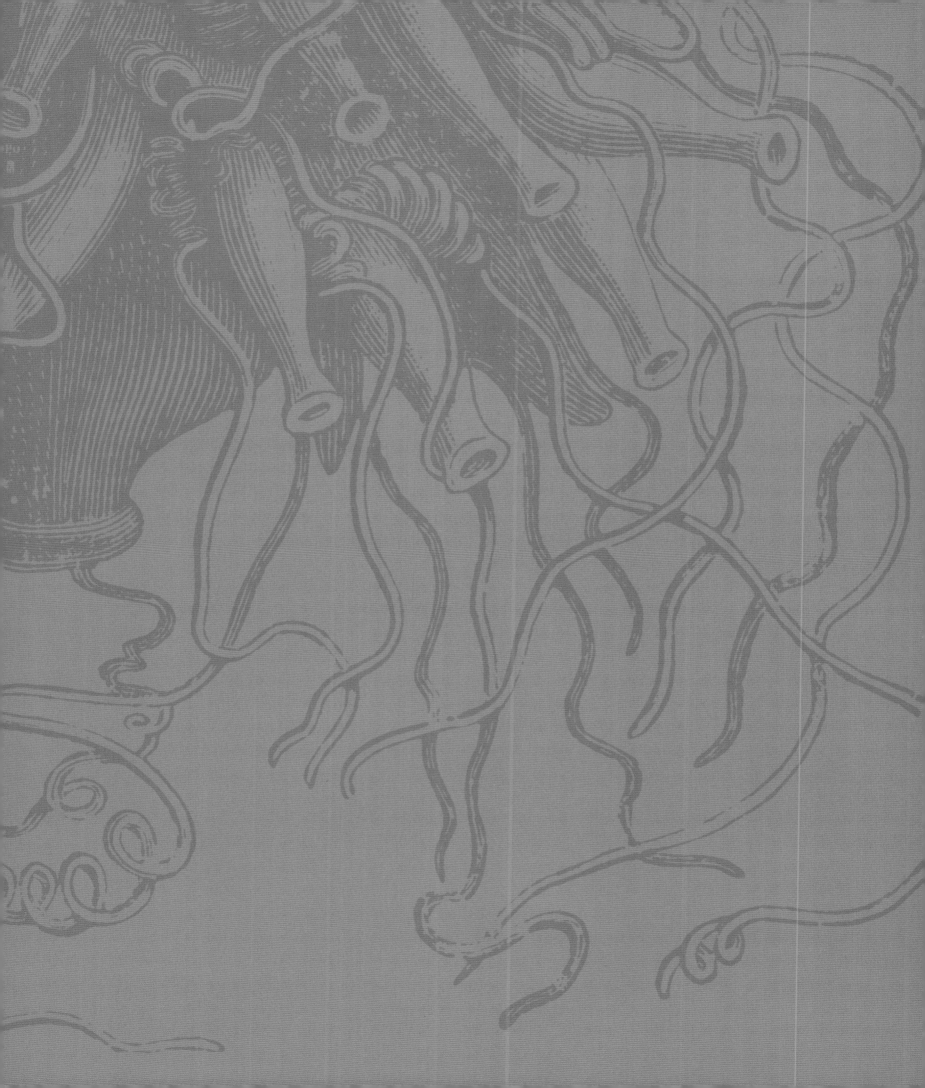

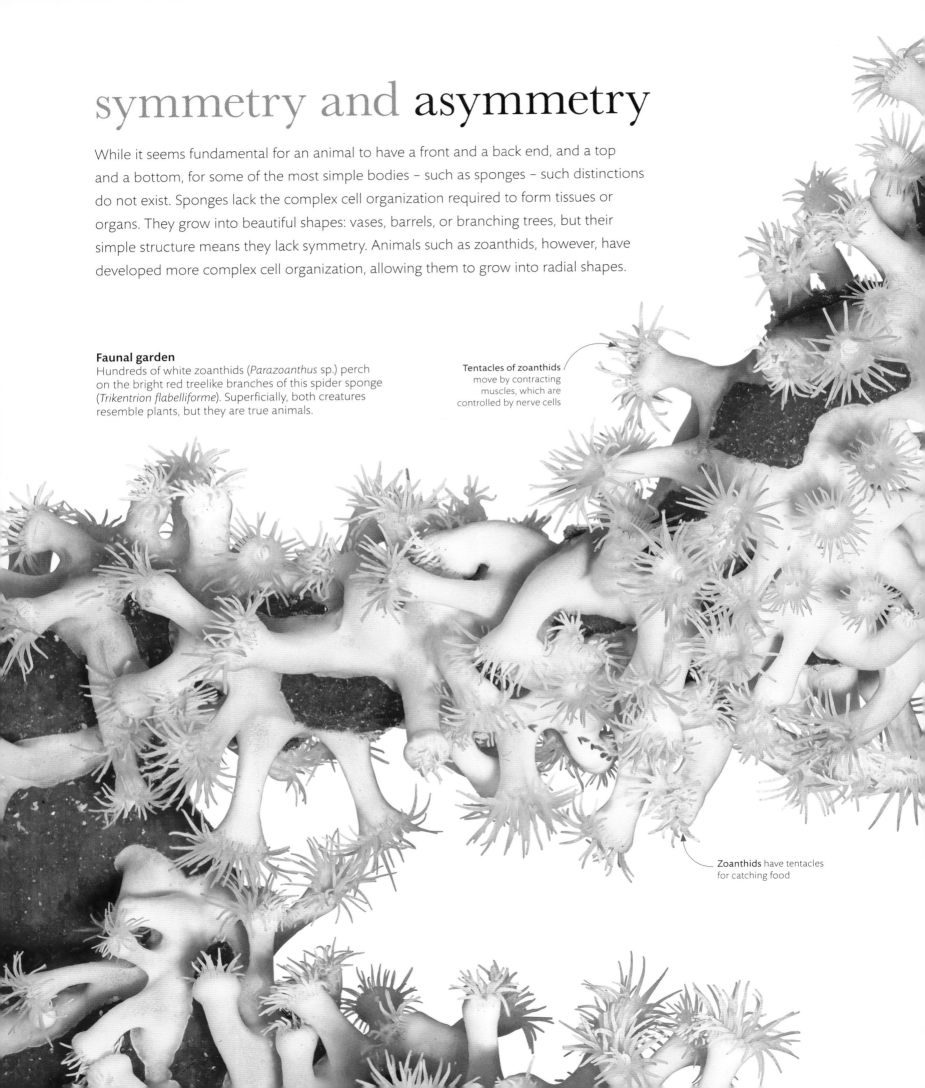

symmetry and **asymmetry**

While it seems fundamental for an animal to have a front and a back end, and a top and a bottom, for some of the most simple bodies – such as sponges – such distinctions do not exist. Sponges lack the complex cell organization required to form tissues or organs. They grow into beautiful shapes: vases, barrels, or branching trees, but their simple structure means they lack symmetry. Animals such as zoanthids, however, have developed more complex cell organization, allowing them to grow into radial shapes.

Faunal garden
Hundreds of white zoanthids (*Parazoanthus* sp.) perch on the bright red treelike branches of this spider sponge (*Trikentrion flabelliforme*). Superficially, both creatures resemble plants, but they are true animals.

Tentacles of zoanthids move by contracting muscles, which are controlled by nerve cells

Zoanthids have tentacles for catching food

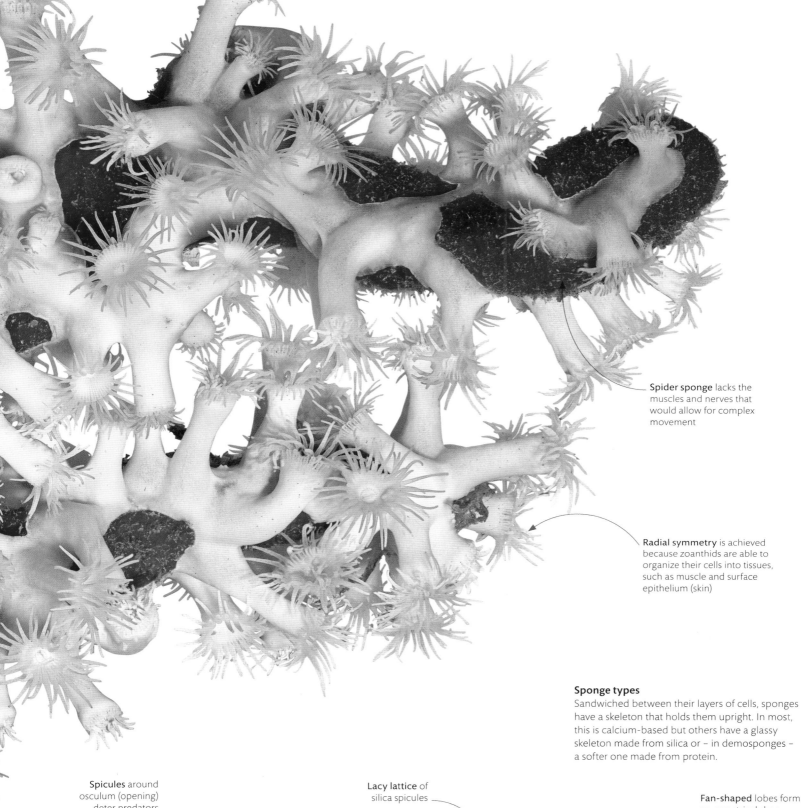

Spider sponge lacks the muscles and nerves that would allow for complex movement

Radial symmetry is achieved because zoanthids are able to organize their cells into tissues, such as muscle and surface epithelium (skin)

Sponge types

Sandwiched between their layers of cells, sponges have a skeleton that holds them upright. In most, this is calcium-based but others have a glassy skeleton made from silica or – in demosponges – a softer one made from protein.

Spicules around osculum (opening) deter predators

Lacy lattice of silica spicules

Fan-shaped lobes form asymmetrical shape

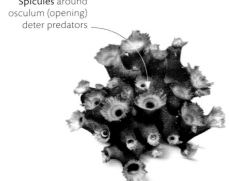

CALCAREOUS SPONGE

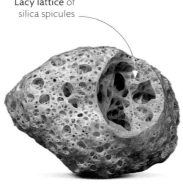

GLASS SPONGE

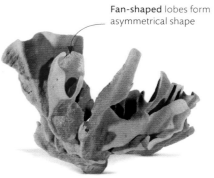

DEMOSPONGE

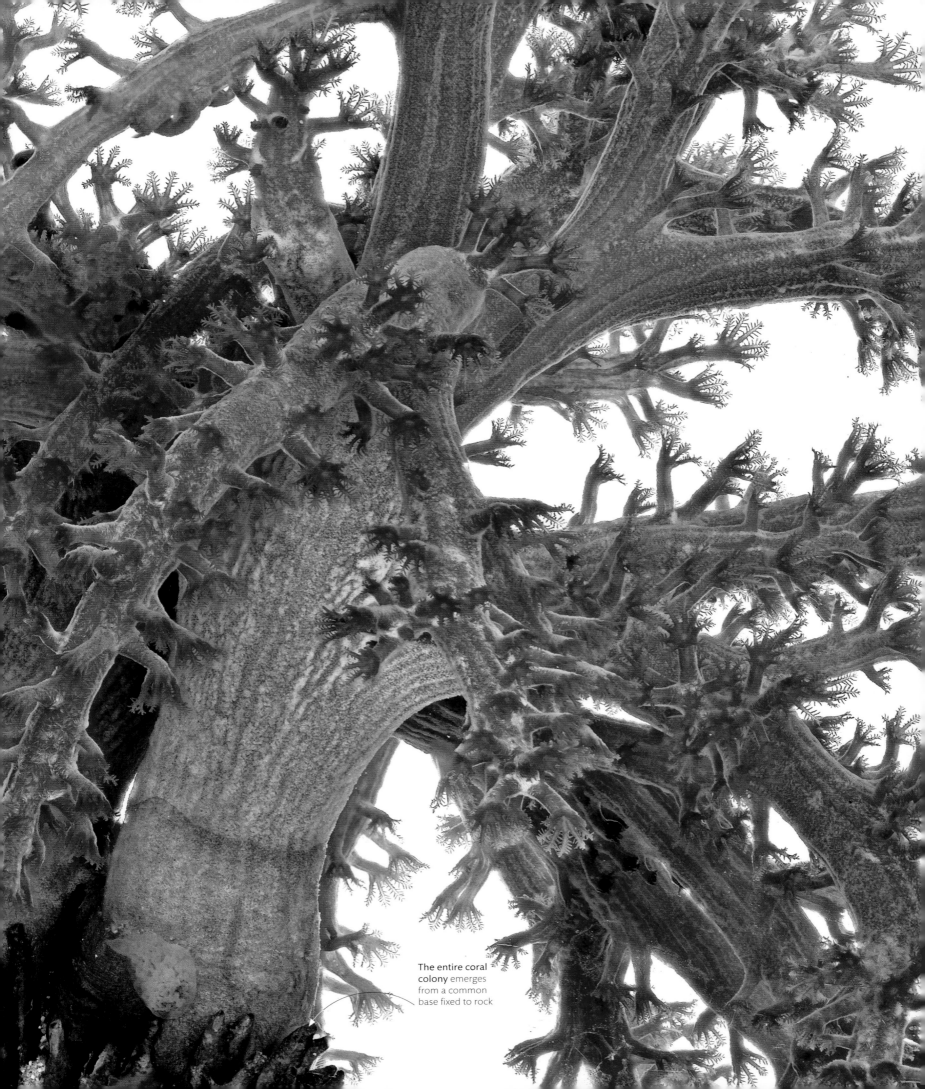

The entire coral colony emerges from a common base fixed to rock

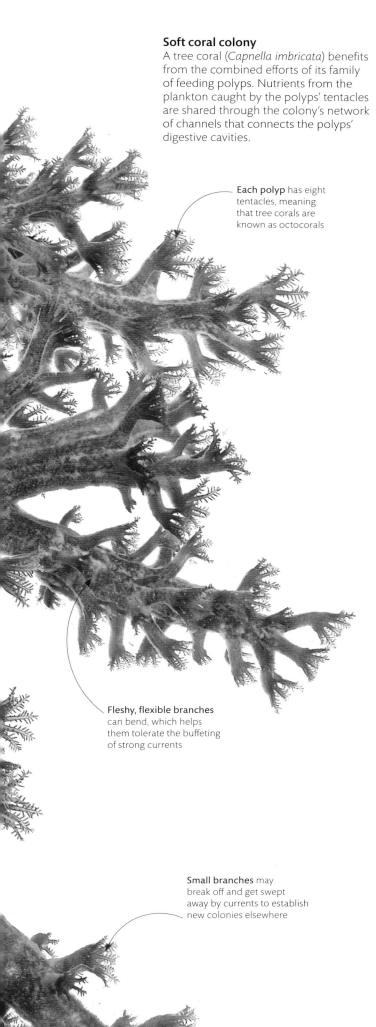

Soft coral colony

A tree coral (*Capnella imbricata*) benefits from the combined efforts of its family of feeding polyps. Nutrients from the plankton caught by the polyps' tentacles are shared through the colony's network of channels that connects the polyps' digestive cavities.

Each polyp has eight tentacles, meaning that tree corals are known as octocorals

Fleshy, flexible branches can bend, which helps them tolerate the buffeting of strong currents

Small branches may break off and get swept away by currents to establish new colonies elsewhere

COLONY DEVELOPMENT

A colony can be any group of animals that live close together, but for corals the connections are especially intimate. Each colony germinates from a single fertilized egg, so all its branches and polyps end up as the genetically identical parts of a massively branching individual.

Coenenchyme (shared colonial tissue) covered with surface epidermis (skin)

Single polyp

Rocky substrate

LATERAL COLONY GROWTH

Hydrocaulus – the branching stem of the colony

Axial (leading) polyp

Lateral (side) polyp

VERTICAL COLONY GROWTH

forming colonies

An animal that relies on its tentacles to gather food gets more nourishment by sweeping more of the water around it. Many anemone-like creatures sprout branches to do just that, forming a colony of animals called polyps. Some branch upwards like trees, others spread sideways into mats, while stony corals lay down rocky foundations (see pp.64–65) that are the basis of coral reefs.

Specialized polyps

Some colonial animals, such as the microscopic hydrozoan *Obelia*, produce polyps that are specialized for different tasks. Some of the animal's polyps are dedicated to feeding, while others produce eggs and sperm for reproduction.

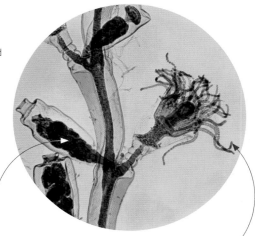

Flasklike gonozooids are reproductive polyps, producing stacks of eggs

Retractable gastrozooids are feeding polyps, with tentacles that catch plankton

SIMPLEST NERVOUS SYSTEM

Lacking any sign of a head or tail, the nervous system of a cnidarian is not centralized into a brain as in other animals. Instead, nerve cell fibres are spread throughout the body as a simple net between the animal's gut and outer body wall. As in other animals, sensors communicate signals to muscle cells using electrical nerve impulses in order to coordinate behaviour.

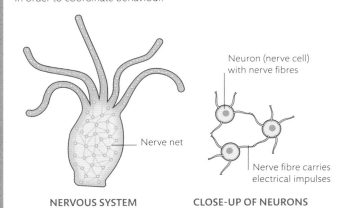

Neuron (nerve cell) with nerve fibres

Nerve net

Nerve fibre carries electrical impulses

NERVOUS SYSTEM   CLOSE-UP OF NEURONS

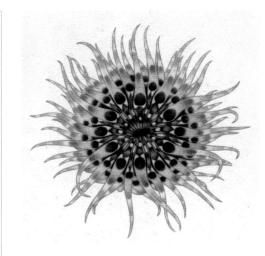

View down the axis

The body parts of a sea anemone are arranged around a central axis. The axis is marked by its gut, which has a single upward-facing opening in the middle of the tentacles. This opening both receives prey and expels waste.

Changing body form

An adult magnificent sea anemone (*Heteractis magnifica*) is arranged around a central point. However, like many other animals with radial symmetry, it begins life as a larva with bilateral symmetry.

Tentacles, moved by muscular contractions, transfer prey to the mouth or retract when danger threatens

Tentacle tips contain specialized stinging cells called nematocysts, which paralyze small prey

radial symmetry

Bodies encircled by tentacles were common among early animals found in prehistoric seas half a billion years ago, and they are still present in many animals living today. Cnidarians – a group of animals that includes anemones, corals, and jellyfish – have no front and back, but have radial bodies that sense the world from all directions at once. All of them are aquatic, and rely on tentacles that wave in water to trap passing prey.

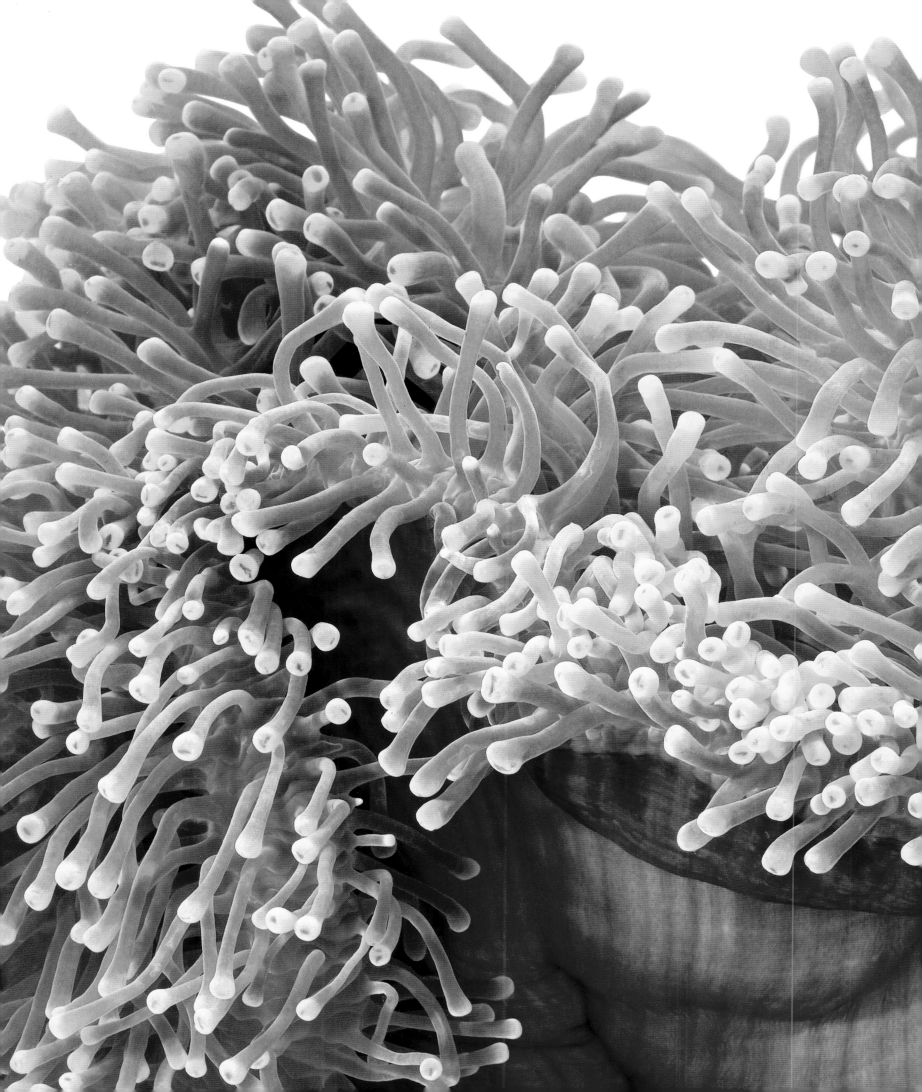

symmetry in motion

An animal that can break free of the ocean floor and live in open water is surrounded by opportunities – but it faces new challenges, too. Jellyfish are free-swimming relatives of anemones; most have rings of tentacles that hang down from a soft floating bell. Lacking firm anchorage, jellyfish risk being swept away, but they can use muscles in their jelly to move against the currents to feast on the planktonic food coming from every direction.

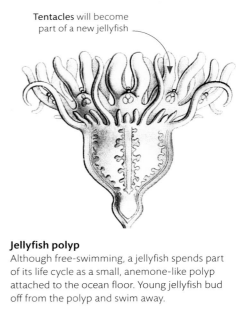

Tentacles will become part of a new jellyfish

Free-swimming predator
The pulsating bell of a Pacific sea nettle (*Chrysaora fuscescens*) pushes the jellyfish up through the water column, trailing its stinging tentacles. The jellyfish lacks the sensory equipment to pursue prey actively, so it relies on the long reach of its tentacles to ensnare tiny fish and shrimp.

Jellyfish polyp
Although free-swimming, a jellyfish spends part of its life cycle as a small, anemone-like polyp attached to the ocean floor. Young jellyfish bud off from the polyp and swim away.

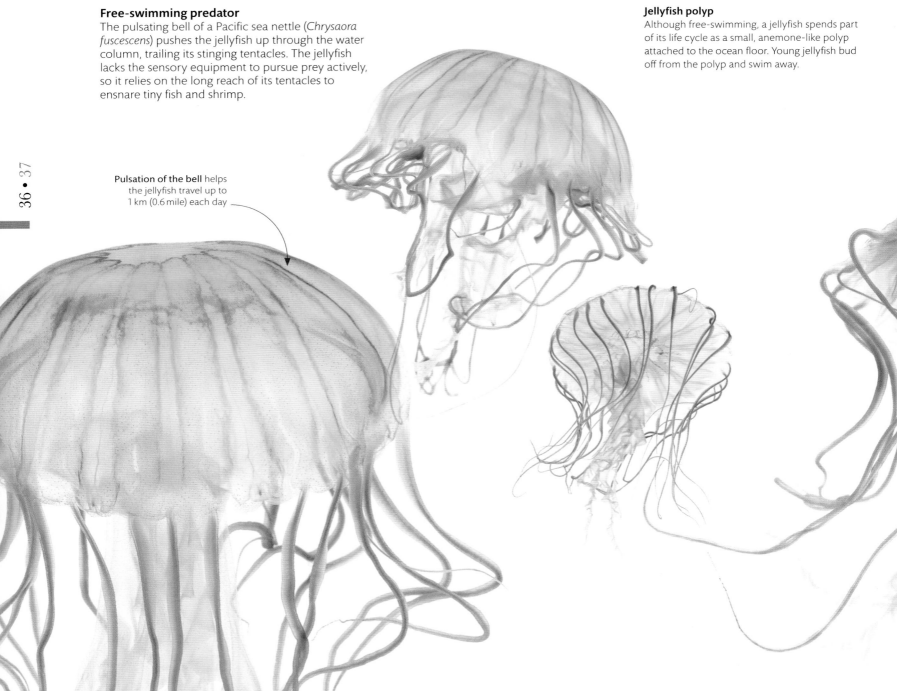

Pulsation of the bell helps the jellyfish travel up to 1 km (0.6 mile) each day

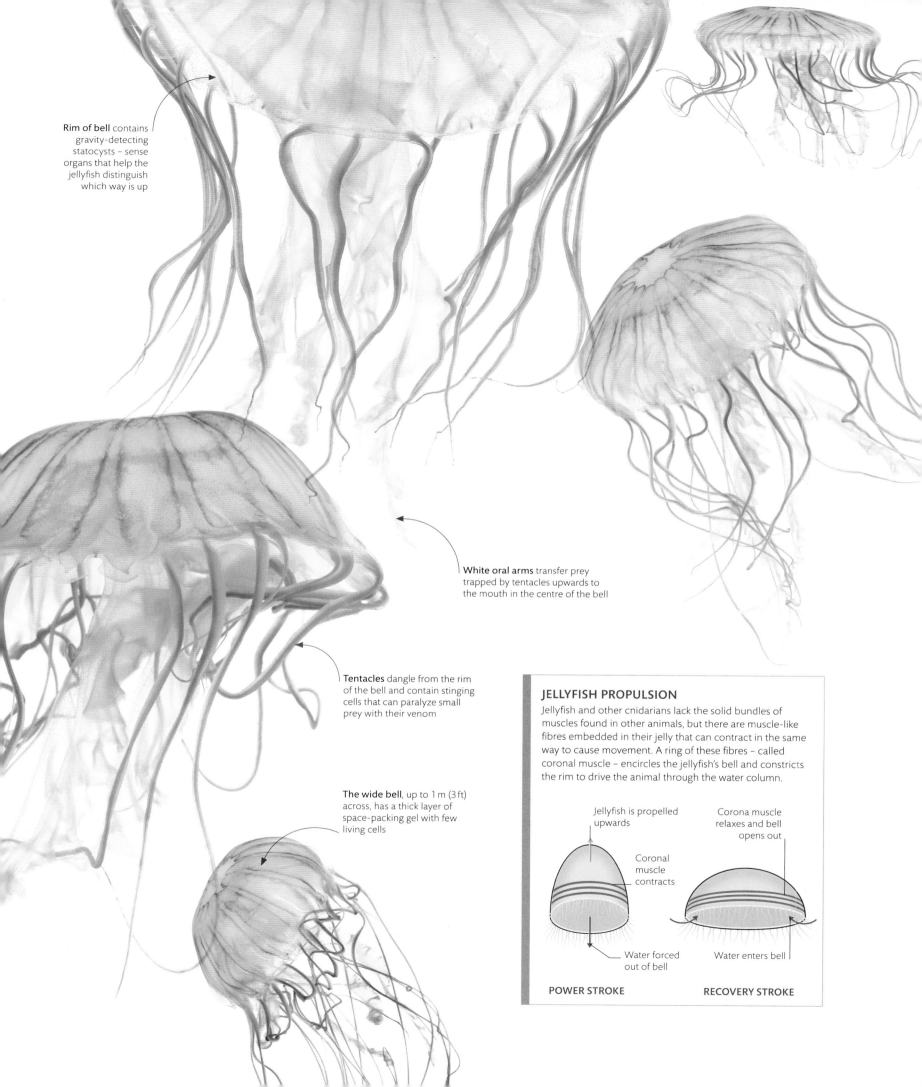

Rim of bell contains gravity-detecting statocysts – sense organs that help the jellyfish distinguish which way is up

White oral arms transfer prey trapped by tentacles upwards to the mouth in the centre of the bell

Tentacles dangle from the rim of the bell and contain stinging cells that can paralyze small prey with their venom

The wide bell, up to 1 m (3 ft) across, has a thick layer of space-packing gel with few living cells

JELLYFISH PROPULSION

Jellyfish and other cnidarians lack the solid bundles of muscles found in other animals, but there are muscle-like fibres embedded in their jelly that can contract in the same way to cause movement. A ring of these fibres – called coronal muscle – encircles the jellyfish's bell and constricts the rim to drive the animal through the water column.

Jellyfish is propelled upwards

Corona muscle relaxes and bell opens out

Coronal muscle contracts

Water forced out of bell

Water enters bell

POWER STROKE

RECOVERY STROKE

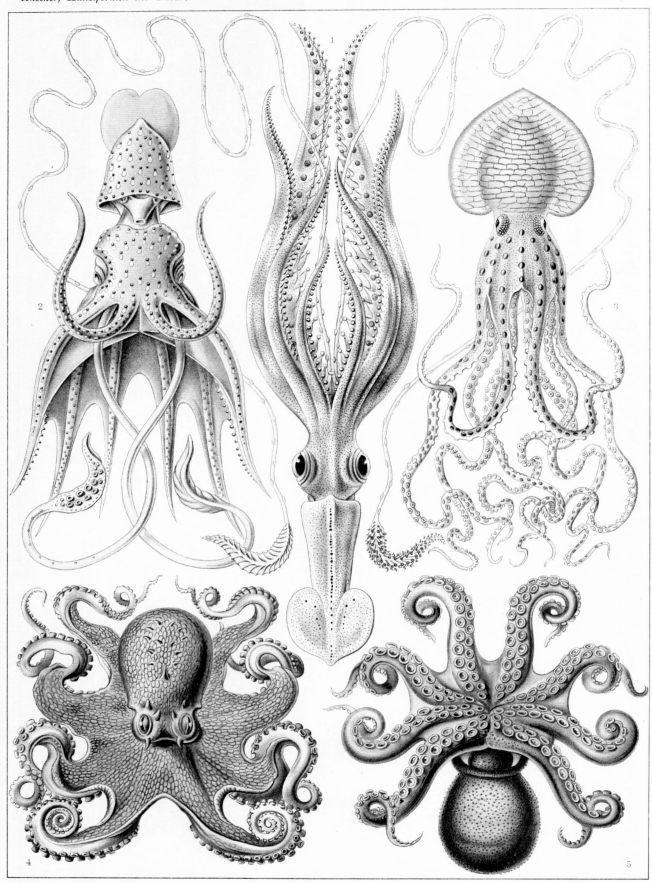

Gamochonia. — Trichterkraken.

Cephalopods

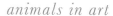

Combining science with artistry, the illustrations of biologist Ernst Heinrich Haeckel (1834–1919), such as these exquisitely arranged gamochonia (cephalopods), included thousands of new species. The evolutionary theorist's major 1904 work *Kunstformen der Natur* (*Art Forms of Nature*) explored symmetry and levels of organization in body forms.

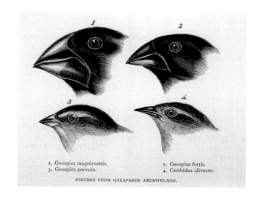

Darwin's finches (1845)

John Gould's observation that Darwin's diverse bird specimens – collected from different Galápagos islands – all belonged to the same family of finches, became a mainspring of Darwin's theory of natural selection.

animals in art

the darwinists

For wealthy young gentlemen of independent means in the 19th century, the study of natural history was a route to self-improvement and a rationale for global travel. Their interest was fuelled by Charles Darwin's accounts of zoological exploration aboard the H.M.S. *Beagle* and, ultimately, by his groundbreaking theories on the origins of humankind.

Although most natural history societies and organizations were run by amateurs, scientific methodology had gained traction in the 100 years since Linnaeus introduced a global classification system for plants (1753) and animals (1758). Under his influence, the British Royal Navy included naturalists on voyages to collect and record flora and fauna, and Charles Darwin was asked to join H.M.S. *Beagle*.

Intellectual rigour called for a new approach to art: zoological drawing had to be accurate and formulaic to enable comparison of anatomy between species. Over the course of the 19th century, the London Zoological Society commissioned hundreds of drawings and paintings.

Its artists included John Reeves, a tea merchant with the East India company who travelled to China in 1812 and spent 19 years in Macao working with Chinese artists on plant and animal pictures.

Ornithologist and taxidermist John Gould was a major illustrator of specimens for Darwin's *The Voyage of the Beagle* (1839), and offered insights through his drawing. Another contributor to Darwin's work was Reverend Leonard Jenyns, part of a group of Cambridge-educated naturalists engaged in a golden age of zoology, botany, and geology. Jenyns used local illustrators for his own *Notebook of the Fauna of Cambridgeshire* and noted that, such was the appetite for rare specimens, persons in the "lower classes" were supplementing their living collecting and selling insects from the fens.

In the footsteps of Darwin came Ernst Haeckel, a German polymath who mapped a genealogical tree, coined biological terms such as "ecology" and "phylogeny", and produced eye-opening works on invertebrates. The remarkable artistry of his biological illustrations was an inspiration to art nouveau artists.

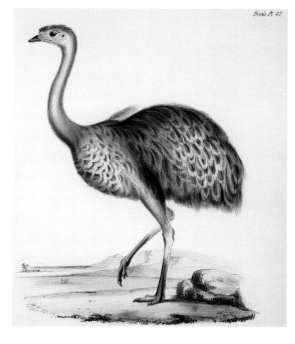

Small rhea (1841)

English ornithologist and artist John Gould identified and illustrated many of the birds Darwin encountered on the voyage of H.M.S. *Beagle*, including a small species of rhea that he named *Rhea darwinii*.

> 66 The affinities of all the beings of the same class have sometimes been represented by a great tree. I believe this simile largely speaks the truth. 99

CHARLES DARWIN, *ON THE ORIGIN OF THE SPECIES*, 1859

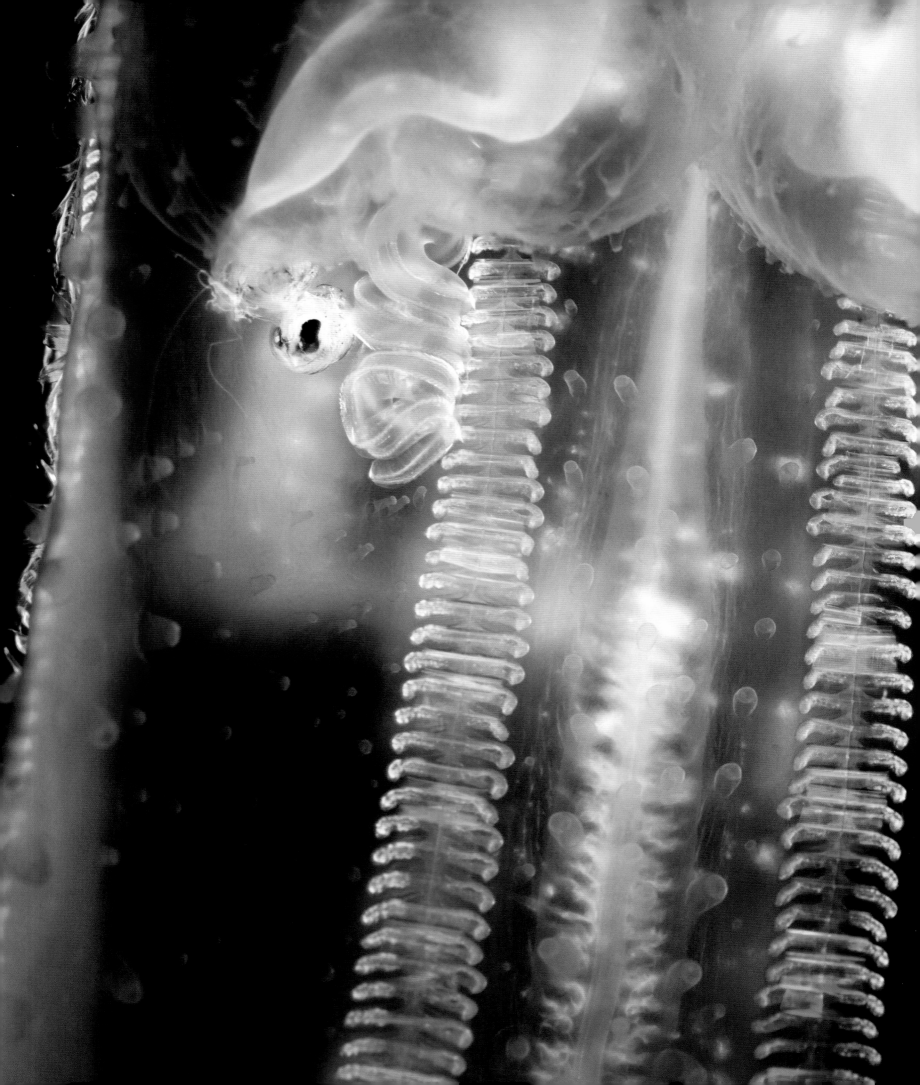

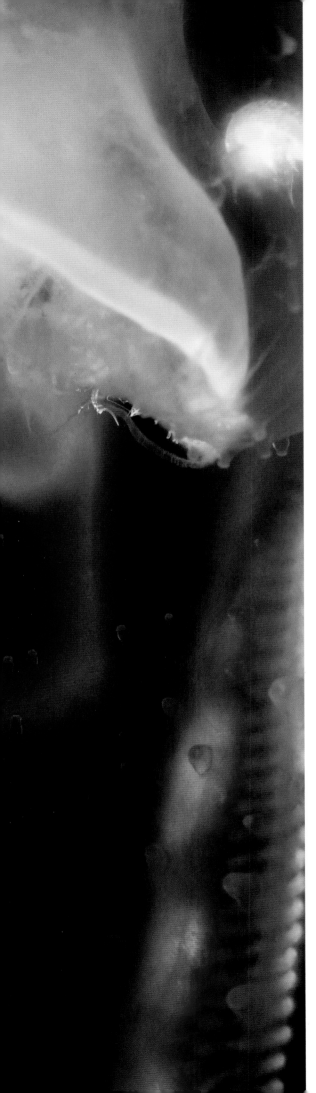

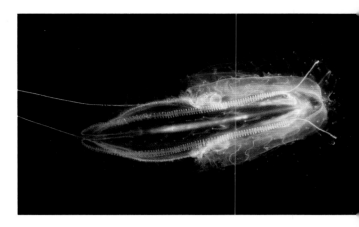

Tiny voyager
Just 10–18 cm (4–7 in) long, the warty comb jelly is usually found in the upper layers of open seas from the Atlantic Ocean and the Mediterranean Sea to subtropical waters.

spotlight species

comb jellies

Drifting through the oceans in sizes that range from the microscopic to 2 m (6 ft), comb jellies, or ctenophores, are often mistaken for jellyfish (see pp.36–37). While both gelatinous invertebrates have existed for at least 500 million years, and share many features, the two are very different.

The word ctenophore means "comb-bearing" and describes the hairlike cilia that these animals use for propulsion – the largest known creatures to do so. These cilia are arranged in comblike groups called ctenes, which are fused at the base and arranged in 8 rows along each side of the animal.

Like jellyfish, comb jellies' bodies are 95 per cent water, covered in a thin cell layer – the ectoderm. An internal layer, or endoderm, lines the animals' gut cavity. The middle, transparent jellylike layer is called the mesoglea, which in comb jellies contains three cell types: contractile or muscular, nerve, and mesenchymal. Mesenchyme develops into a range of tissues in complex organisms, but in comb jellies, it is simply connective tissue.

Comb jellies are found in all oceans, from equatorial to polar waters. There are only 187 known species and they come in myriad shapes, from lobed to long and beltshaped. Some even have retractable fringed tentacles covered in cells that exude a gluelike substance to catch prey. They are hermaphroditic, reproducing by releasing both sperm and eggs, and rely on currents to bring them into contact with others. Comb jellies are rapacious carnivores, with a mouth at one end and two anal pores that release waste at the other. Individuals of some species can eat 500 copepods (tiny crustaceans) an hour, and have been known to wipe out fish populations, as they leave nothing for any surviving fish larvae or fry to eat.

Light show
Adult warty comb jellies (*Leucothea multicornis*) develop two transparent lobes, as well as the combs (or ctenes) that generate rainbowlike colours as they refract and reflect light.

bodies with simple heads

The ability to move forwards with a brain that leads from the front was a strategic improvement over the way radial animals seemed to flounder in open water. Flatworms were among the first animals to be able to move from place to place in this way. They have distinct front and back ends, with a body symmetry that means their left and right sides are mirror images. Despite the limitations of shape, they have evolved into large, flamboyant forms.

Delicate beauty
Facing forwards with their wide heads, flatworms like this Hyman's polyclad (*Pseudobiceros hymanae*) glide over rocky reefs. The name "polyclad", meaning "many branches", refers to how its branched gut disperses food through its paper-thin body without the need for circulating blood.

Body is so thin oxygen can reach all cells just by seeping in through surface of animal

Flatworm is tapered at the rear

LEADING FROM THE FRONT
Forward-moving animals have more sense organs at the front, or head, where they are most needed. A flatworm has the simplest centralized nervous system of any animal. A collection of nerve cells, or neurons, in its head process incoming information, then communicate with the rest of its body via nerve cords and nerves that contain lines of neurons.

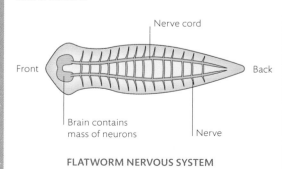

Nerve cord

Front

Back

Brain contains mass of neurons

Nerve

FLATWORM NERVOUS SYSTEM

Slippery mucus released by glandular cells on the underside help flatworm glide smoothly across the sea bed

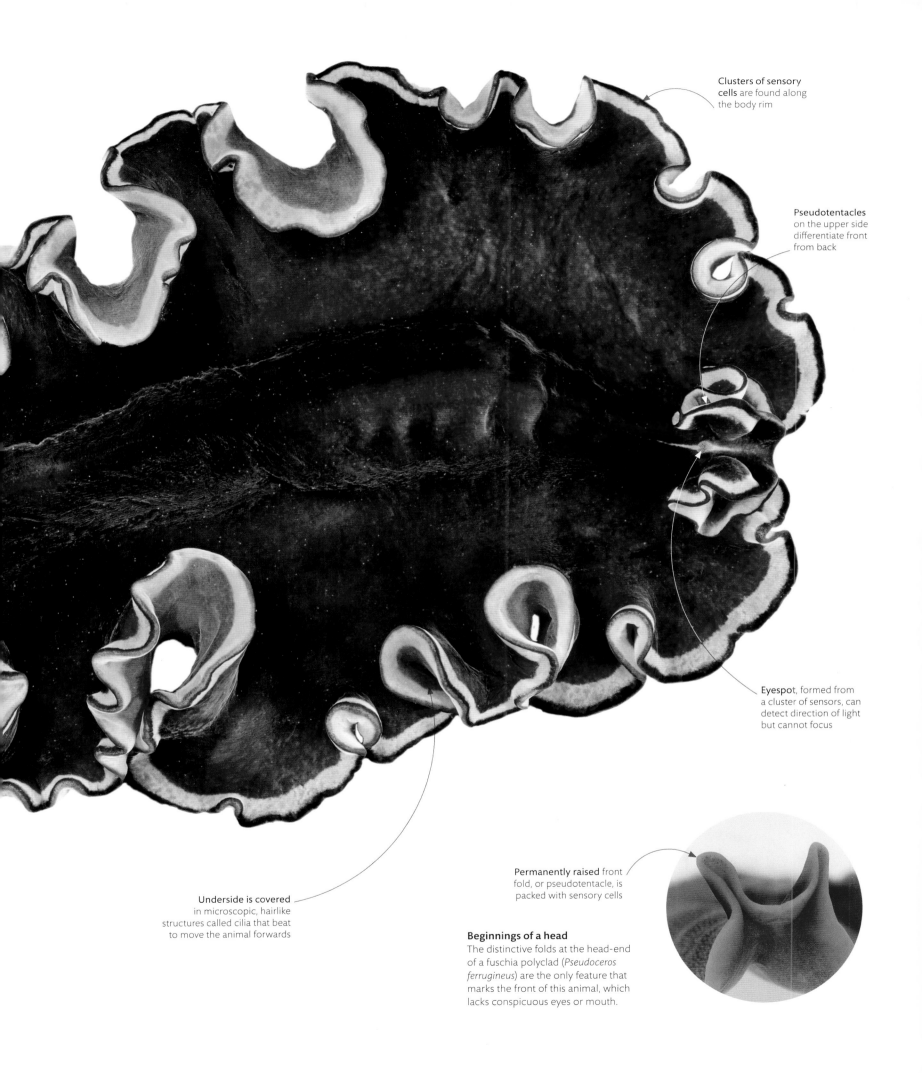

Clusters of sensory cells are found along the body rim

Pseudotentacles on the upper side differentiate front from back

Eyespot, formed from a cluster of sensors, can detect direction of light but cannot focus

Underside is covered in microscopic, hairlike structures called cilia that beat to move the animal forwards

Permanently raised front fold, or pseudotentacle, is packed with sensory cells

Beginnings of a head
The distinctive folds at the head-end of a fuschia polyclad (*Pseudoceros ferrugineus*) are the only feature that marks the front of this animal, which lacks conspicuous eyes or mouth.

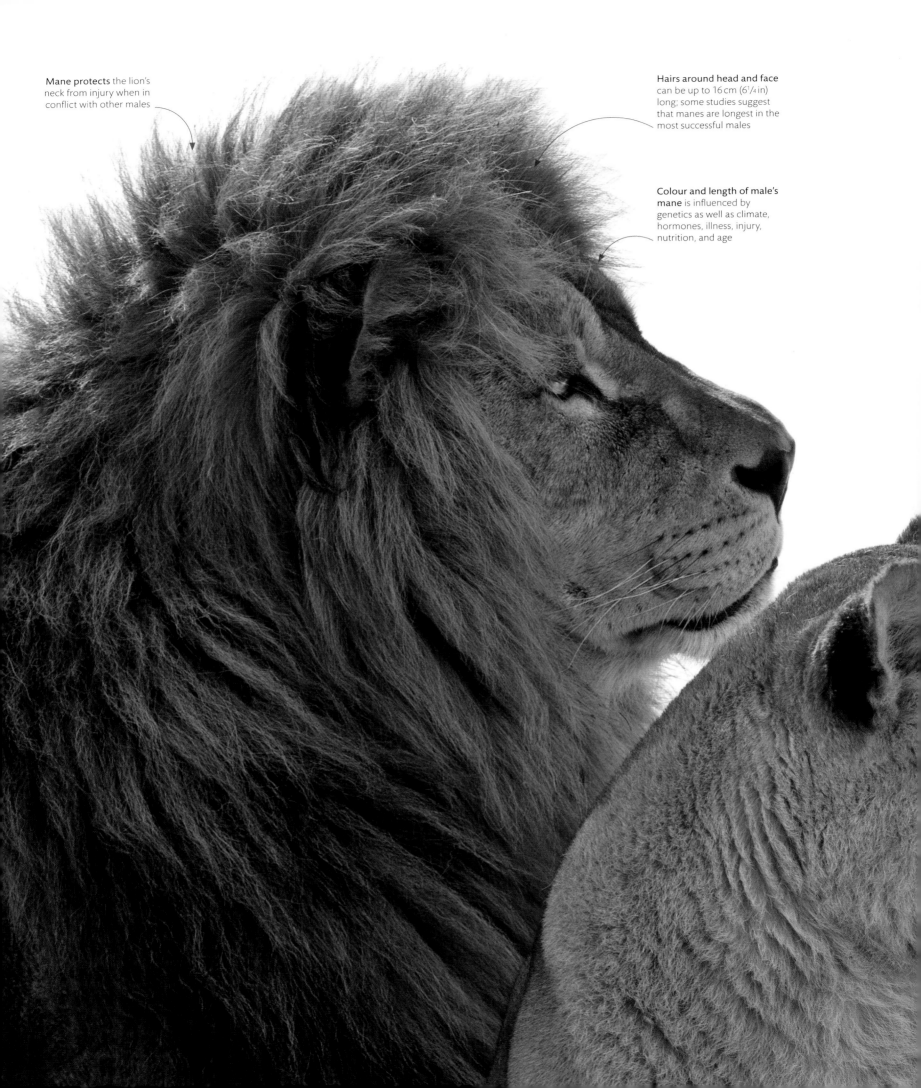

Mane protects the lion's neck from injury when in conflict with other males

Hairs around head and face can be up to 16 cm (6¼ in) long; some studies suggest that manes are longest in the most successful males

Colour and length of male's mane is influenced by genetics as well as climate, hormones, illness, injury, nutrition, and age

sexual differences

Most animals develop into one of two sexes, determined by genes or the environment. The differences in sexually mature individuals advertise that they are ready to produce offspring of their own. In some animals this sexual dimorphism is reflected in body size. In many mammals the males are bigger than females. In addition, the larger the male, the more likely he is to succeed with choosy females. In other species, such as some fish, a larger female may produce more eggs or bigger offspring, increasing their chances of survival.

Adults in waiting
All lion cubs look similar, and a male only starts to develop a mane at 12 months, but sexual development is determined by genetics before birth. Like all mammals, male cubs have XY chromosomes, while the females have XX. The Y chromosome masculinizes the body, while its absence feminizes it.

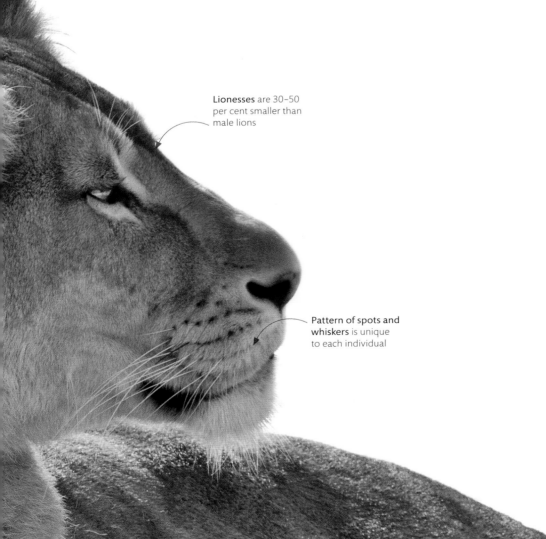

Male and female
The mane of a male lion (*Panthera leo*) is a clear indicator of his fitness to sire a lioness's cubs as its thickness correlates with his testosterone level. A fuller mane means higher fertility and a more assertive temperament, which can be passed on to the lioness's cubs.

Lionesses are 30–50 per cent smaller than male lions

Pattern of spots and whiskers is unique to each individual

GENDER EXTREMES

Sexual dimorphism is driven to extremes in some deep-sea anglerfish species, where the male is 10 times smaller than the female. Tiny males attach parasitically to females, an act that brings both sexes – perhaps uniquely among animals – to sexual maturity. It is only after the large-bodied female has been fertilized that she can produce her eggs.

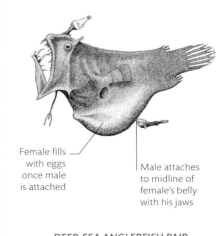

Female fills with eggs once male is attached

Male attaches to midline of female's belly with his jaws

DEEP-SEA ANGLERFISH PAIR
Linophryne argyresca

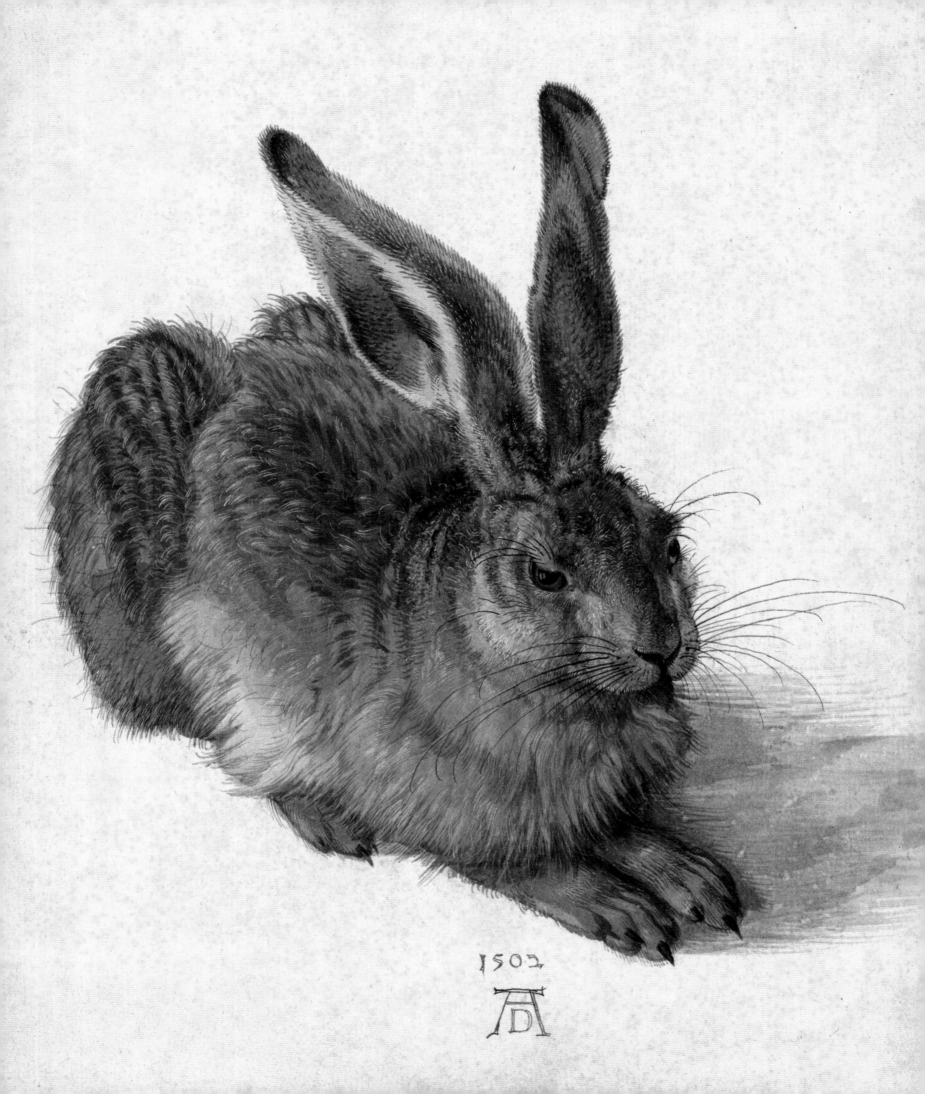

1502

Head of a Bear (c.1480)
Leonardo da Vinci's sketchbooks reveal his fascination with bears, many of which lived wild in the mountains of Tuscany and Lombardy. This small-scale observation in metalpoint on prepared paper was probably of an animal in captivity.

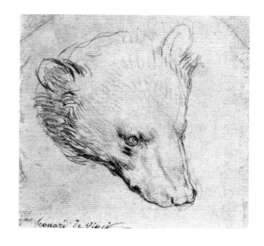

animals in art

the renaissance eye

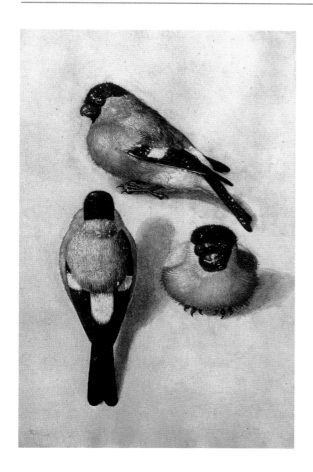

In the quiet spaces between their great works, artists of the Renaissance produced watercolours and sketches of animals that revealed a profound sensitivity to the natural world. Nature was the new religion of the Renaissance, fuelled by retranslations of ancient Greek texts that advocated an ethical and scientific approach to the study of animals.

The genius of the Renaissance, Leonardo da Vinci, prepared for his major works with anatomical studies of animals and sketches that tap into animal behaviour. Following in his footsteps in the north, Albrecht Dürer's preparatory watercolours of flora and fauna were destined to enhance the grand visions of sacred stories in his paintings, woodcuts, and engravings. They reveal a deep engagement with nature.

A new sensitivity to animals had emerged from humanist rediscovery of the works of Greek philosophers such as Aristotle. His reverence for animal life sat in sharp contrast to medieval notions of unfeeling beasts with satanic qualities. The result was an explosion in the study of natural history. Swiss naturalist Conrad Gessner mirrored Aristotle's work with his five-volume *Historia Animalium* (1551–58), which set out to catalogue every known animal and mythical creature. His vast collection of illustrations included Dürer's woodcut of a rhinoceros as a metallic, armour-plated beast. At the time, no rhino had been seen in Europe: the only specimen, a gift from the King of Portugal to Pope Leo X, drowned in a shipwreck in 1616.

Three Studies of a Bullfinch (1543)
Sophisticated perspective and precise anatomical observation are encapsulated in Albrecht Dürer's watercolours, which were usually preparatory studies for major works.

The Young Hare (1502)
The almost tangible softness of the fur and the spark of life in the eyes of this young hare has made Albrecht Dürer's study one of his most revered works. He painted his hare in gouache and watercolour in his workshop in Nuremburg, most likely combining observation of hares in the field with observation of a taxidermist's specimen in his studio.

" Every kind of animal will reveal to us something natural and something beautiful. "

ARISTOTLE, *ON THE PARTS OF ANIMALS BOOK 1*, 350 BCE

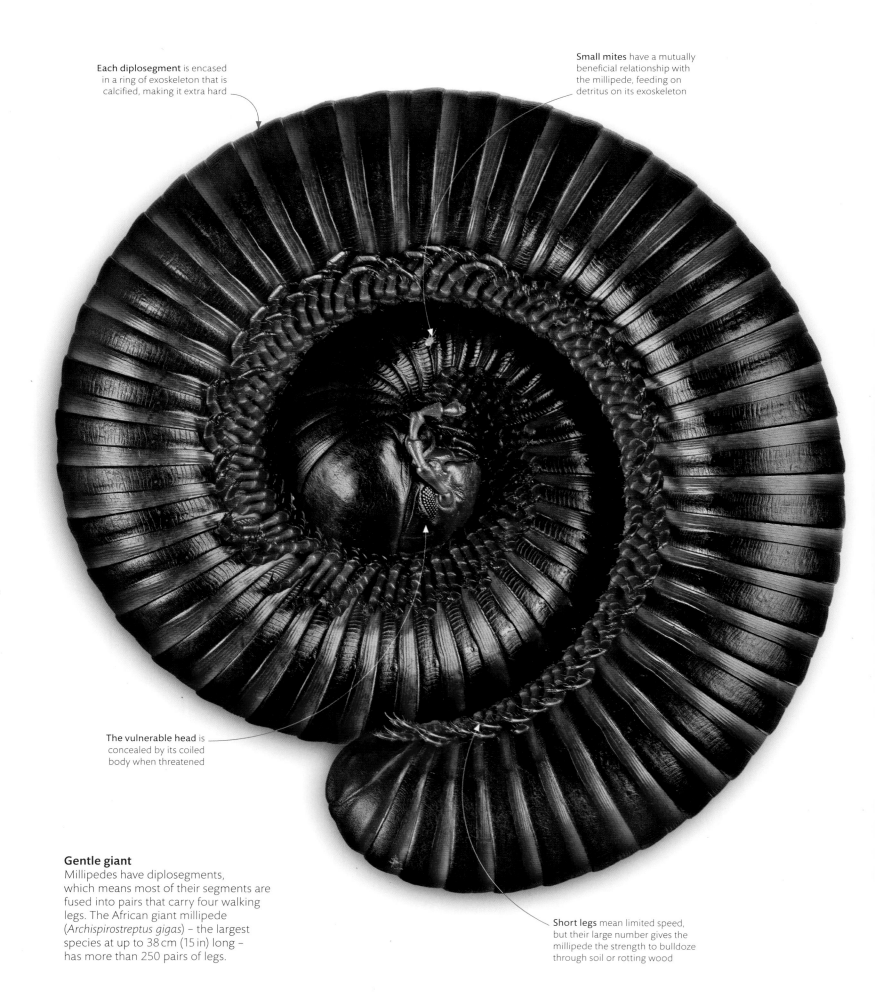

Each diplosegment is encased in a ring of exoskeleton that is calcified, making it extra hard

Small mites have a mutually beneficial relationship with the millipede, feeding on detritus on its exoskeleton

The vulnerable head is concealed by its coiled body when threatened

Short legs mean limited speed, but their large number gives the millipede the strength to bulldoze through soil or rotting wood

Gentle giant
Millipedes have diplosegments, which means most of their segments are fused into pairs that carry four walking legs. The African giant millipede (*Archispirostreptus gigas*) – the largest species at up to 38 cm (15 in) long – has more than 250 pairs of legs.

segmented bodies

Some animals get bigger during their development by repeatedly duplicating segments of their body. In the evolution of worms, this kind of segmentation meant body sections could be moved independently (see pp.66–67). Hard-bodied arthropods, including millipedes and centipedes, inherited this segmented body plan from wormlike ancestors, but the addition of jointed legs has made their locomotion even more effective.

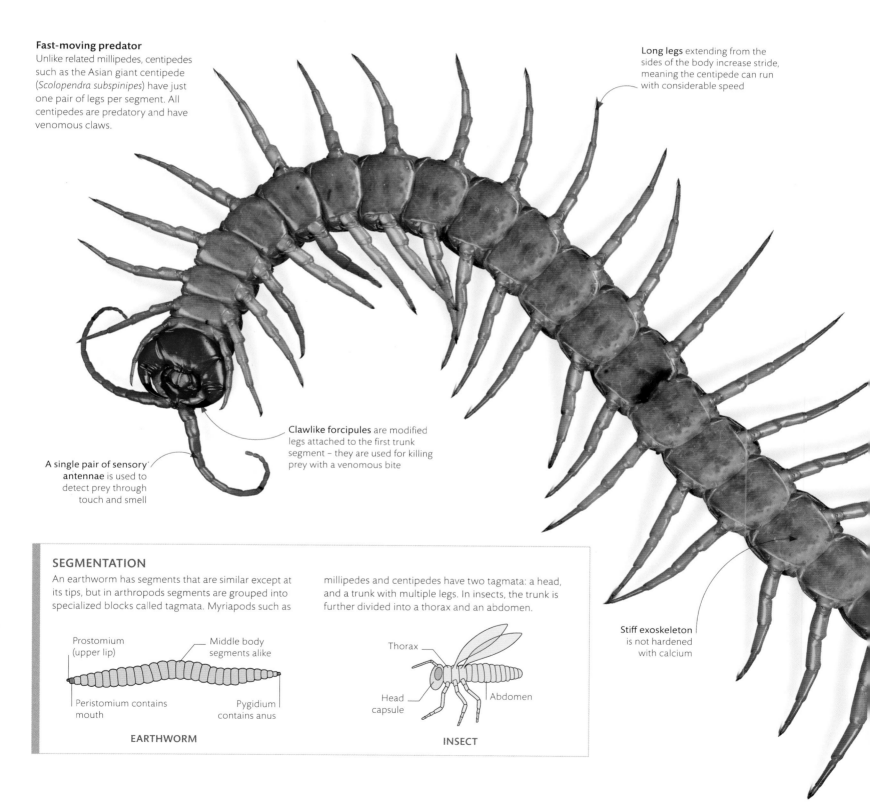

Fast-moving predator
Unlike related millipedes, centipedes such as the Asian giant centipede (*Scolopendra subspinipes*) have just one pair of legs per segment. All centipedes are predatory and have venomous claws.

Long legs extending from the sides of the body increase stride, meaning the centipede can run with considerable speed

Clawlike forcipules are modified legs attached to the first trunk segment – they are used for killing prey with a venomous bite

A single pair of sensory antennae is used to detect prey through touch and smell

Stiff exoskeleton is not hardened with calcium

SEGMENTATION

An earthworm has segments that are similar except at its tips, but in arthropods segments are grouped into specialized blocks called tagmata. Myriapods such as millipedes and centipedes have two tagmata: a head, and a trunk with multiple legs. In insects, the trunk is further divided into a thorax and an abdomen.

Prostomium (upper lip)

Middle body segments alike

Peristomium contains mouth

Pygidium contains anus

EARTHWORM

Thorax

Head capsule

Abdomen

INSECT

vertebrate bodies

A spine, or vertebral column, provides a solid support for the body but is flexible so the body can bend. This is possible because the column is built from smaller articulating blocks, or vertebrae, made of cartilage or bone. The power for bending comes from muscle blocks called myotomes that flank the column. The spines of the first vertebrates – fish – swung sinuously from side to side; this movement made them a success in water and later helped their four-legged descendants to crawl on land.

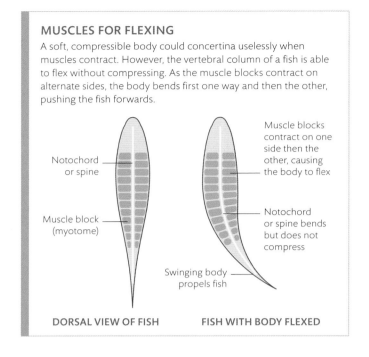

MUSCLES FOR FLEXING

A soft, compressible body could concertina uselessly when muscles contract. However, the vertebral column of a fish is able to flex without compressing. As the muscle blocks contract on alternate sides, the body bends first one way and then the other, pushing the fish forwards.

Notochord or spine

Muscle block (myotome)

Muscle blocks contract on one side then the other, causing the body to flex

Notochord or spine bends but does not compress

Swinging body propels fish

DORSAL VIEW OF FISH

FISH WITH BODY FLEXED

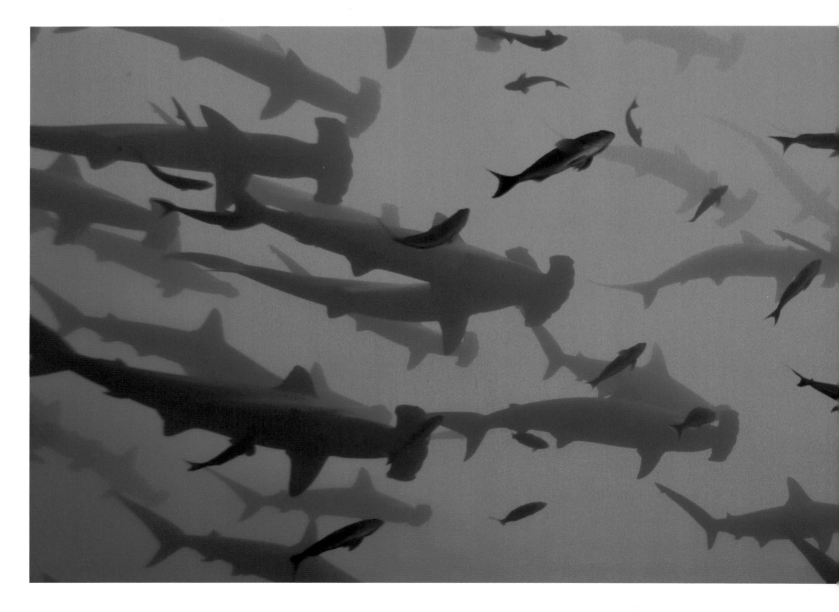

Lancelet

The fishlike lancelet (*Branchiostoma lanceolatum*) has a rubbery notochord supporting its back: the evolutionary precursor of a bony or cartilaginous spine. A notochord is present in vertebrate embryos; it is replaced by a vertebral column during development.

The back is supported by a notochord sandwiched between rows of myotomes

Muscle blocks (myotomes) are visible as faint chevrons in the body wall

Intestine digests tiny food particles filtered from the water

Sharks on the move

Large schools of mainly female scalloped hammerheads (*Sphyrna lewini*) gather around the Galapagos Islands by day, before dispersing to hunt alone at night – propelled by the same sideways sweep of a tail that drove the first vertebrates half a billion years ago.

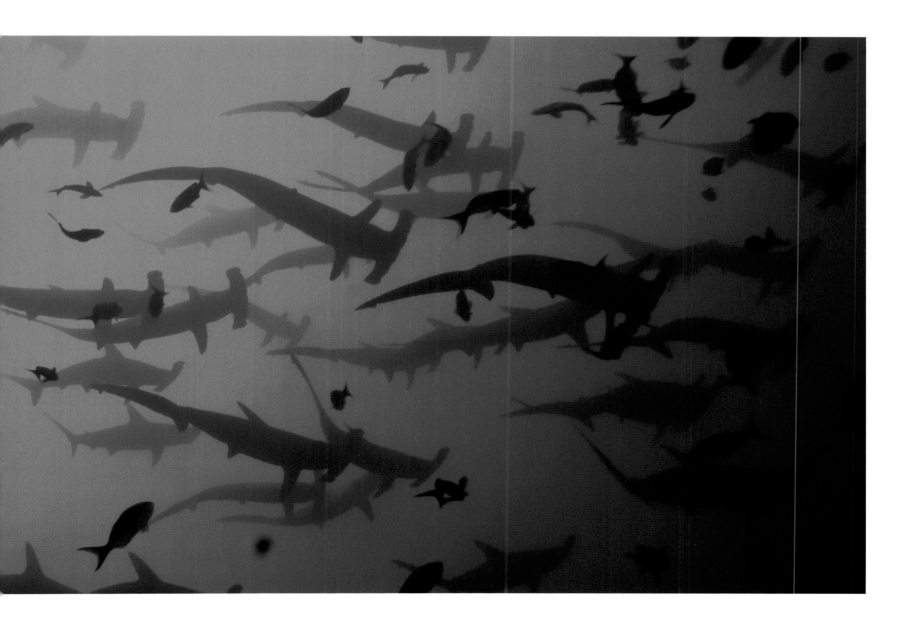

Frog evolution

Frogs' body form evolved partly by reducing the number of vertebrae in their spine to nine or eight. The few living frogs with nine vertebrae today belong to the "ancient frogs", or Archaeobatrachia, two of which also appear to retain a tail, although this is actually a boneless mating organ. Mesobatrachians, an intermediate evolutionary stage, and the diverse Neobatrachians ("new frogs") acquired more typical frog traits – new frogs have a tongue and a voice.

Tail is part of the male cloaca, used for internal fertilization

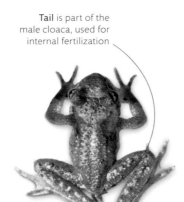

ARCHAEOBATRACHIA
Tailed frog
Ascaphus truei

Mouth contains no movable tongue, as in many mesobatrachians

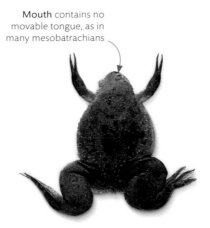

MESOBATRACHIA
African clawed frog
Xenopus laevis

Eardrum present, as in most neobatrachians

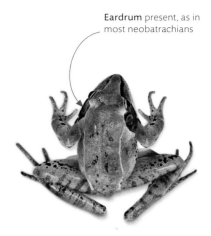

NEOBATRACHIA
European common frog
Rana temporaria

Lifestyles

Despite their reliance on keeping their skin moist, frogs and toads have met the demands of a range of habitats. Most require a water source to reproduce (see pp.316–17), but can be found even in deserts, where burrowing species sit out dry periods underground, often sealed in cocoons to retain moisture. Many ambush prey on forest floors, while others are active tree-climbing hunters.

Sticky disc on each finger (and toe) aids climbing

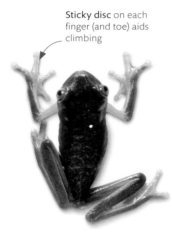

FLY-CATCHER
Red-eyed tree frog
Agalychnis callidryas

Markings mimic leaf litter where frog hides during the day

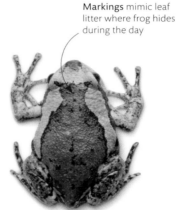

WORM-EATER
Banded bullfrog
Kaloula pulchra

Wide mouth suits a sit-and-wait predator with no teeth, which must swallow large prey whole

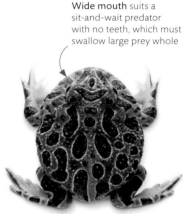

MOUSE-CATCHER
Argentinian horned frog
Ceratophrys ornata

Size variations

Frog and toad sizes are as diverse as their numbers, and the discovery of new miniature species constantly moves the lower boundary of the size range. The smallest known living frog is *Paedophryne amauensis* of Papua New Guinea, which measures just 7.7 mm (¼ in) long; the largest, at 30 cm (13 ½ in) long and weighing up to 3.3 kg (7 lb 3 oz), is the West African goliath frog (*Conraua goliath*).

Body colour changes to white with yellow, black-rimmed spots in the sun

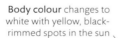
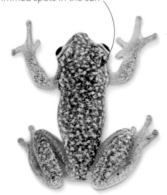

3–3.3 CM (1 ⅕ IN)
Madagascan reed frog
Heterixalus alboguttatus

Long "neck" section allows frog to move head from side to side

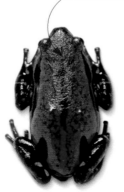

4.1–6.2 CM (1 ¾–2 ⅓ IN)
West African rubber frog
Phrynomantis microps

Skin controls evaporation so frog can adjust to seasonally wet or dry habitats

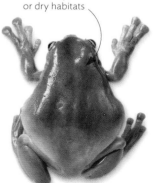

7–11.5 CM (2 ¾–4 ½ IN)
White's tree frog
Ranoidea caerulea

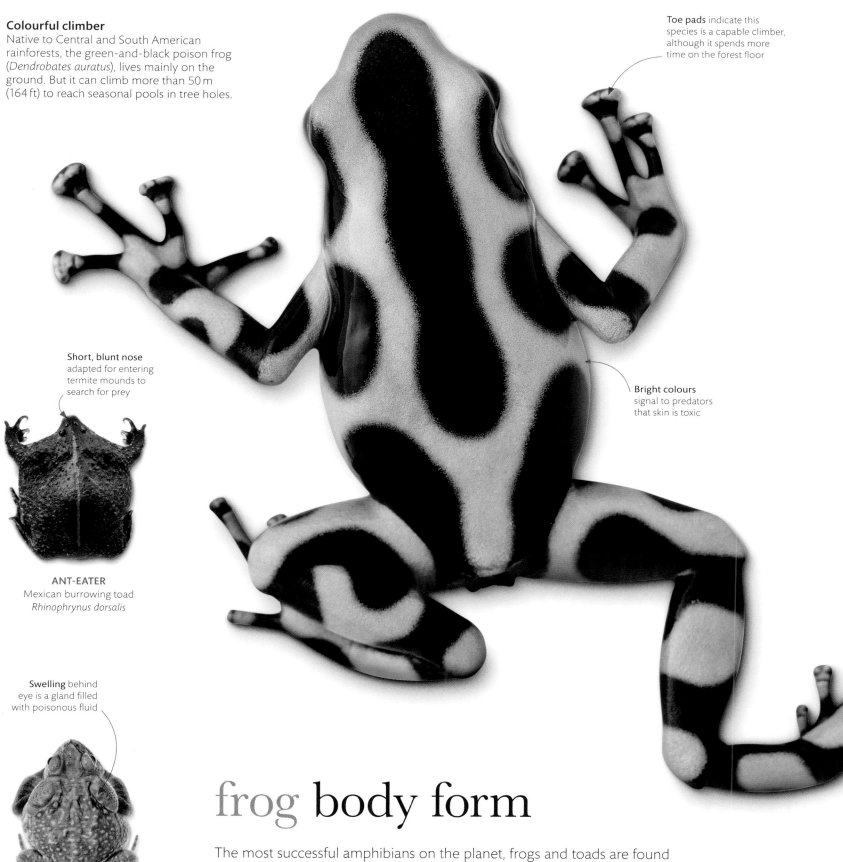

Colourful climber
Native to Central and South American rainforests, the green-and-black poison frog (*Dendrobates auratus*), lives mainly on the ground. But it can climb more than 50 m (164 ft) to reach seasonal pools in tree holes.

Toe pads indicate this species is a capable climber, although it spends more time on the forest floor

Short, blunt nose adapted for entering termite mounds to search for prey

Bright colours signal to predators that skin is toxic

ANT-EATER
Mexican burrowing toad
Rhinophrynus dorsalis

Swelling behind eye is a gland filled with poisonous fluid

UP TO 22 CM (8⅔ IN)
American cane toad
Rhinella marina

frog body form

The most successful amphibians on the planet, frogs and toads are found on every continent except Antarctica. Their body shape has remained largely unchanged for 250 million years. Their heads are attached by a single neck vertebra to truncated bodies, they have elongated hindlimbs, and, in adults, the tail is replaced by fused vertebrae known as a urostyle.

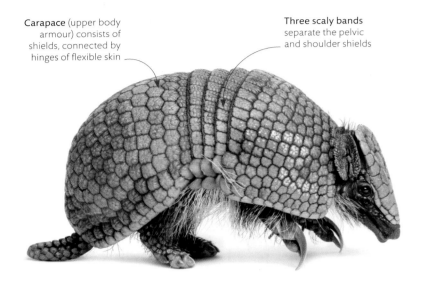

Carapace (upper body armour) consists of shields, connected by hinges of flexible skin

Three scaly bands separate the pelvic and shoulder shields

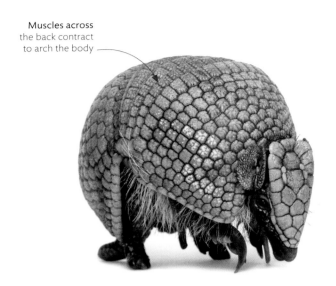

Muscles across the back contract to arch the body

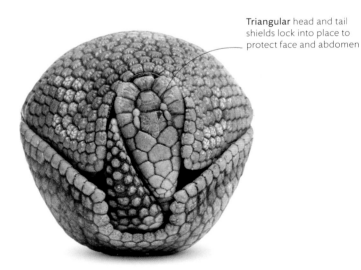

Triangular head and tail shields lock into place to protect face and abdomen

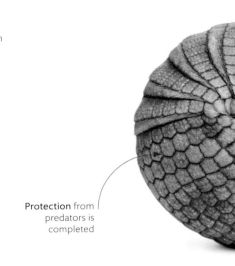

Protection from predators is completed

shape shifters

Strong muscles and a degree of flexibility are all it takes for any animal to change shape as it moves, but some animals manage especially dramatic transformations when confronted with danger. All armadillos are already protected by bony plates that cover the upper part of their bodies and, when hazards threaten, most lie flat on the ground with their limbs withdrawn. However, two species of three-banded armadillos have the flexibility it takes to roll up into a ball in self-defence.

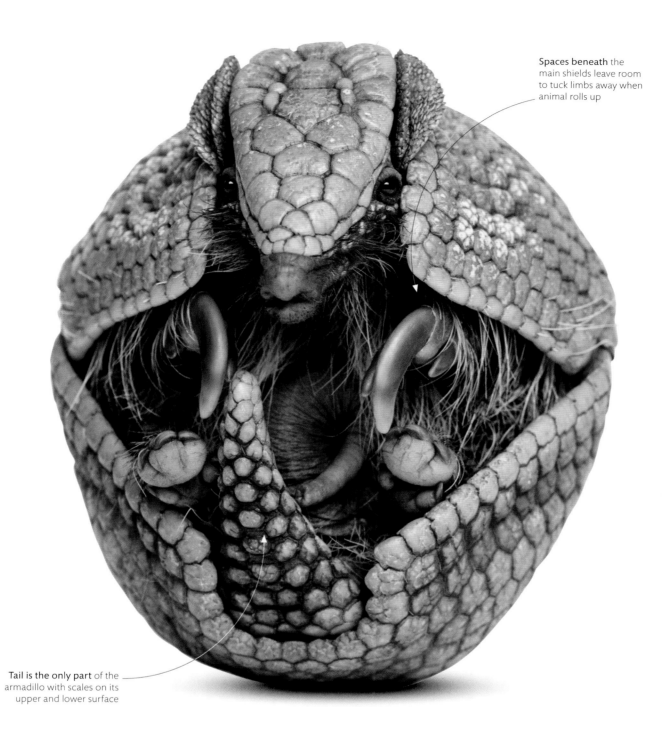

Spaces beneath the main shields leave room to tuck limbs away when animal rolls up

Tail is the only part of the armadillo with scales on its upper and lower surface

Armour-plated protection
The carapace of the southern three-banded armadillo (*Tolypeutes matacus*) is made up of plates of bone, or osteoderms, overlaid with a horny epidermis. The plates are only partially attached to the body, which leaves space inside for it to withdraw its limbs completely and so roll into a ball.

LIFE ON THE EDGE

An elephant's sheer size buffers it in an unpredictable environment. It moves slowly and has large reserves that allow it to survive on very little nutrition for weeks. But the tiniest warm-blooded animals, such as shrews, which can weigh as little as 2 g (³⁄₄ oz), have no energy reserves and lose a far greater proportion of their energy as radiated body heat, so need to eat almost continually just to prevent starvation.

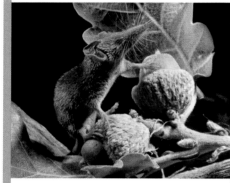

SHREW (SORICIDAE FAMILY)

Ears of the savanna elephant can be 2 m (6 ft 6 in) from top to bottom

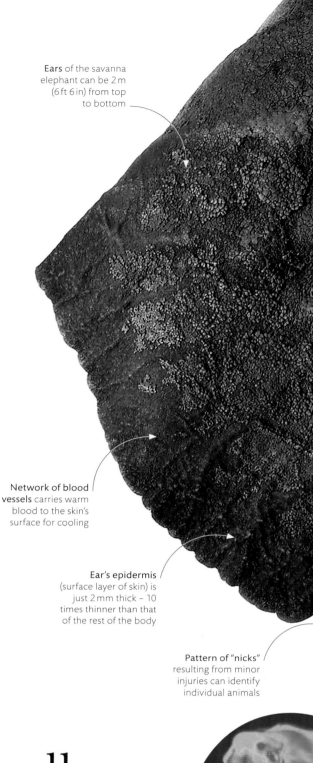

Network of blood vessels carries warm blood to the skin's surface for cooling

Adapted for life in the tropics

The skin of an African savanna elephant (*Loxodonta africana*) – the world's largest land animal – is sparsely haired to help compensate for its massive heat-generating bulk. While it lacks the oil-producing glands of other mammals, mud-baths condition its skin, and wrinkles help to lock in moisture.

Ear's epidermis (surface layer of skin) is just 2 mm thick – 10 times thinner than that of the rest of the body

Pattern of "nicks" resulting from minor injuries can identify individual animals

large and small

The largest animals alive today – elephants and whales – are vast compared with the tiniest invertebrates that can perch on a pinhead. Their size brings benefits – bigger animals can repel predators and intimidate competitors – but they require more food and oxygen. The pull of gravity also subjects their bodies to formidable stresses and strains, so they need very strong skeletons and powerful muscles to be able to move.

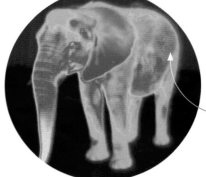

Thermal image shows that the bulk of the elephant's body is warm (red) but the tips of its ears are cool (blue)

Staying cool

Mammals generate body heat to maintain vital functions. Overheating could be a problem for this warm-blooded giant, but elephants can lose heat by flapping their ears to cool the blood pumped through them.

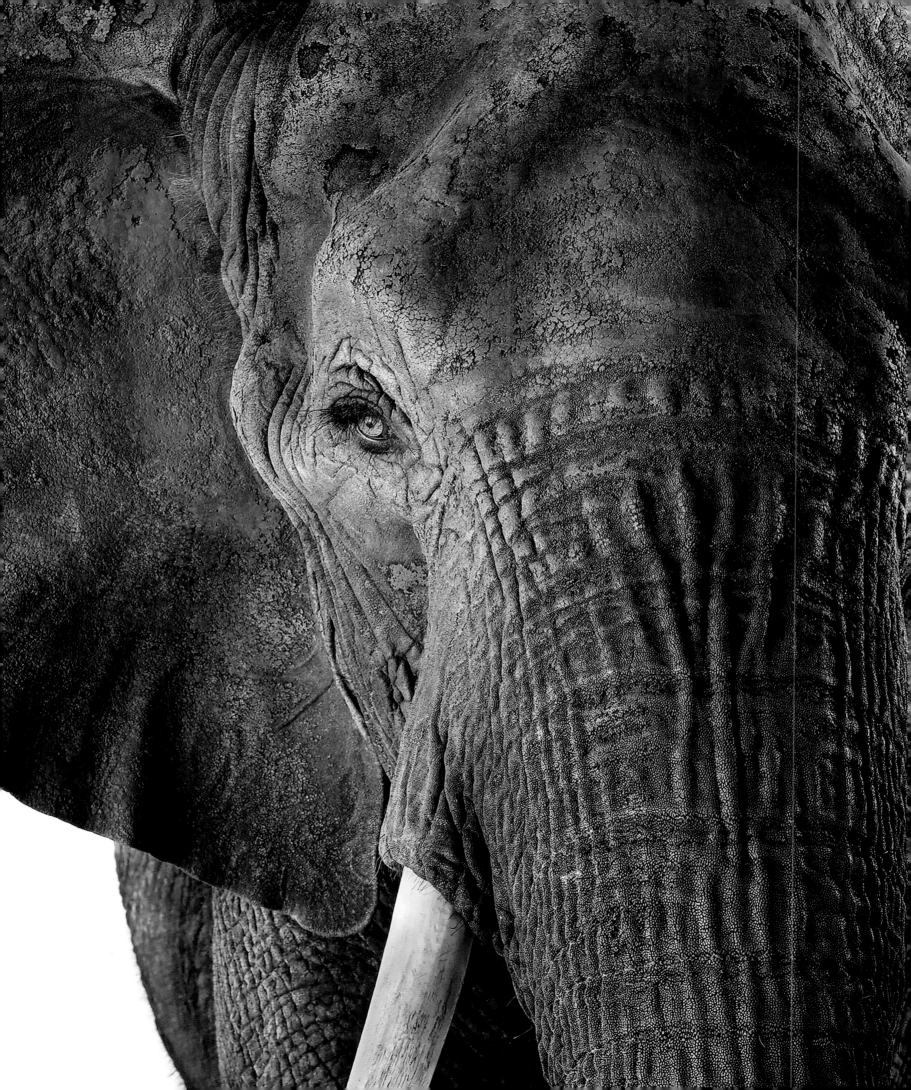

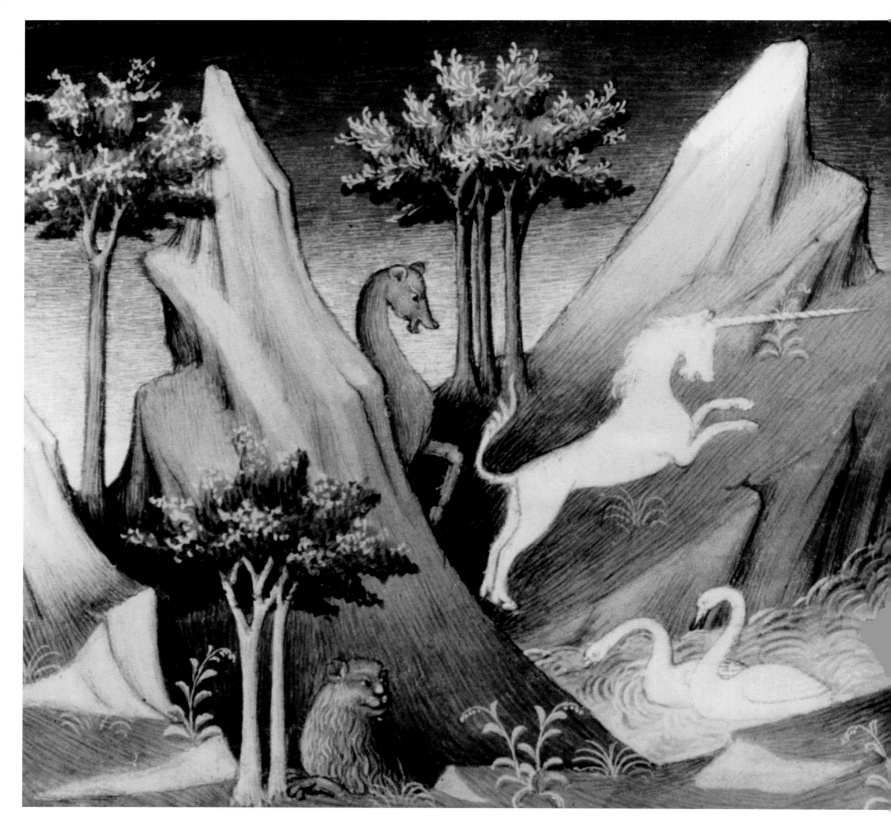

Malabar unicorn
Italian explorer Marco Polo claimed an encounter
with a unicorn occurred during his 24-year journey
across Asia, but his description of an ugly animal
wallowing in the mud suggests it was probably a
rhino. For the 15th-century edition of Polo's *Livre des
Merveilles du Monde* (*Book of Wonders of the World*),
the illustrator opted for a more traditional unicorn
surrounded by indigenous animals of Malabar, India.

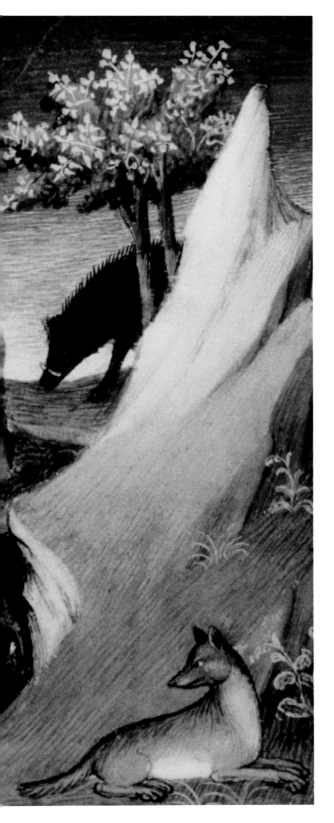

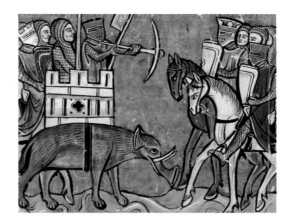

Recreating elephants
Elephants in the 13th-century *Rochester Bestiary* were given a variety of colours, trumpet-shaped trunks, and boarlike features. This war elephant is shown bearing a turreted howdah of soldiers.

animals in art

fantasy beasts

Bestiaries were medieval catalogues of creatures based on the belief that the beasts, birds, fish, and even rocks of the Earth had God-given qualities and characteristics. Stories about the good and evil inherent in the world's wildlife were easy-to-grasp tutorials for the medieval mind. Many bestiaries were lavish, illuminated works with animal pictures that would linger in the imagination of illiterate audiences.

Many of the European bestiaries, written mainly in Latin or in vernacular French in France, originate from the *Physiologus* – an ancient Alexandrian text, featuring 50 animals, which was probably written between the 2nd and 4th century CE. Here, nature's innocent beasts were afforded behaviours and traits that linked them to Christ or the devil. The pelican, for example, was believed to revive its dead young with its own blood, making it a powerful symbol of the Resurrection.

In 13th-century bestiaries, the whale is shown as a huge multi-finned fish that mimics an island to lure ships onto its back. When the unwitting sailors build a fire on its sandy skin, the hot, angry whale drags them into the deep: a reminder of the devil's temptations and sinners' descent into hell. Hedgehogs are depicted rolling on grapes to spear them and carry them to their young. A roaring panther is shown with sweet smelling breath that draws all animals to him.

Most of the illustrations were the work of artistically untrained monks forced to base their attempts at exotic animals on verbal accounts and carvings. A crocodile with a doglike head seems to be drawn as much from the imagination as a unicorn.

Even the first-hand observations of the 13th-century explorer Marco Polo added little to medieval understanding of wildlife in distant lands. The Venetian explorer dictated his recollections of his 24-year journey to Rustichello da Pisa, an inmate incarcerated with him in a Genoese prison. What became a bestseller in Europe was illustrated with fanciful miniatures of animals and birds, but Polo's descriptions of "everything different… and excelling in both size and beauty" were treated with scepticism.

   To tame the unicorn so it can be captured, a virgin girl is placed in its path.

ROCHESTER BESTIARY, 13TH CENTURY

tall animals

A high head gives an animal a better view of danger and enables it to reach food inaccessible to competitors. In this respect, no other living animal can match the giraffe. Its stature comes from an elongated neck and its especially long lower leg bones. But this height also demands a stronger heart to pump blood so far against gravity, and double the typical mammalian blood pressure to keep it moving.

Heads above the rest
Giraffes (*Giraffa camelopardalis*) are the tallest living animals – males can be 6 m (20 ft) from ground to crown. Several factors have favoured their evolution: the height provides for better vigilance and higher browsing, and their shape provides a greater surface area from which to radiate heat, enabling them to remain cool.

NECK CIRCULATION

When a giraffe lowers its head, a network of elastic-walled blood vessels at the base of the brain – the rete mirabile – expands to help absorb the rush of blood that might otherwise damage it. A series of one-way valves along the jugular vein close to prevent blood flowing backwards under pressure of gravity.

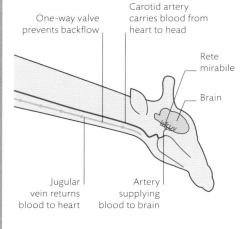

One-way valve prevents backflow

Carotid artery carries blood from heart to head

Rete mirabile

Brain

Jugular vein returns blood to heart

Artery supplying blood to brain

BLOOD FLOW IN LOWERED HEAD

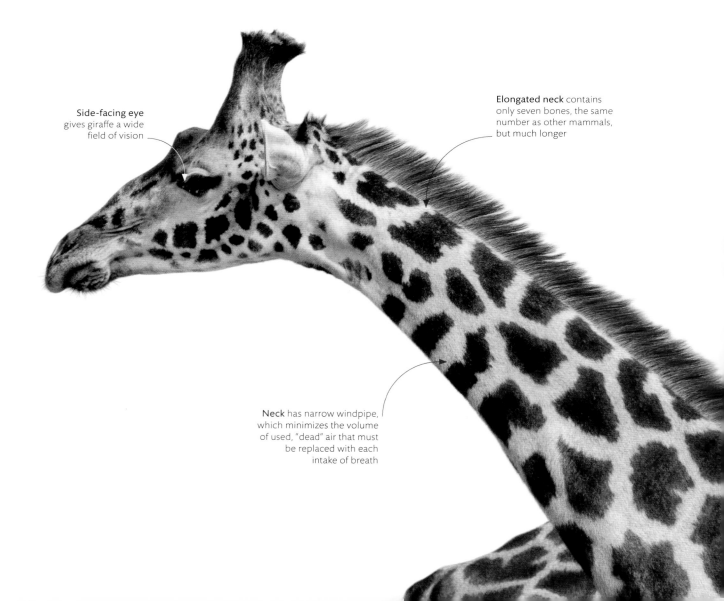

Side-facing eye gives giraffe a wide field of vision

Elongated neck contains only seven bones, the same number as other mammals, but much longer

Neck has narrow windpipe, which minimizes the volume of used, "dead" air that must be replaced with each intake of breath

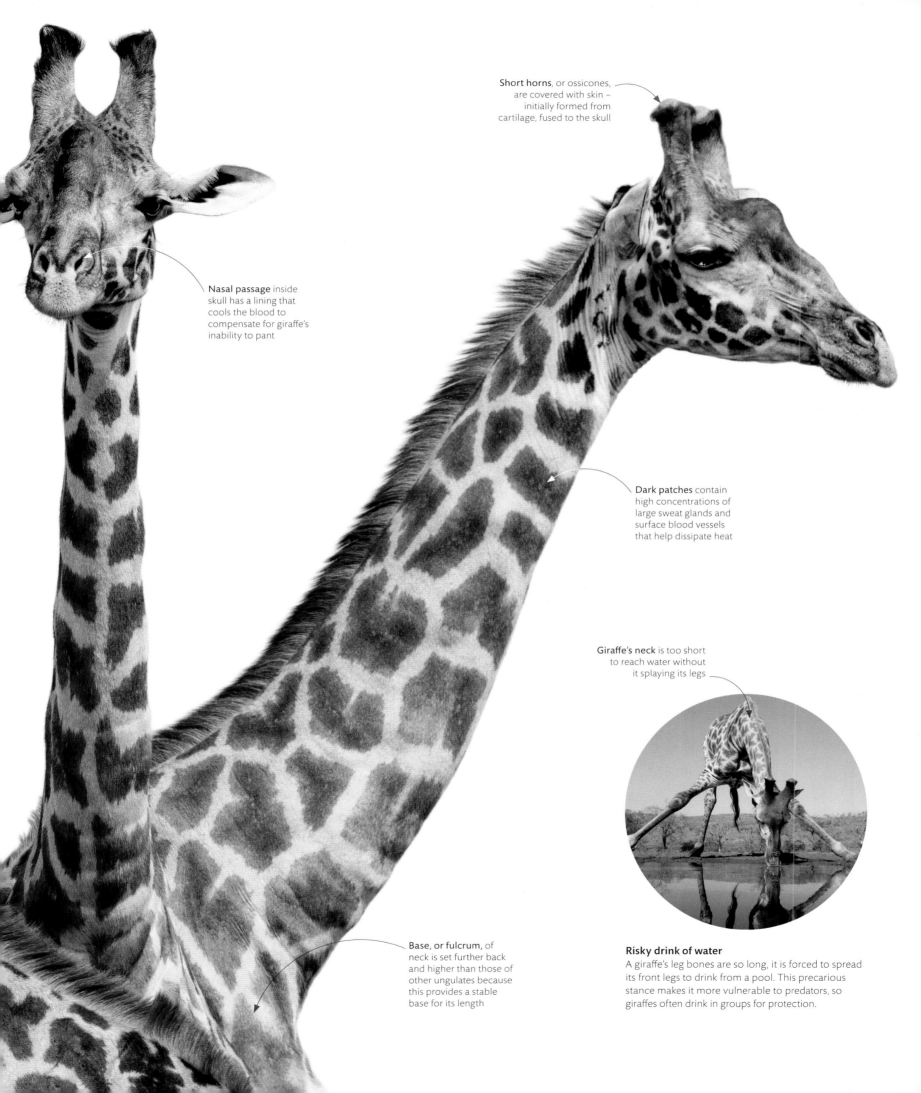

Short horns, or ossicones, are covered with skin – initially formed from cartilage, fused to the skull

Nasal passage inside skull has a lining that cools the blood to compensate for giraffe's inability to pant

Dark patches contain high concentrations of large sweat glands and surface blood vessels that help dissipate heat

Giraffe's neck is too short to reach water without it splaying its legs

Base, or fulcrum, of neck is set further back and higher than those of other ungulates because this provides a stable base for its length

Risky drink of water

A giraffe's leg bones are so long, it is forced to spread its front legs to drink from a pool. This precarious stance makes it more vulnerable to predators, so giraffes often drink in groups for protection.

skeletons

skeleton. an internal or external framework – often of a stiff material, such as bone or cartilage – that supports the shape and movement of an animal.

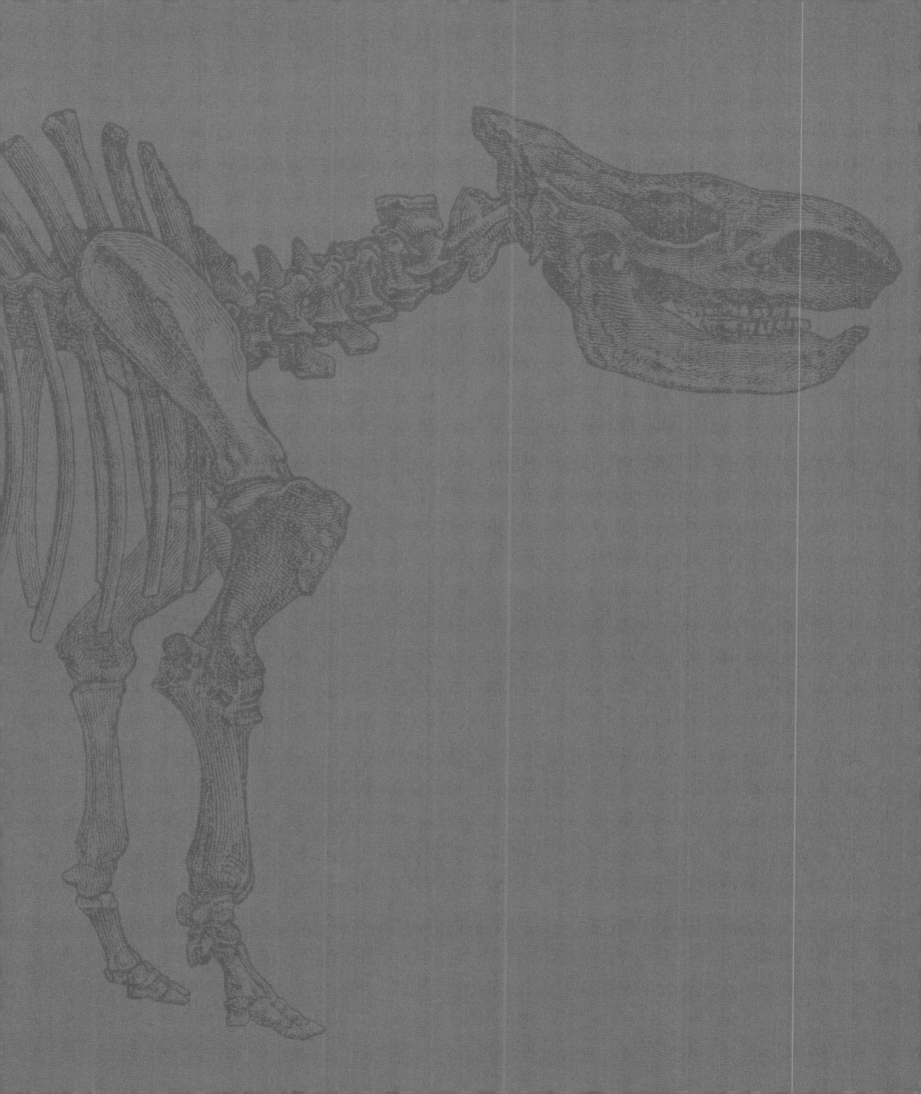

Sea fans

The silver tree gorgonian (*Muricea* sp.) is reinforced with hollow pipes of a horny material called gorgonin. It is strong enough to resist the buffeting of ocean currents. The fanlike shape of this yellow sea fan (*Annella mollis*) also helps it withstand the current.

Communal skeleton of the gorgonian helps the colony grow erect branches, exposing the polyps to more plankton

Gorgonians with dense, featherlike colonies are more easily uprooted by strong currents, so grow in calmer, deeper waters

Polyps

The skeleton of a coral or gorgonian is a lifeless substructure but supports living polyps on the surface. Each polyp sits within its own hard cup, called a corallite, which protects its soft body.

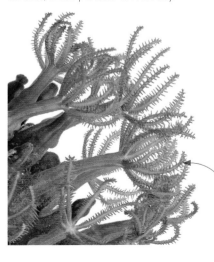

Each tentacle is covered with stinging cells that enable it to stun its prey

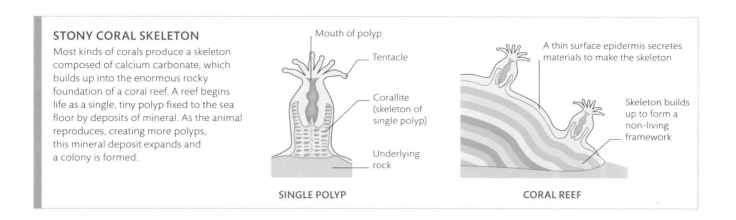

communal skeletons

A skeleton can support a sprawling colony just as effectively as a woody trunk and branches support a tree's foliage. Corals and their relatives excel at making massive skeletons to carry thousands of tiny polyps (see pp.32–33). Built from hornlike protein, rocky mineral, or chitin (the same material found in the shells of crabs) these skeletons can grow thick beneath the thin layer of skin that connects all the polyps living on their surface.

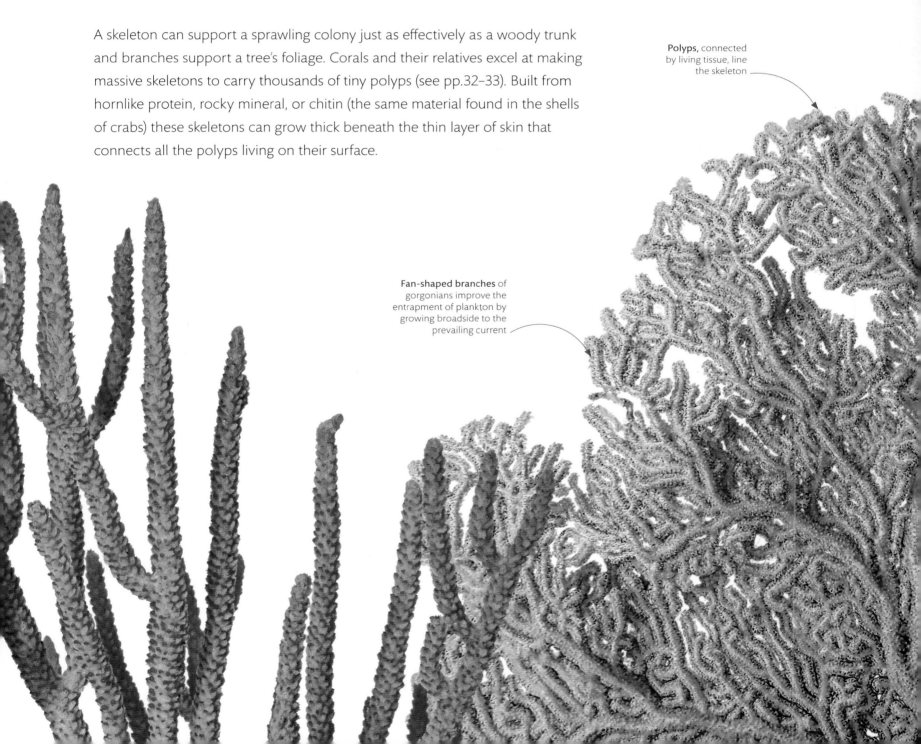

Polyps, connected by living tissue, line the skeleton

Fan-shaped branches of gorgonians improve the entrapment of plankton by growing broadside to the prevailing current

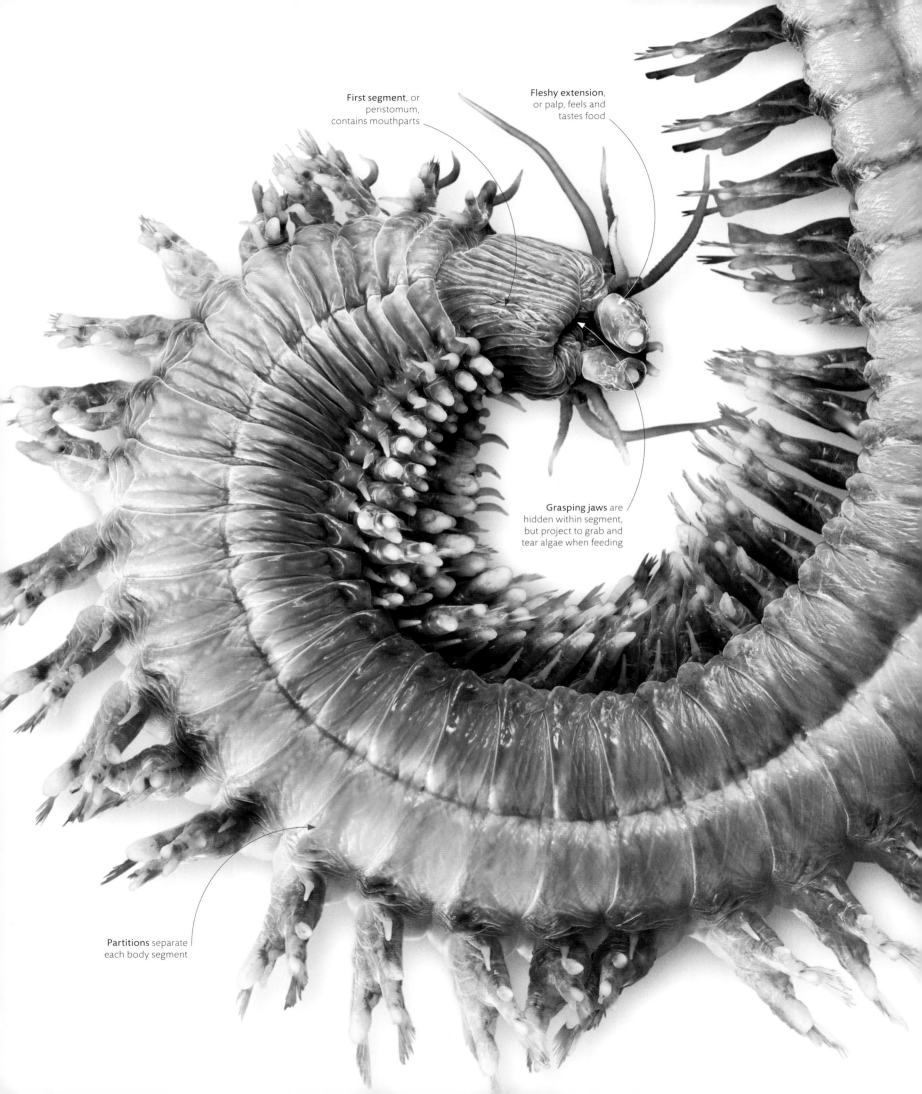

First segment, or peristomum, contains mouthparts

Fleshy extension, or palp, feels and tastes food

Grasping jaws are hidden within segment, but project to grab and tear algae when feeding

Partitions separate each body segment

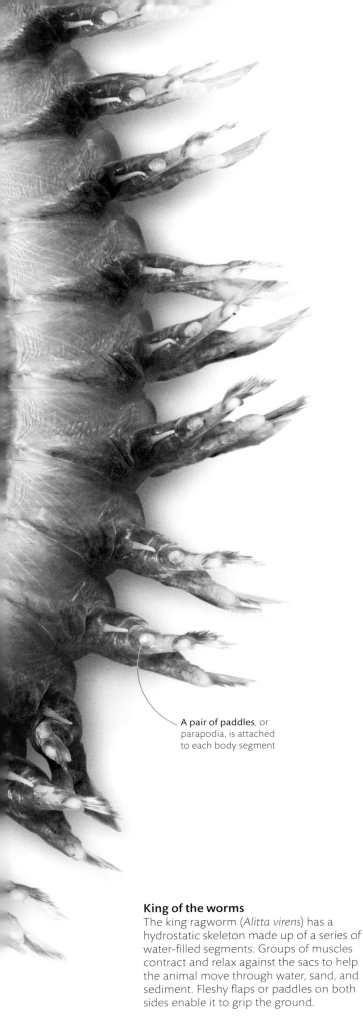

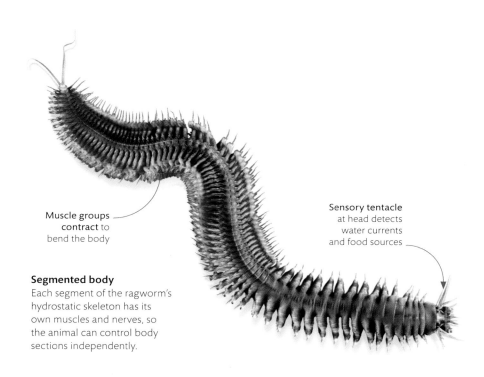

Muscle groups contract to bend the body

Sensory tentacle at head detects water currents and food sources

Segmented body
Each segment of the ragworm's hydrostatic skeleton has its own muscles and nerves, so the animal can control body sections independently.

water skeletons

Annelids – the group of animals that includes earthworms – have bodies that are not supported by hard bones or solid armour. Instead they have flexible, water-filled (or hydrostatic) skeletons. Water can be the perfect skeletal substance, as it cannot be compressed and it flows to fill any shape. As a result, by coordinating their efforts, the animal's muscles can squeeze against the water-filled parts of the body to move it forward.

A pair of **paddles**, or parapodia, is attached to each body segment

King of the worms
The king ragworm (*Alitta virens*) has a hydrostatic skeleton made up of a series of water-filled segments. Groups of muscles contract and relax against the sacs to help the animal move through water, sand, and sediment. Fleshy flaps or paddles on both sides enable it to grip the ground.

MOVING IN WAVES
By contracting alternate groups of muscles along either side of the body, while those on the opposite side relax and lengthen, a ragworm creates S-shaped transverse waves that pass along its body, helping it to move along the sand or swim through water.

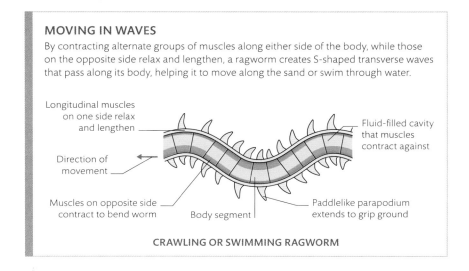

Longitudinal muscles on one side relax and lengthen

Direction of movement

Muscles on opposite side contract to bend worm

Body segment

Fluid-filled cavity that muscles contract against

Paddlelike parapodium extends to grip ground

CRAWLING OR SWIMMING RAGWORM

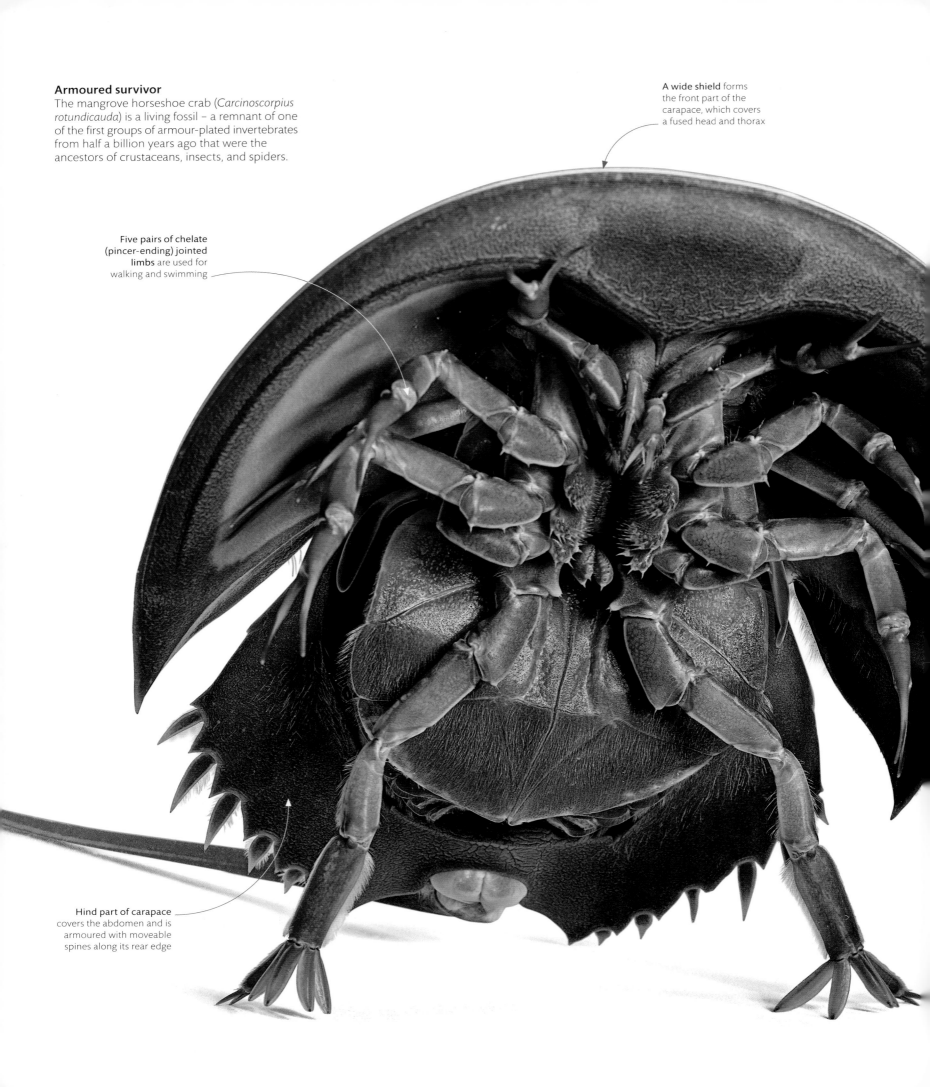

Armoured survivor
The mangrove horseshoe crab (*Carcinoscorpius rotundicauda*) is a living fossil – a remnant of one of the first groups of armour-plated invertebrates from half a billion years ago that were the ancestors of crustaceans, insects, and spiders.

A wide shield forms the front part of the carapace, which covers a fused head and thorax

Five pairs of chelate (pincer-ending) jointed limbs are used for walking and swimming

Hind part of carapace covers the abdomen and is armoured with moveable spines along its rear edge

False crab
Despite their name, horseshoe crabs are more closely related to arachnids than to true crabs. Like them, they lack antennae and the body is divided into two parts: the prosoma (fused head and thorax) and the opisthosoma (or abdomen).

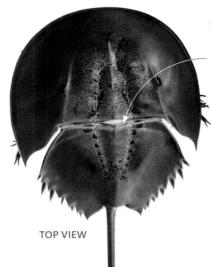

A flexible hinge in the carapace helps the animal bend in the middle

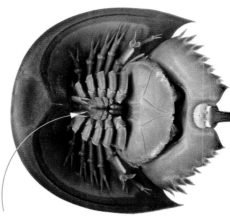

Chelicerae are used for capturing prey and moving it into the mouth

TOP VIEW

UNDERSIDE VIEW

The stiff, tail-like telson helps the animal to steer and contains sensors that detect light

external skeletons

Many aquatic animals with soft, flabby bodies are supported by water, both externally and internally. But a hard skeleton provides a more rigid framework that keeps the animal's shape. Animals supported by an external casing – known as an exoskeleton – have better control over their movements, can move faster, and can also grow bigger. Exoskeletons develop around the body like a suit of armour, meaning that they must be moulted periodically to make room underneath for the animal to grow.

The horseshoe crab's exoskeleton, unlike those of many other aquatic arthropods, does not contain calcium carbonate

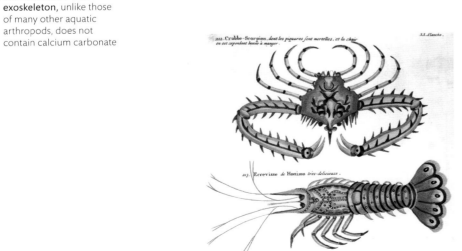

Reinforced exoskeletons
Arthropods have a rigid exoskeleton made of a substance called chitin. The exoskeletons of many arthropods that live in water – especially crustaceans – are also reinforced with minerals such as calcium carbonate, which make the exoskeleton stronger, but also heavier. However, the water surrounding aquatic arthropods supports the weight of its heavy skeleton.

exoskeletons on land

The animals that colonized land for the first time half a billion years ago had a hard external skeleton, or exoskeleton, that articulated around flexible joints. This armour protected them from physical harm and even – by evolving a waxy layer – helped prevent dehydration. But there was a big drawback for these land animals, because without the buoyancy of water, armour is heavy. Today the largest armoured invertebrates are still found in the sea (see pp.68–69). Those that have invaded land, such as woodlice, are limited to small size. But the spiders and, above all, the insects – with their better waterproofing and breathing pores (see box) – make up for this in numbers and diversity.

INSECT EXOSKELETONS

Skeletal modifications helped insects become the dominant arthropods on land. Not only are they better waterproofed, but their outer "shells" are punctured by a system of breathing pores (spiracles) that deliver oxygen directly to muscles through microscopic air-filled tubes.

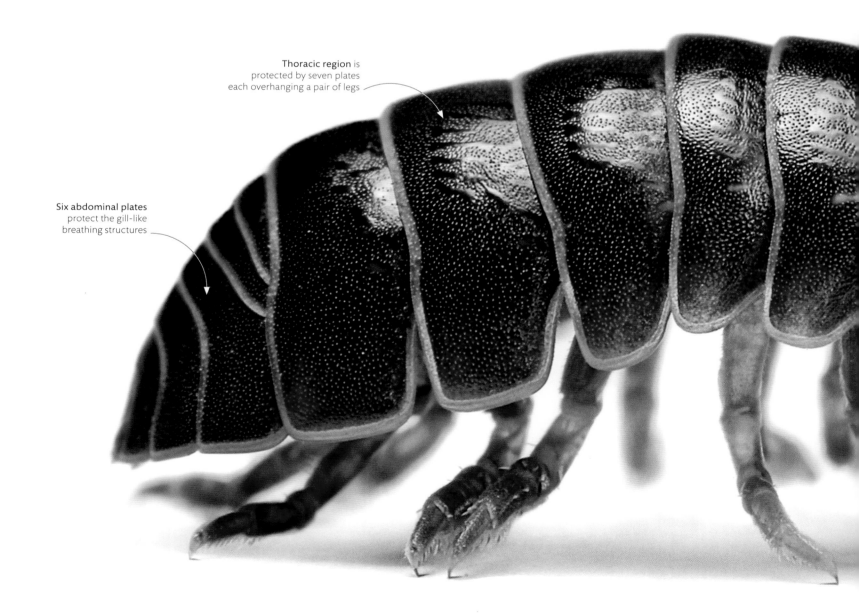

Spiracle, or breathing hole

Cuticle surface impregnated with waterproofing wax

Sensory bristle (seta)

Cuticle, hardened at surface

Epidermis

Trachea, or breathing tube

Glandular cell secretes substances that form cuticle

Bristle-forming cell, connects to nerve cells in epidermis

PROTECTIVE OUTER LAYER OF AN INSECT

Thoracic region is protected by seven plates each overhanging a pair of legs

Six abdominal plates protect the gill-like breathing structures

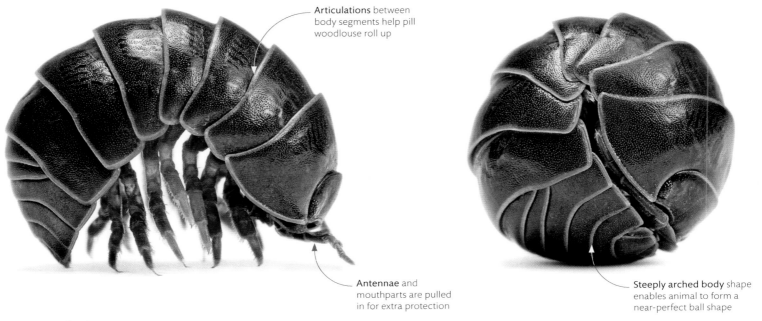

Articulations between body segments help pill woodlouse roll up

Antennae and mouthparts are pulled in for extra protection

Steeply arched body shape enables animal to form a near-perfect ball shape

Defence mechanism
Unlike other species of woodlice, pill woodlice have the extra defensive strategy of being able to roll into a ball – something that helps protect them from predators.

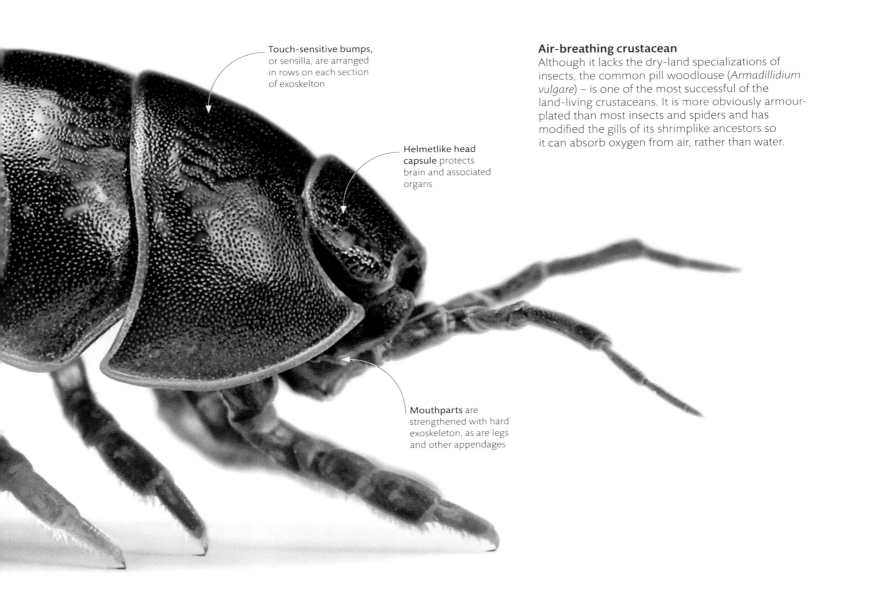

Touch-sensitive bumps, or sensilla, are arranged in rows on each section of exoskelton

Helmetlike head capsule protects brain and associated organs

Mouthparts are strengthened with hard exoskeleton, as are legs and other appendages

Air-breathing crustacean
Although it lacks the dry-land specializations of insects, the common pill woodlouse (*Armadillidium vulgare*) – is one of the most successful of the land-living crustaceans. It is more obviously armour-plated than most insects and spiders and has modified the gills of its shrimplike ancestors so it can absorb oxygen from air, rather than water.

Waving spines

Having a skeleton that is made up of plates joined into a test (shell) means reduced mobility. But in urchins such as the white-dotted long-spined urchin (*Diadema setosum*), muscles can move the spines to repel intruders, while beneath its body, suckerlike tube feet (see pp.212–13) pull the animal over the sea bed.

The base of each spine has a flexible connection with the rest of the skeleton, which allows the spine to move from side to side

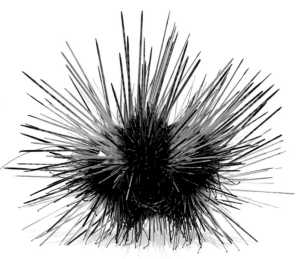

SIDE VIEW

chalky skeletons

Starfish and urchins belong to a group of animals called echinoderms – a word meaning "spiny skin". The name refers to their unique chalky skeleton, which is made from crystals of calcium carbonate that are packed loosely into the animal's rough skin (as in starfish), or joined together into a shell, or "test" (as in sea urchins). Despite their starlike radial symmetry, these animals are among the closest living relatives of backboned animals.

Long sucker-tipped tubes called tube feet reach beyond the spines and are used for attaching to the sea bed and crawling along

Hard-tipped defensive spines are hollow and brittle, releasing a mild venom when broken

PENTARADIAL BODY PLAN

Most echinoderms have pentaradial symmetry, meaning that their body is arranged in five parts around a central point. The symmetry is seen most clearly in the five arms of a typical starfish, but it occurs in the shell-like plates visible on a dead urchin that has lost its spines, as well as in related animals, such as brittle stars (see pp.212–13) and sea cucumbers.

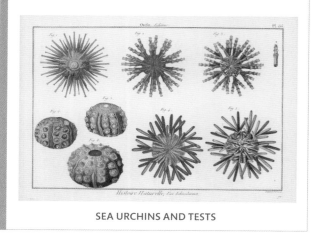

SEA URCHINS AND TESTS

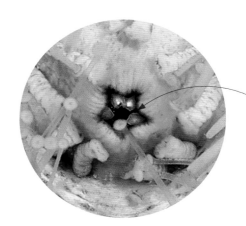

JAW APPARATUS

Five teeth protrude from the jaw, helping the sea urchin rasp algae from rocks

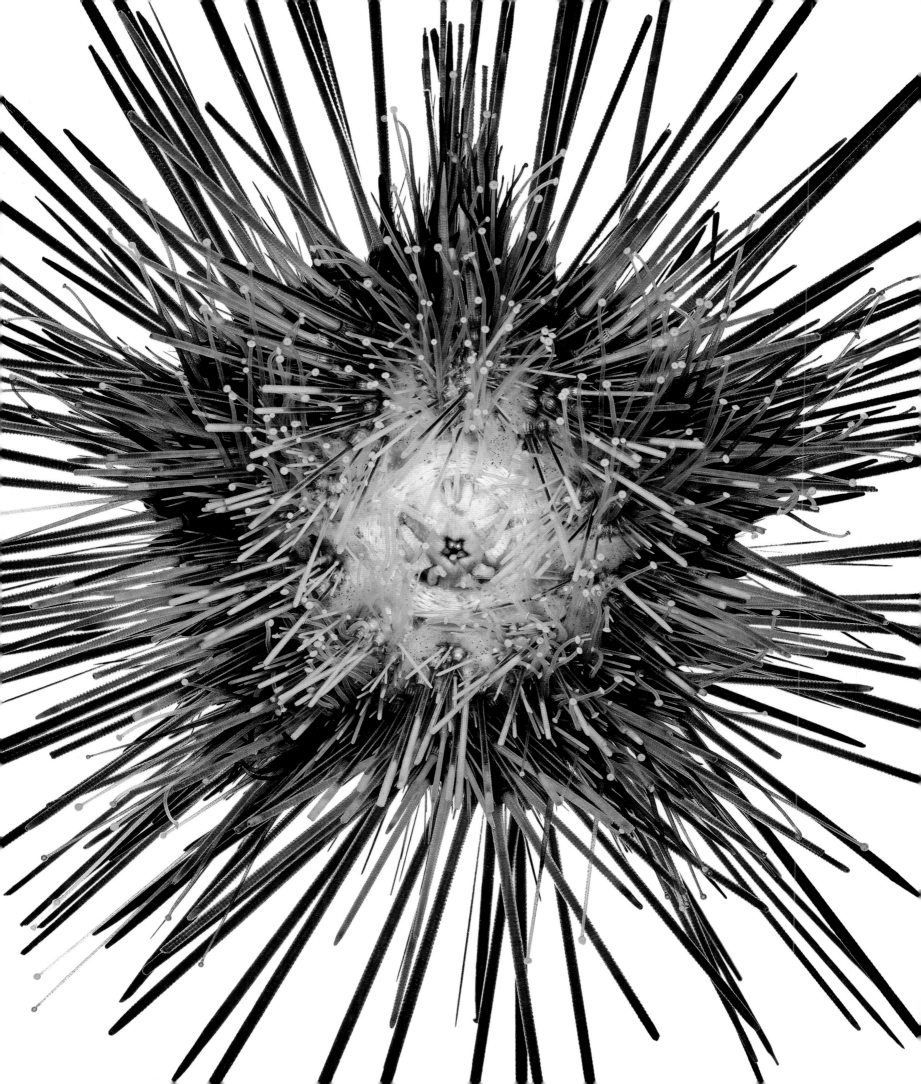

internal skeletons

Backboned animals – vertebrates – have a solid, jointed frame inside their body, surrounded by muscle. Unlike an invertebrate exoskeleton, this endoskeleton is built from within. It grows with the body, so it never needs to be moulted. This is possible because of the evolution of hard bone and flexible cartilage: living tissues that shape and reshape during development.

Dyed components
Like most vertebrates, this preserved skeleton of an adult clingfish (*Gobiesox* sp.) – a coastal fish named for its sucker-like pelvic fins – is mainly bone (here stained violet). But it also retains parts made from softer cartilage (blue) that exclusively supported the embryo.

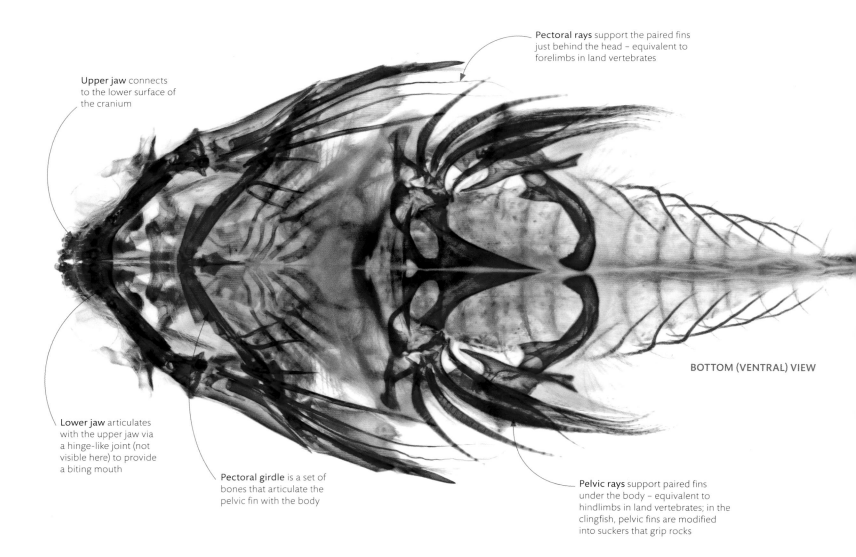

The vertebral column, a repeating sequence of bony elements called vertebrae, provides a flexible supporting axis for the body's muscles

Pectoral rays support the paired fins just behind the head – equivalent to forelimbs in land vertebrates

Upper jaw connects to the lower surface of the cranium

Lower jaw articulates with the upper jaw via a hinge-like joint (not visible here) to provide a biting mouth

Pectoral girdle is a set of bones that articulate the pelvic fin with the body

Pelvic rays support paired fins under the body – equivalent to hindlimbs in land vertebrates; in the clingfish, pelvic fins are modified into suckers that grip rocks

BOTTOM (VENTRAL) VIEW

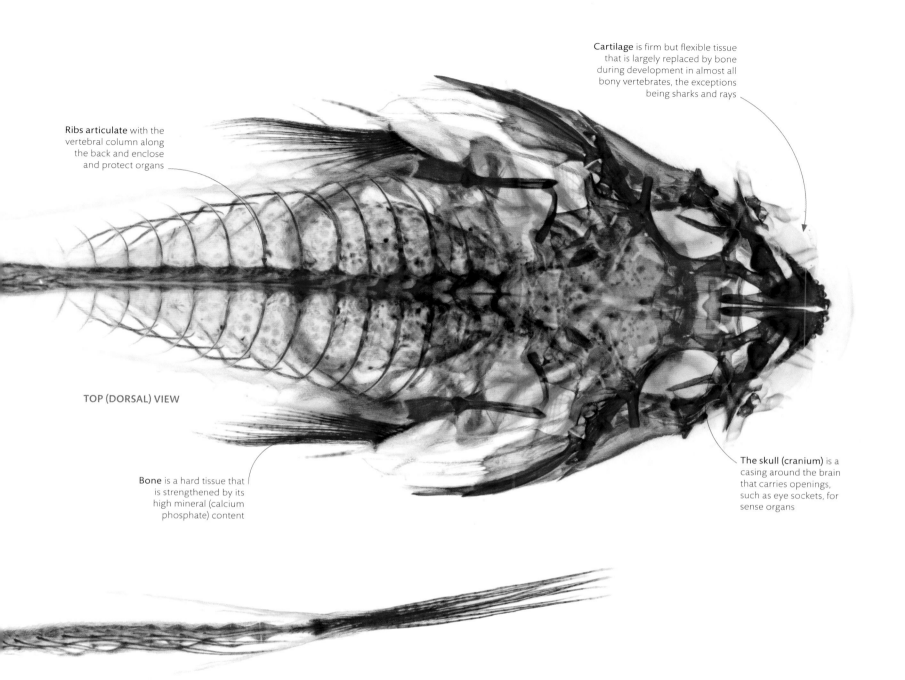

Cartilage is firm but flexible tissue that is largely replaced by bone during development in almost all bony vertebrates, the exceptions being sharks and rays

Ribs articulate with the vertebral column along the back and enclose and protect organs

TOP (DORSAL) VIEW

Bone is a hard tissue that is strengthened by its high mineral (calcium phosphate) content

The skull (cranium) is a casing around the brain that carries openings, such as eye sockets, for sense organs

VERTEBRATE SKELETON

The vertebral column, also called the backbone or spine, supports the length of a vertebrate's body and encloses the spinal cord, while the skull protects the brain. Together, these make up the axial skeleton. Articulating with the axial skeleton are parts that aid locomotion: the so-called appendicular skeleton. The earliest vertebrates – which were all fish – had fins, but they later evolved walking limbs.

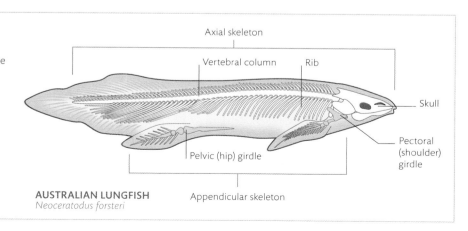

Axial skeleton

Vertebral column   Rib

Skull

Pectoral (shoulder) girdle

Pelvic (hip) girdle

Appendicular skeleton

AUSTRALIAN LUNGFISH
Neoceratodus forsteri

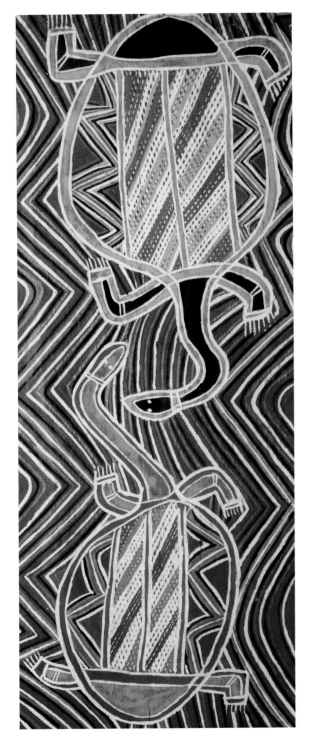

aboriginal insights

The painted walls of the rocky outcrops that sheltered Australian aboriginal clans for millennia tell the story of the native fish and animals these groups relied on for food. Many of these prehistoric artists used X-ray effects that reveal the inner working and structures of the animals they hunted and their complete understanding of every part of them.

The first human migrants are thought to have arrived on Australian soil more than 50,000 years ago, marking the beginning of the oldest continuous culture in history. The earliest records of their civilization include charcoal drawings (c.20,000 BCE) of now extinct animals such as the Tasmanian tiger, but it is in the more recent galleries of rock art in the Ubirr caves of Arnhem Land, Northern Australia, that the continuum of aboriginal life and belief is on full view.

Across the walls of the caves' interiors, the rich variety of animals and fish from the nearby East Alligator River and Nadab flood plain is painted in an X-ray style that dates back 8,000 years. Most were painted in the "freshwater period", spanning the past 2,000 years, and depict an abundance of fish, mussels, water fowl, wallabies, goannas, and echidnas.

Long-necked turtles

A bark painting of two long-necked turtles uses a variety of aboriginal techniques including traditional cross-hatching, known as "rarrk", which is believed to endow the turtles with spiritual power.

The artists' colours were derived from charcoal and ochre, a local hard clay found in red (the most enduring colour), pink, white, yellow, and, occasionally, blue. It was ground to a powder and mixed with egg, water, pollen, or animal fat or blood to make paint. The bones and organs, as familiar to the hunter as the living animal, were mapped out inside the lifelike outlines of each creature. Over the centuries, aboriginal art diversified across different regions of Australia with symbolic dot paintings prevalent in central and western desert regions; X-ray art in Northern Territory; and a style of fine cross-hatching particular to "rarrk" paintings in Arnhem Land. These are often rendered on the inside of a piece of bark. Artists use hairlike bristles from reeds or human hair for painstaking line-filling within the outlines of their animal portraits. They believe that the technique bestows spirituality on the subject.

The symbolic associations of Australia's animals sit at the heart of Dreamtime. This creation myth is based on the belief that the rivers, streams, land, hills, rocks, plants, animals, and people were created by spirits, who gave tools, land, totems, and dreaming to every clan.

Sacred rules for clan conduct, moral laws, and beliefs were passed down through storytelling, dance, painting, and song. In many cave galleries, ancestral art has been painted over across the centuries but the core messages remain the same.

> **Cave... he never move. No one can shift that cave, because it dream. It story, it law.**

BIG BILL NEIDJIE, BUNIDJ CLAN

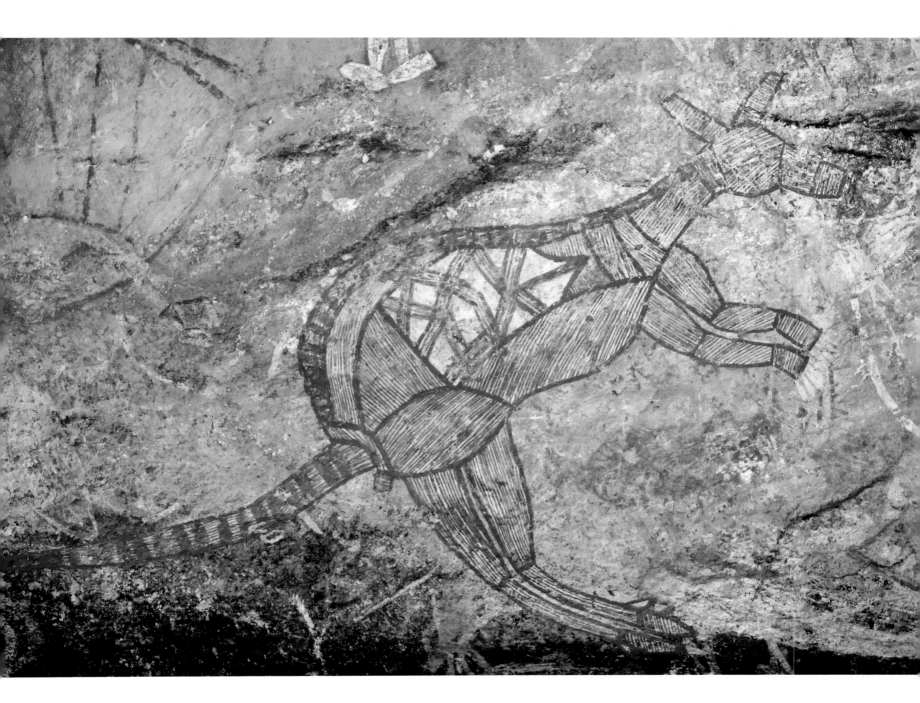

Rock art wallaby

A wallaby with its tailbones and spine visible decorates the cave walls at Ubirr, Kakadu, Northern Territory, one of the oldest continuously inhabited shelters. Animal food sources were painted here for thousands of years.

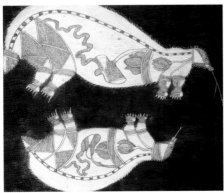

Inside-out anteaters

Two anteaters, painted on the inside of a strip of bark, offer a lesson in anatomy. Heart, stomach, and intestines are rendered in detail in a 20th-century X-ray style, cross-hatched painting from Western Arnhem Land, Australia.

Skeletal adaptations

As part of their internal skeleton, some vertebrates have developed unique extra structures, such as protective helmets (casques), horns, and plates. In other animals the bone shape itself has been extended or altered.

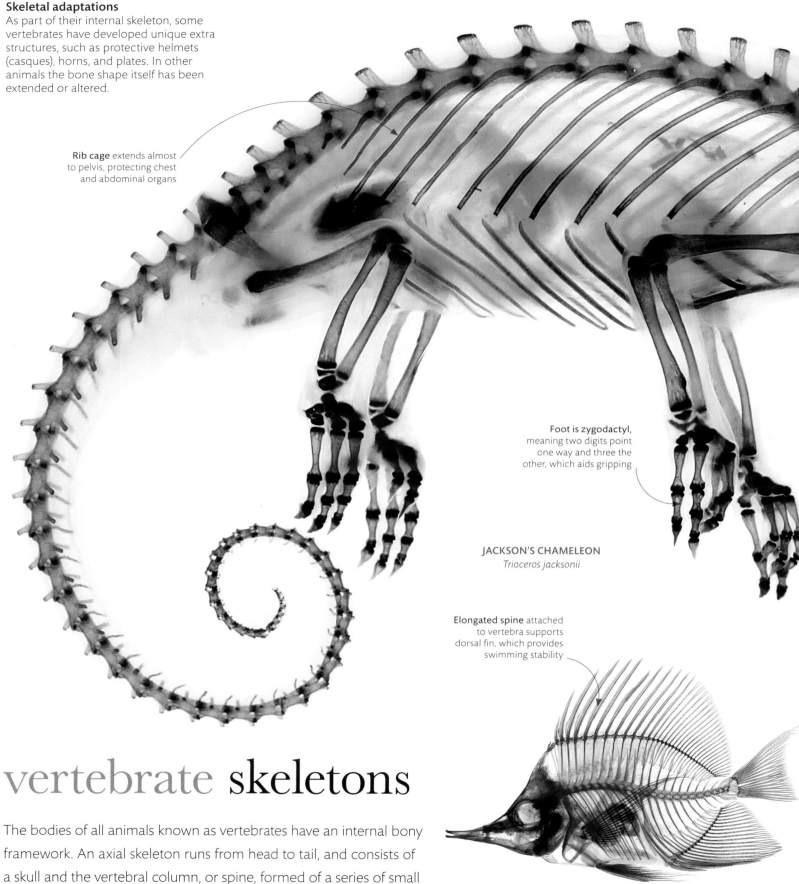

Rib cage extends almost to pelvis, protecting chest and abdominal organs

Foot is zygodactyl, meaning two digits point one way and three the other, which aids gripping

JACKSON'S CHAMELEON
Trioceros jacksonii

Elongated spine attached to vertebra supports dorsal fin, which provides swimming stability

YELLOW LONGNOSE BUTTERFLYFISH
Forcipiger flavissimus

vertebrate skeletons

The bodies of all animals known as vertebrates have an internal bony framework. An axial skeleton runs from head to tail, and consists of a skull and the vertebral column, or spine, formed of a series of small vertebrae. Attached to this is the appendicular skeleton that supports limbs in four-legged animals, fins in fish, and legs and wings in birds.

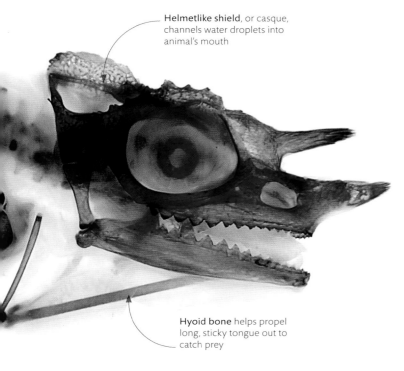

Helmetlike shield, or casque, channels water droplets into animal's mouth

Hyoid bone helps propel long, sticky tongue out to catch prey

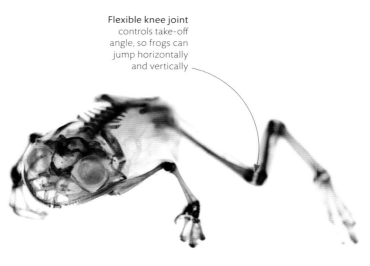

Flexible knee joint controls take-off angle, so frogs can jump horizontally and vertically

JAPANESE TREE FROG
Hyla japonica

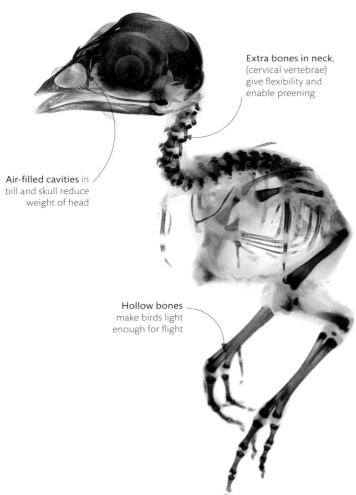

Extra bones in neck, (cervical vertebrae) give flexibility and enable preening

Air-filled cavities in bill and skull reduce weight of head

Hollow bones make birds light enough for flight

JAPANESE QUAIL
Coturnix japonica

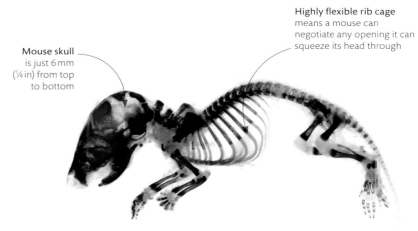

Highly flexible rib cage means a mouse can negotiate any opening it can squeeze its head through

Mouse skull is just 6 mm (¼ in) from top to bottom

HOUSE MOUSE
Mus musculus

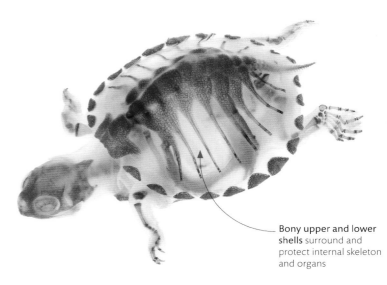

Bony upper and lower shells surround and protect internal skeleton and organs

JAPANESE POND TURTLE
Mauremys japonica

Shell from bone
Like other chelonians, the upper shell (carapace) and lower shell (plastron) of an Indian star tortoise (*Geochelone elegans*) are made up of interlocking plates that correspond with the dermal bones that grow within the skin. Above this is a hornlike layer of keratin, impregnated with pigments.

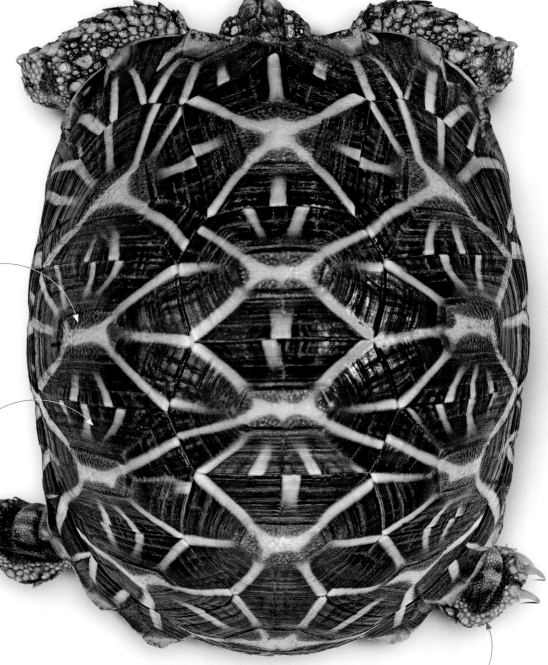

Each scute –
a hornlike layer covering
a plate of bone – sports
a starlike pattern

Carapace (upper shell)
has a high dome, as in
other terrestrial tortoises

Thick, elephantine
limbs carry claws that
help grip the ground

vertebrate shells

The shells of chelonians (turtles and tortoises) provide a unique form of protection among backboned animals: much of their body is enveloped in bone. This provides impressive protection against predators but compromises mobility because of its weight and rigidity. The necks of chelonians have to be longer and more flexible than in other reptiles to reach out for food, while extra-strong limb muscles provide propulsion on land or in water.

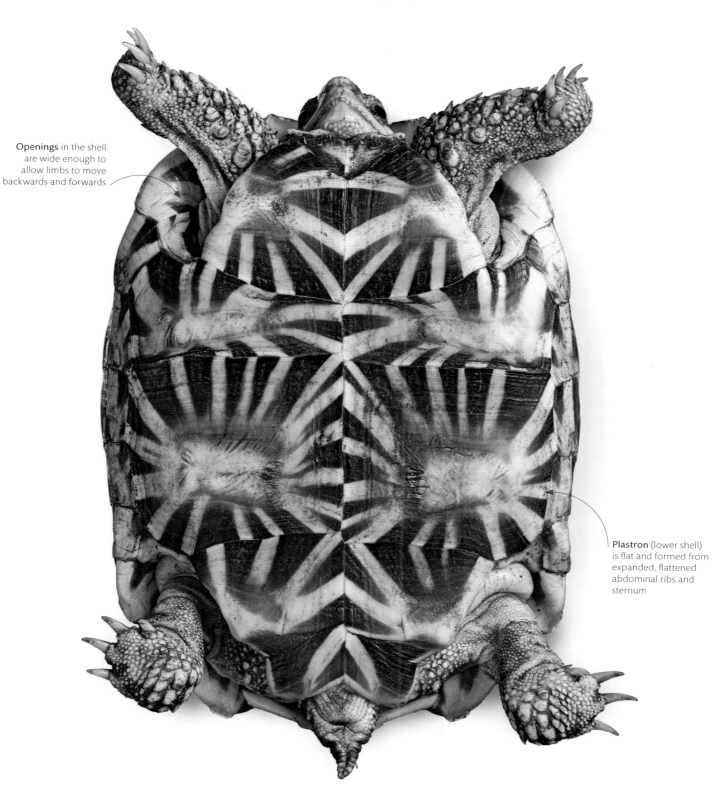

Openings in the shell are wide enough to allow limbs to move backwards and forwards

Plastron (lower shell) is flat and formed from expanded, flattened abdominal ribs and sternum

BODY ARMOUR

With spine and ribs fused to the shell – and shoulder and pelvic bones uniquely within the ribcage – chelonians enjoy almost total body protection; many can even retract their limbs and head inside their shell. However, conventional breathing by moving the ribcage is impossible. Instead, the shoulder girdle swings back and forth to ventilate the lungs.

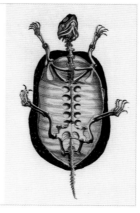

**TORTOISE SKELETON
(PLASTRON REMOVED)**

Neck bends sideways so that the head rests under the edge of the shell

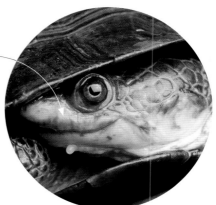

Retracting the neck

Some chelonians pull their head back into the shell by curving the neck into a S-shape; others, such as this gibba turtle (*Mesoclemmys gibba*), fold their head and neck to the side.

bird skeletons

Despite the wealth of species – more than 10,000 – the body plan of most birds is remarkably similar, largely due to the physical constraints of flight. They have two legs used for walking and perching, their forelimbs are fashioned as wings, and some bones of the spine, collar, and pelvis are fused to take the stresses of jumping and landing. Most of the bones are pneumatic, or air-filled, which minimizes weight (see box), so less energy is spent on becoming airborne and parents can sit on their eggs without breaking them.

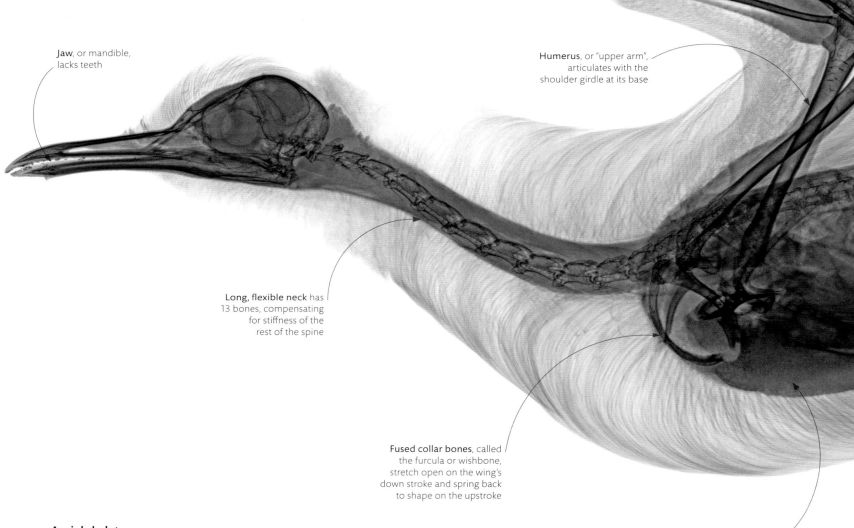

Jaw, or mandible, lacks teeth

Humerus, or "upper arm", articulates with the shoulder girdle at its base

Long, flexible neck has 13 bones, compensating for stiffness of the rest of the spine

Fused collar bones, called the furcula or wishbone, stretch open on the wing's down stroke and spring back to shape on the upstroke

Large keel bone, or carina, of the breastbone, serves as an attachment point for flight muscles

Aerial skeleton
This taxidermied specimen of a Mediterranean gull (*Larus melanocephalus*) shows how wings pivot around the bones of the pectoral (shoulder) girdle. Compared with that of their reptilian ancestors, the body skeleton of a bird is shorter and more compact, with centre of gravity shifted over the hind limbs.

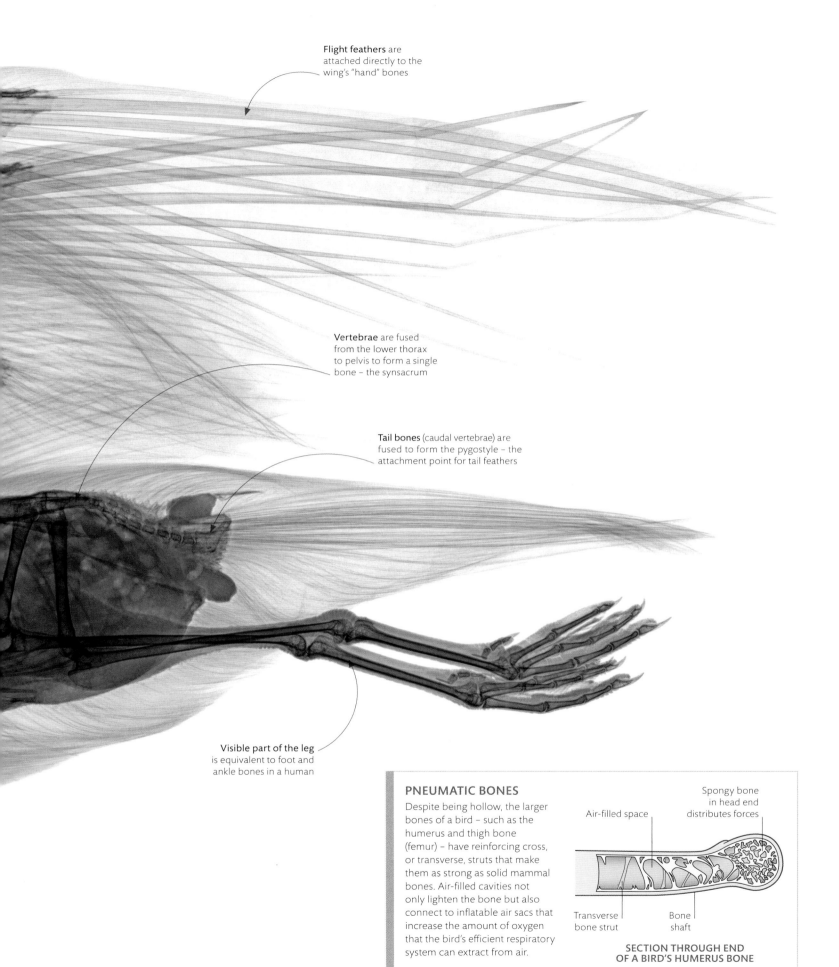

Flight feathers are attached directly to the wing's "hand" bones

Vertebrae are fused from the lower thorax to pelvis to form a single bone – the synsacrum

Tail bones (caudal vertebrae) are fused to form the pygostyle – the attachment point for tail feathers

Visible part of the leg is equivalent to foot and ankle bones in a human

PNEUMATIC BONES

Despite being hollow, the larger bones of a bird – such as the humerus and thigh bone (femur) – have reinforcing cross, or transverse, struts that make them as strong as solid mammal bones. Air-filled cavities not only lighten the bone but also connect to inflatable air sacs that increase the amount of oxygen that the bird's efficient respiratory system can extract from air.

Air-filled space

Spongy bone in head end distributes forces

Transverse bone strut

Bone shaft

SECTION THROUGH END OF A BIRD'S HUMERUS BONE

cheetah

Many carnivores rely on ambush or the cooperaton of the pack, but the cheetah (*Acinonyx jubatus*) uses sudden bursts of speed to catch its prey. With huge, bounding strides facilitated by its flexible spine, the cheetah can reach 102 kph (63 mph), making this big cat the fastest animal on four legs.

Native to Africa and a few locations in Iran, cheetahs inhabit a wide range of habitats, from dry forest and scrubland to grasslands and even desert. They typically hunt small antelopes, such as Thomson's gazelle (*Eudorcas thomsonii*): animals that, having co-evolved with hunters, are also fast – and vigilant, too. A herd is quick to spot a predator and will keep it in view to deny it the element of surprise; a cheetah must strike at the right moment or waste precious energy on a fruitless pursuit.

Relying on stealth and camouflage, the cheetah approaches slowly, crouching low and freezing mid-stride if seen. When it gets near enough to its prey – ideally less than 50 m (165 ft) away – it explodes into a sprint. Within seconds, the cheetah has reached 60 kph (37 mph) and more than doubled its breathing rate as it homes in on the bolting gazelle. As the cheetah draws level, it swipes its paw and hooks the quarry off balance with its dew claw.

Explosive speed is no guarantee of a meal. Sprinting for prey is an energy-sapping hunting strategy, and the cheetah tires sooner than the gazelle. Should the target not be felled within 300 m (985 ft), the cheetah will give up the chase.

If the hunt is a success, the cheetah suffocates the victim by clamping its jaws over the animal's throat, panting through enlarged nostrils as it recovers. Even then, a meal is not certain – many cheetahs lose their kills to leopards, lions, or hyenas (the same predators that take a high toll on cheetah cubs). Dragging the carcass to cover is a priority, as is eating the meat as rapidly as possible: a cheetah can devour up to 14 kg (30 lb) of flesh in one sitting.

Closing in for the kill
Typically, half of all sprints result in a kill, but by targeting fawns, such as this Thomson's gazelle, the success rate can reach 100 per cent.

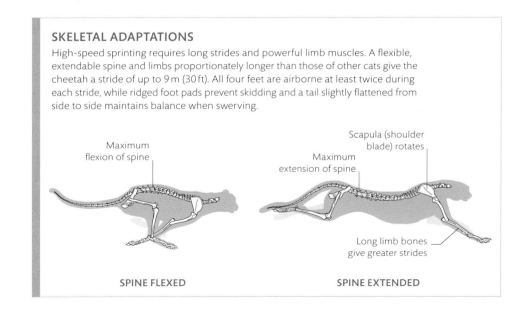

SKELETAL ADAPTATIONS
High-speed sprinting requires long strides and powerful limb muscles. A flexible, extendable spine and limbs proportionately longer than those of other cats give the cheetah a stride of up to 9 m (30 ft). All four feet are airborne at least twice during each stride, while ridged foot pads prevent skidding and a tail slightly flattened from side to side maintains balance when swerving.

Maximum flexion of spine

Scapula (shoulder blade) rotates

Maximum extension of spine

Long limb bones give greater strides

SPINE FLEXED

SPINE EXTENDED

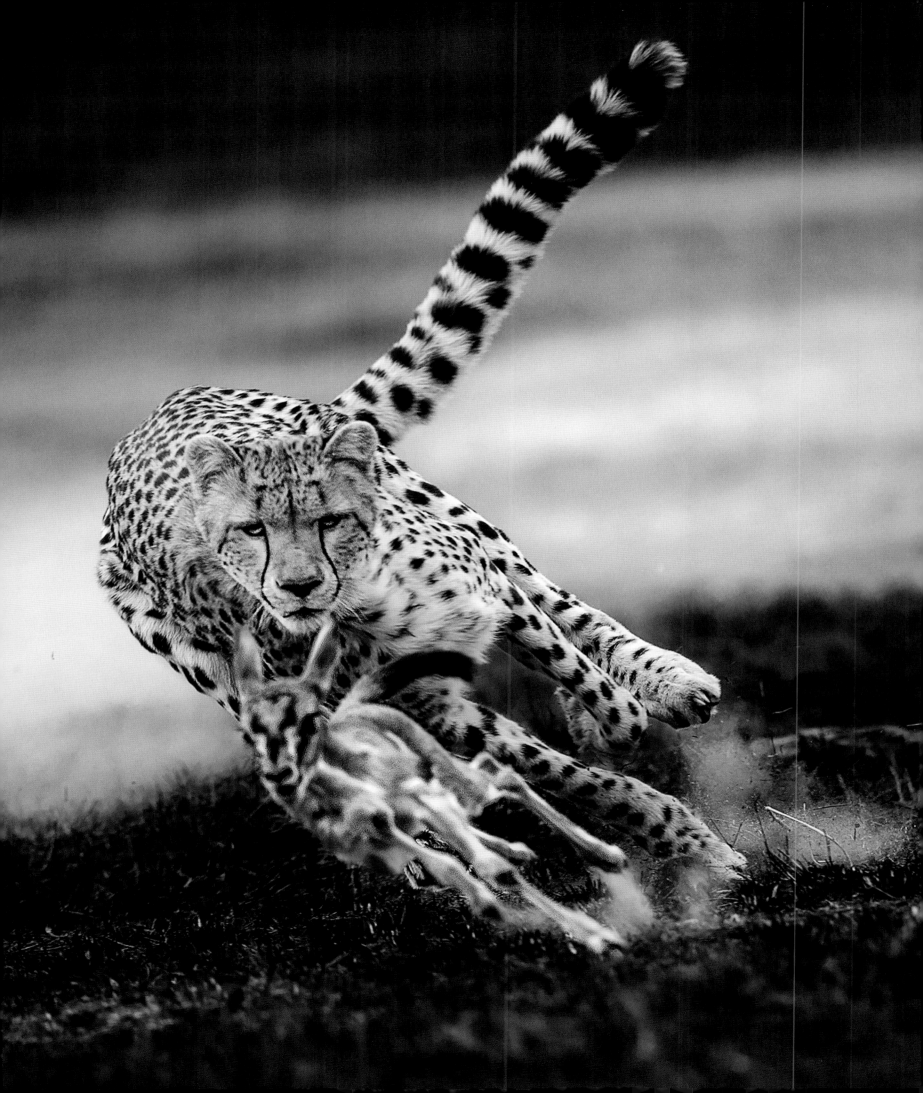

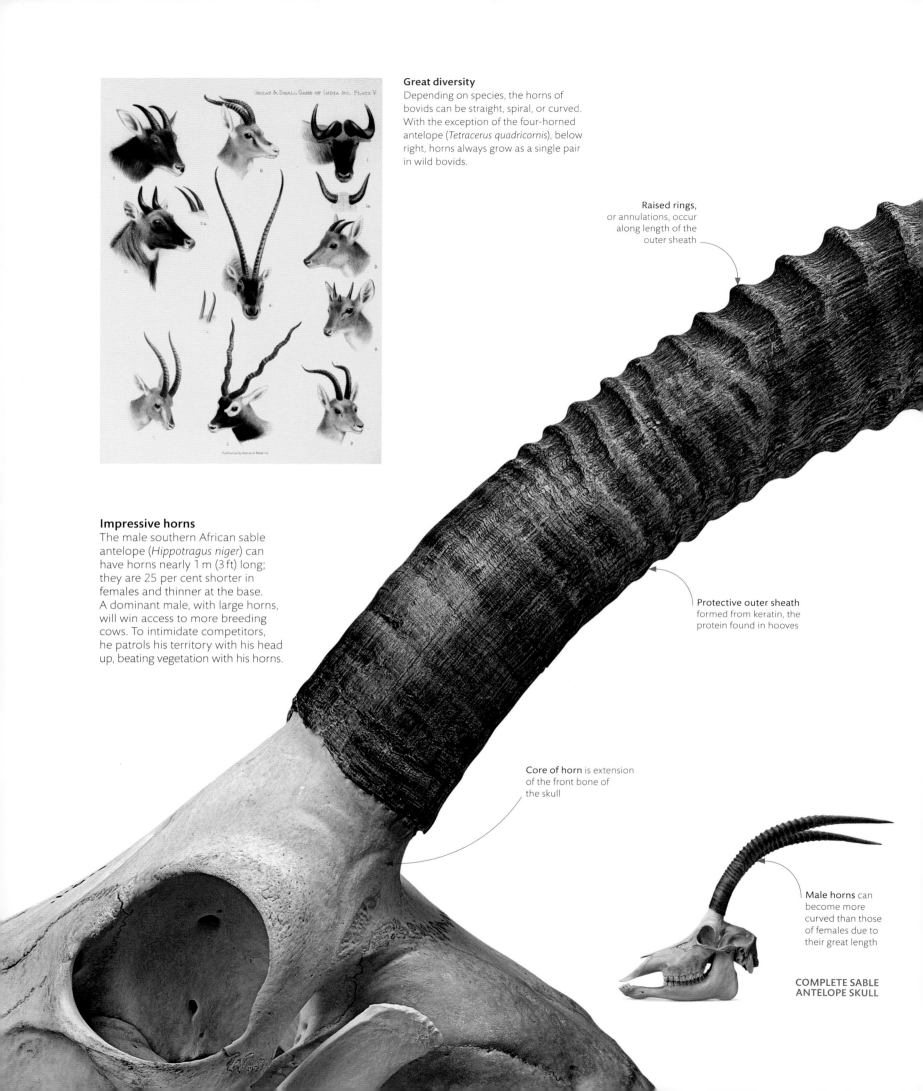

Great diversity

Depending on species, the horns of bovids can be straight, spiral, or curved. With the exception of the four-horned antelope (*Tetracerus quadricornis*), below right, horns always grow as a single pair in wild bovids.

Impressive horns

The male southern African sable antelope (*Hippotragus niger*) can have horns nearly 1 m (3 ft) long; they are 25 per cent shorter in females and thinner at the base. A dominant male, with large horns, will win access to more breeding cows. To intimidate competitors, he patrols his territory with his head up, beating vegetation with his horns.

Raised rings, or annulations, occur along length of the outer sheath

Protective outer sheath formed from keratin, the protein found in hooves

Core of horn is extension of the front bone of the skull

Male horns can become more curved than those of females due to their great length

COMPLETE SABLE ANTELOPE SKULL

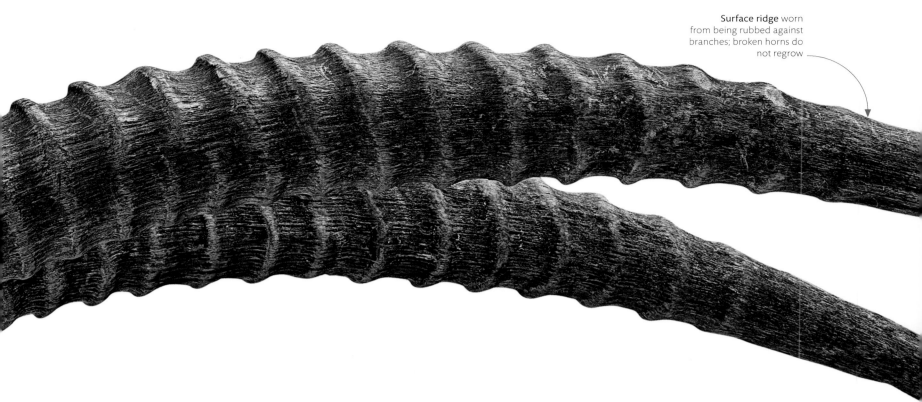

Surface ridge worn from being rubbed against branches; broken horns do not regrow

mammal horns

Many animals have projections on their body – from the Amazonian purple scarab beetle (see p.115) to the hornlike scales of some lizards – but only hoofed mammals, such as antelope, have true horns that are bony extensions of their skulls. Typically bigger in males, and often lacking in females, horns – unlike the branching antlers of male deer (see pp.88–89) – are permanent, unbranched fixtures, used in battles to establish male dominance or as defence against predators.

HORNS AND ANTLERS

True horns are produced by bovids, or Bovidae (which includes cattle, antelope, and goats). Only deer of the Cervidae family have antlers. New antlers are grown each year, nourished by a layer of skin called velvet, and shed at the end of the breeding season. In contrast, horns grow continuously throughout an animal's life and the outer layer has a dried sheath of horny keratin.

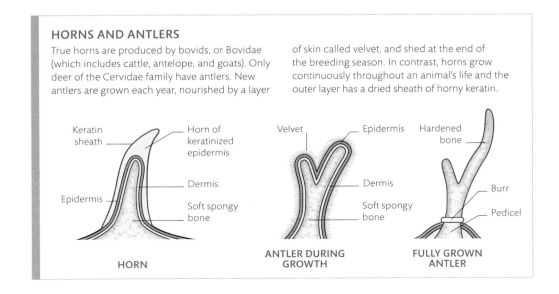

HORN

ANTLER DURING GROWTH

FULLY GROWN ANTLER

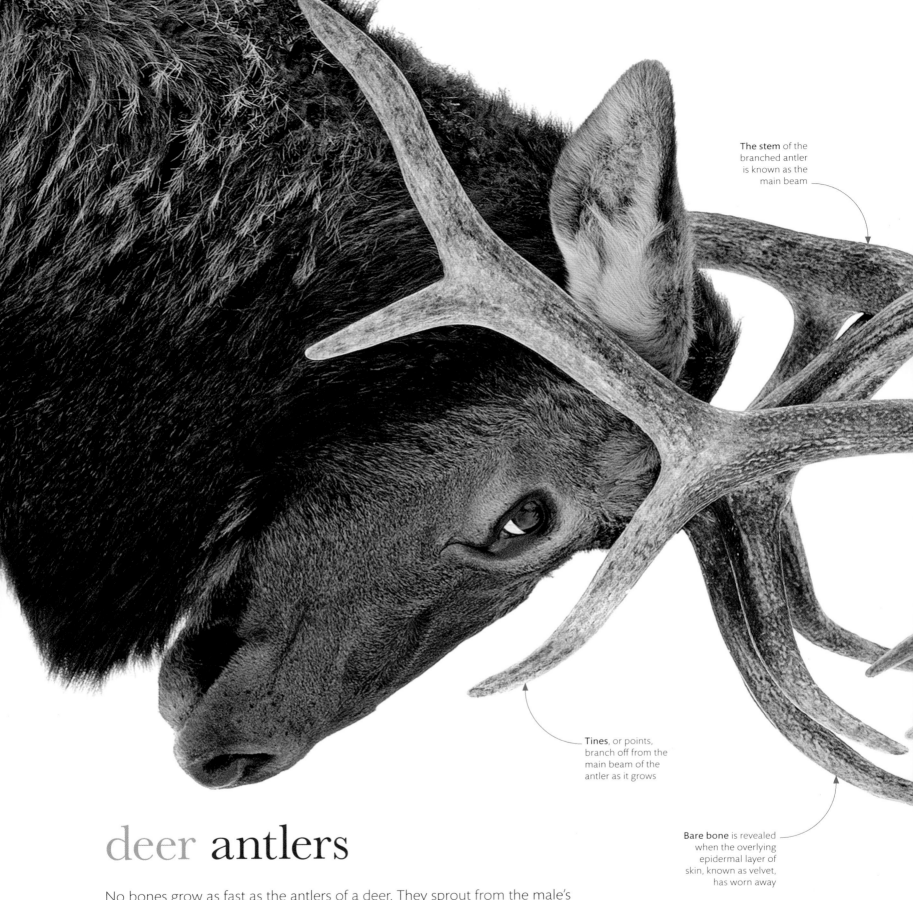

The stem of the branched antler is known as the main beam

Tines, or points, branch off from the main beam of the antler as it grows

Bare bone is revealed when the overlying epidermal layer of skin, known as velvet, has worn away

deer antlers

No bones grow as fast as the antlers of a deer. They sprout from the male's skull, and over several months they branch into a heavy display used in combat over females. Antlers are fed by a rich blood supply within a layer of skin that later shrivels away, exposing the bone. When breeding is over, the antlers fall off – so new ones must grow the next year.

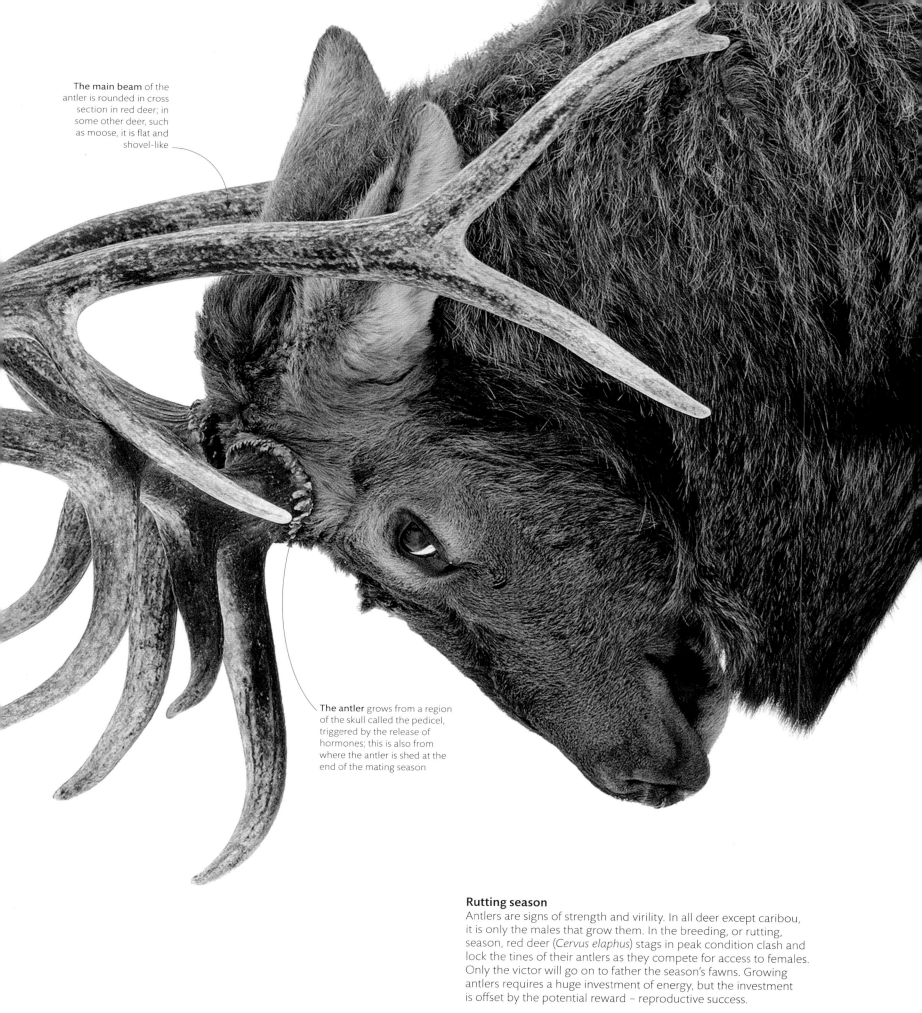

The main beam of the antler is rounded in cross section in red deer; in some other deer, such as moose, it is flat and shovel-like

The antler grows from a region of the skull called the pedicel, triggered by the release of hormones; this is also from where the antler is shed at the end of the mating season

Rutting season

Antlers are signs of strength and virility. In all deer except caribou, it is only the males that grow them. In the breeding, or rutting, season, red deer (*Cervus elaphus*) stags in peak condition clash and lock the tines of their antlers as they compete for access to females. Only the victor will go on to father the season's fawns. Growing antlers requires a huge investment of energy, but the investment is offset by the potential reward – reproductive success.

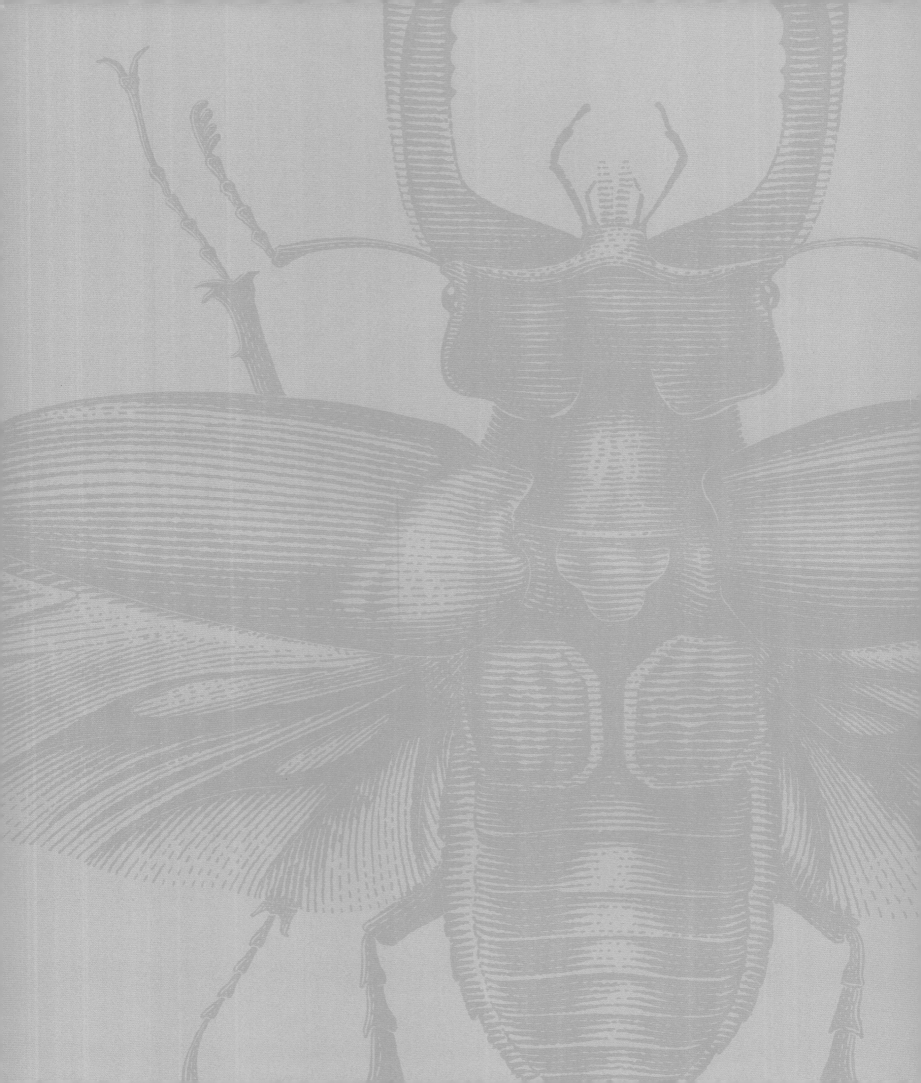

skin, coats, and armour

skin. a thin layer of tissue forming the outer covering of the body, often consisting of two layers: the dermis and epidermis.

coat. the natural covering of an animal, such as fur, feathers, scales, or a test.

armour. a strong, defensive covering that protects the body from injury.

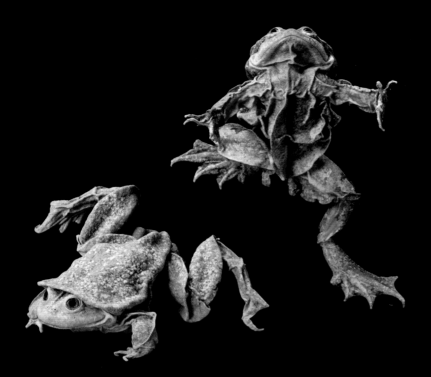

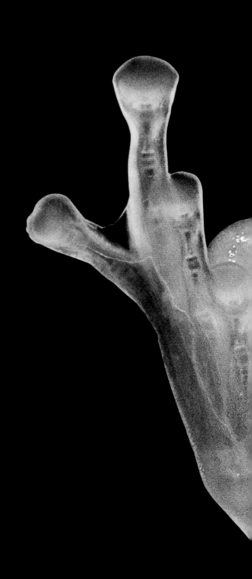

Transparent skin
The two ways in which amphibians gather oxygen are strikingly evident in the body of a reticulated glass frog (*Hyalinobatrachium valerioi*). The frog's transparent, permeable skin accounts for much of its oxygen intake, while the rest of the oxygen it needs is drawn into the blood via the frog's lungs.

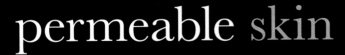

Underwater frog
At lower temperatures, water carries more oxygen. High in the Andes Mountains of South America, the loose-folded skin of the Titicaca water frog (*Telmatobius culeus*) helps it stay in the lake's cool depths by maximizing oxygen uptake without recourse to lungs.

permeable skin

Skin is a protective barrier that separates the delicate living tissues inside the body from the harsh, changeable environment outside. It seals the body from infection and can self-repair when injured. But a completely airtight seal is not wholly beneficial. At least a small quantity of oxygen usually seeps into the surface directly from water or air outside, but for many animals this so called cutaneous gaseous exchange is critical. Amphibians may get more than 50 per cent of their oxygen this way – necessitating a skin that is sufficiently permeable to let oxygen through unimpeded.

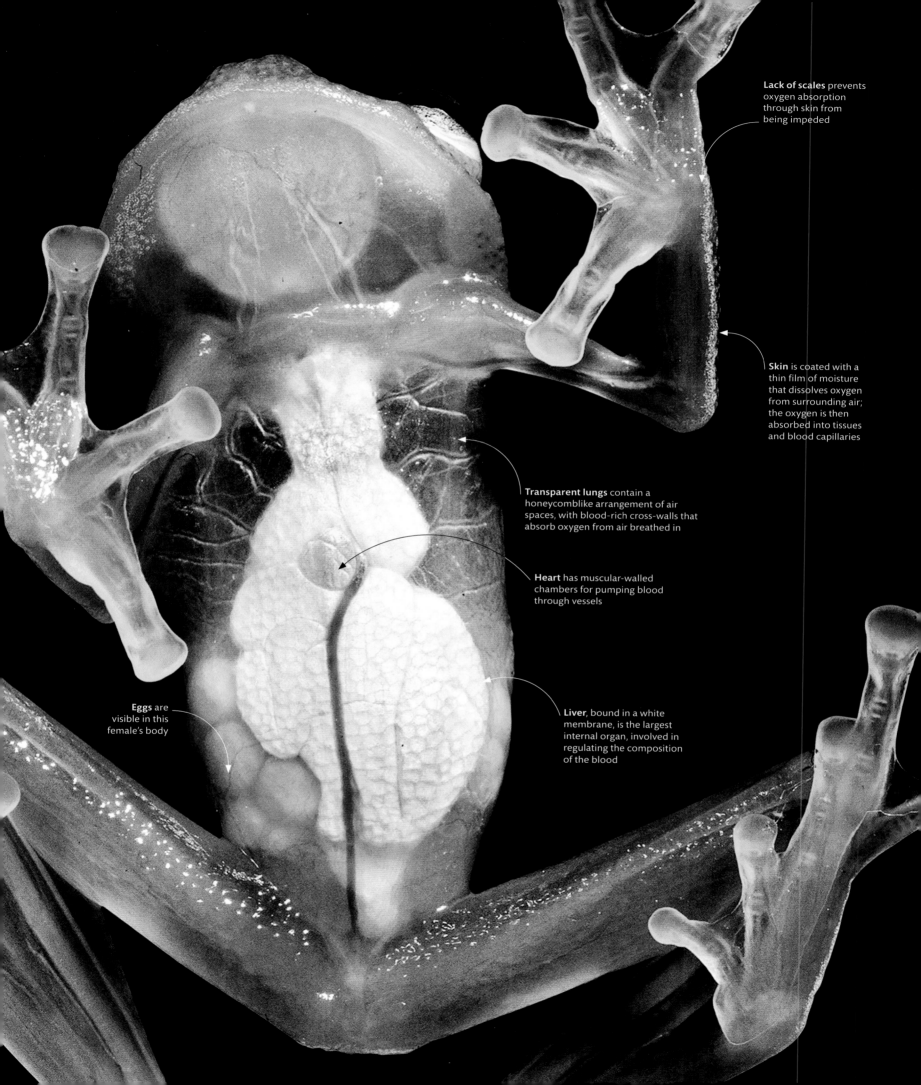

Lack of scales prevents oxygen absorption through skin from being impeded

Skin is coated with a thin film of moisture that dissolves oxygen from surrounding air; the oxygen is then absorbed into tissues and blood capillaries

Transparent lungs contain a honeycomblike arrangement of air spaces, with blood-rich cross-walls that absorb oxygen from air breathed in

Heart has muscular-walled chambers for pumping blood through vessels

Liver, bound in a white membrane, is the largest internal organ, involved in regulating the composition of the blood

Eggs are visible in this female's body

getting oxygen

The chemical processes of respiration release energy from
nutrients, and almost all animals use oxygen to help this happen.
Oxygen is taken into the body from the surroundings, and waste
carbon dioxide escapes out. This happens most effectively across
a large surface area with a thin wall. The simplest route is through
the skin, but that alone usually only satisfies the smallest animals.
Special respiratory organs – gills and lungs – make
gaseous exchange more efficient for larger bodies.

Permeable skin
helps the animal
absorb some
oxygen across its
entire body surface

Spots of pigment help
to camouflage the slug
on the sea bed

RESPIRATORY ORGANS

Gills are extensions of the body that absorb oxygen
under water; lungs are air-filled pockets used for
breathing on land. Both have a lining of thin epithelium
with a large surface area overlying a rich blood supply,
maximizing oxygen uptake and carbon dioxide removal.

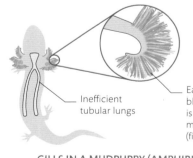

**Inefficient
tubular lungs**

**Each efficient,
blood-rich gill
is made up of
many lamellae
(filaments)**

GILLS IN A MUDPUPPY (AMPHIBIAN)

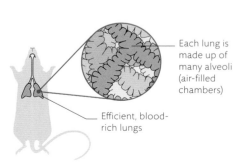

**Each lung is
made up of
many alveoli
(air-filled
chambers)**

**Efficient, blood-
rich lungs**

LUNGS IN A RAT (MAMMAL)

Cerata are projections
in some sea slugs that
contain defensive stinging
cells, as well as helping
gaseous exchange

SPANISH SHAWL
Flabellinopsis iodinea

Double plume of gills
helps sea slug gather
more oxygen

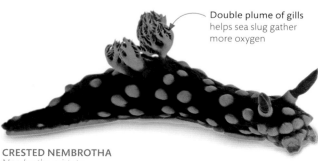

CRESTED NEMBROTHA
Nembrotha cristata

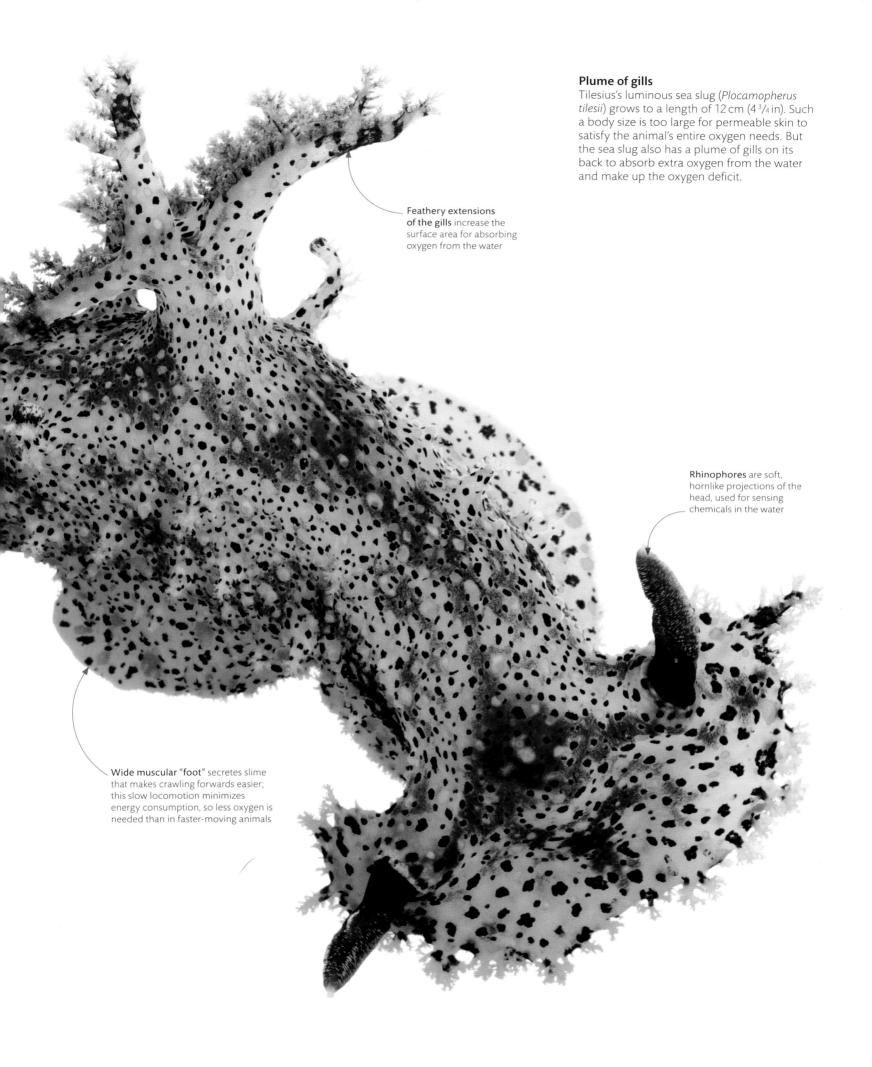

Plume of gills
Tilesius's luminous sea slug (*Plocamopherus tilesii*) grows to a length of 12 cm (4 ³/₄ in). Such a body size is too large for permeable skin to satisfy the animal's entire oxygen needs. But the sea slug also has a plume of gills on its back to absorb extra oxygen from the water and make up the oxygen deficit.

Feathery extensions of the gills increase the surface area for absorbing oxygen from the water

Rhinophores are soft, hornlike projections of the head, used for sensing chemicals in the water

Wide muscular "foot" secretes slime that makes crawling forwards easier; this slow locomotion minimizes energy consumption, so less oxygen is needed than in faster-moving animals

POISON FROGS

South American poison frogs (family Dendrobatidae) have bright colours that warn predators of their especially potent poisons. A few of the 100 species are used by Emberá and Noanamá Indians to poison their blow-dart tips in traditional hunting.

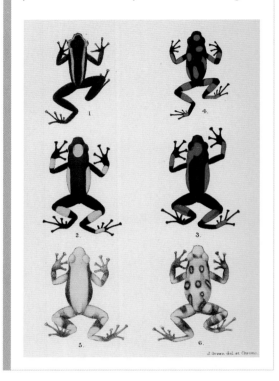

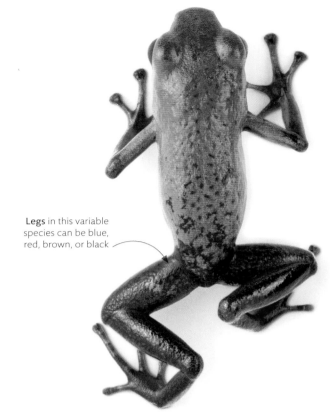

Legs in this variable species can be blue, red, brown, or black

Poison from diet
Like other South American poisonous species, left, this strawberry poison frog (*Oophaga pumilio*) acquires its toxins by eating toxic arthropods, such as ants.

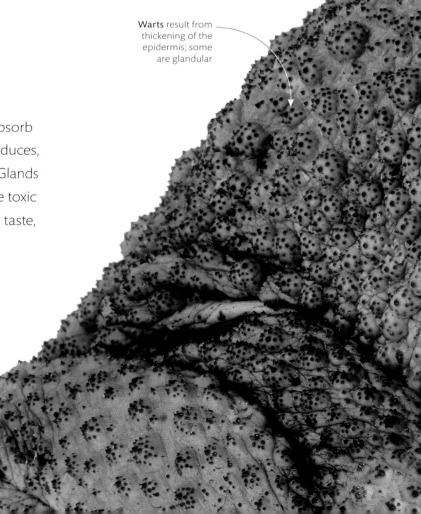

Warts result from thickening of the epidermis; some are glandular

toxic skin

Amphibian skin is thin and lacks scales, because it needs to absorb oxygen across its surface. The skin relies on the poisons it produces, some of which can be devastatingly effective, for protection. Glands in the skin – sometimes swollen into wartlike growths – release toxic fluid, which at the very least deters predators through its bitter taste, but in some species it is fast-acting and lethal.

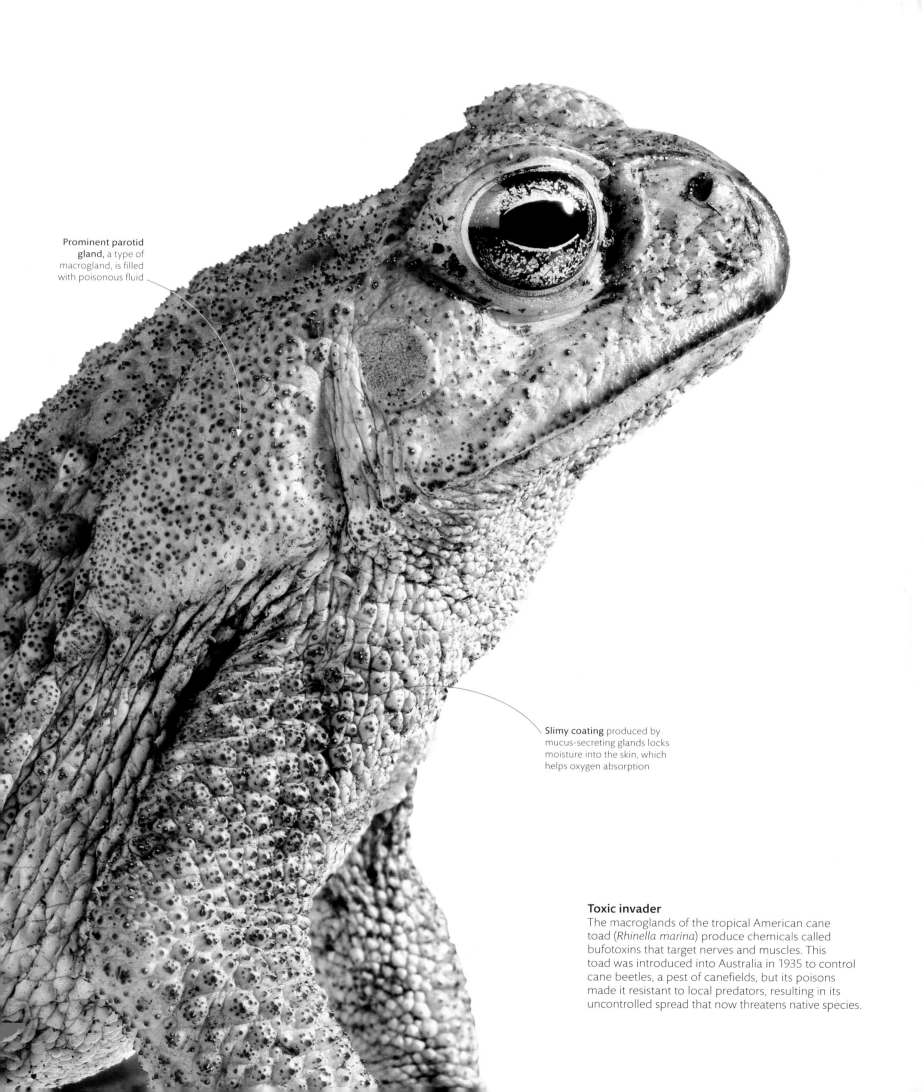

Prominent parotid gland, a type of macrogland, is filled with poisonous fluid

Slimy coating produced by mucus-secreting glands locks moisture into the skin, which helps oxygen absorption

Toxic invader
The macroglands of the tropical American cane toad (*Rhinella marina*) produce chemicals called bufotoxins that target nerves and muscles. This toad was introduced into Australia in 1935 to control cane beetles, a pest of canefields, but its poisons made it resistant to local predators, resulting in its uncontrolled spread that now threatens native species.

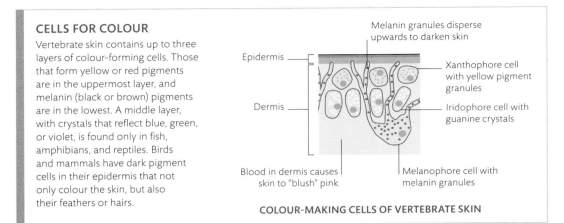

CELLS FOR COLOUR

Vertebrate skin contains up to three layers of colour-forming cells. Those that form yellow or red pigments are in the uppermost layer, and melanin (black or brown) pigments are in the lowest. A middle layer, with crystals that reflect blue, green, or violet, is found only in fish, amphibians, and reptiles. Birds and mammals have dark pigment cells in their epidermis that not only colour the skin, but also their feathers or hairs.

Melanin granules disperse upwards to darken skin

Epidermis

Dermis

Xanthophore cell with yellow pigment granules

Iridophore cell with guanine crystals

Blood in dermis causes skin to "blush" pink

Melanophore cell with melanin granules

COLOUR-MAKING CELLS OF VERTEBRATE SKIN

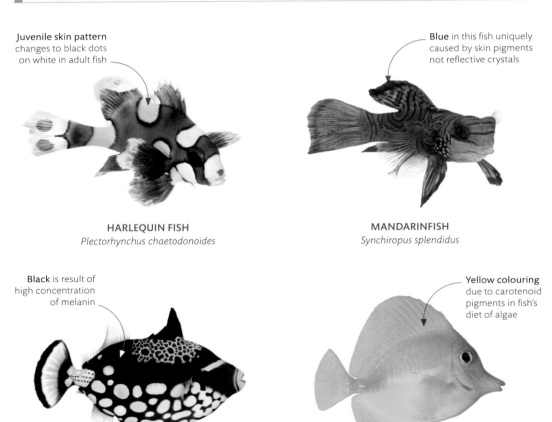

Juvenile skin pattern changes to black dots on white in adult fish

Blue in this fish uniquely caused by skin pigments not reflective crystals

HARLEQUIN FISH
Plectorhynchus chaetodonoides

MANDARINFISH
Synchiropus splendidus

Black is result of high concentration of melanin

Yellow colouring due to carotenoid pigments in fish's diet of algae

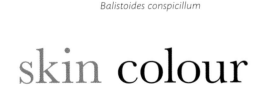

YOUNG CLOWN TRIGGERFISH
Balistoides conspicillum

YELLOW TANG
Zebrasoma flavescens

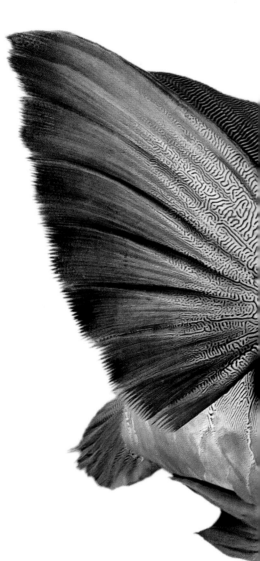

skin colour

Many of the outstanding colours that adorn the bodies of animals come from pigments generated by chemical processes inside skin cells – the biological equivalents of paints and dyes. Brown and black are due to pigments called melanins; yellow, orange, and red come from carotenoids – the same pigments that colour carrots, daffodils, and egg yolks. But greens, blues, and violets usually arise because of the way skin, scales, or feathers bend and reflect light at the body surface.

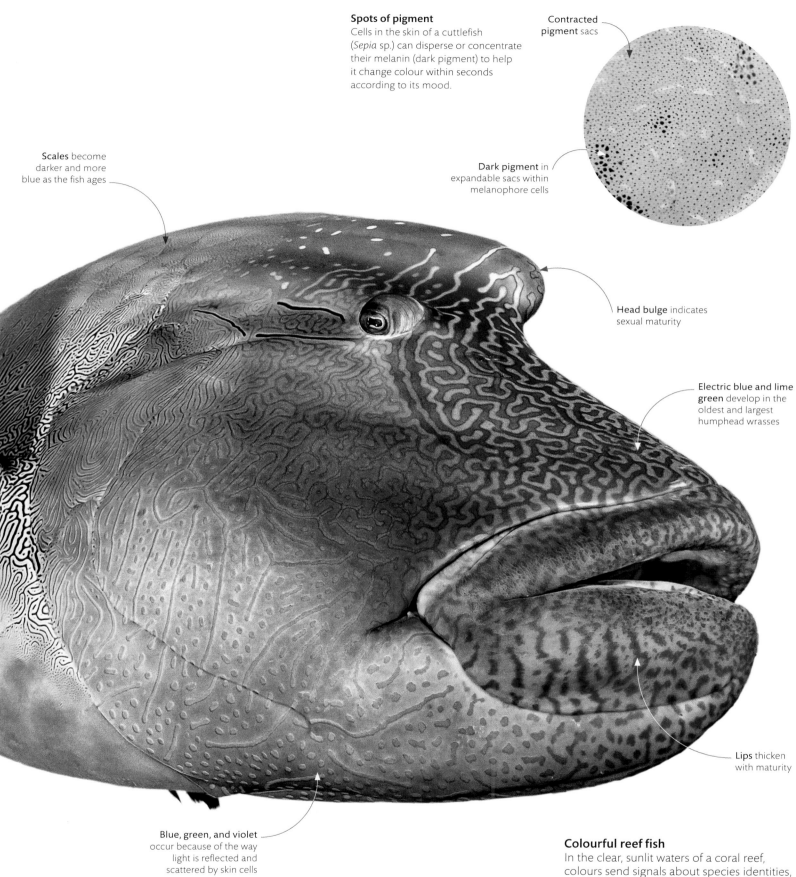

Spots of pigment
Cells in the skin of a cuttlefish
(*Sepia* sp.) can disperse or concentrate
their melanin (dark pigment) to help
it change colour within seconds
according to its mood.

Contracted
pigment sacs

Dark pigment in
expandable sacs within
melanophore cells

Scales become
darker and more
blue as the fish ages

Head bulge indicates
sexual maturity

**Electric blue and lime
green** develop in the
oldest and largest
humphead wrasses

Lips thicken
with maturity

Blue, green, and violet
occur because of the way
light is reflected and
scattered by skin cells

Colourful reef fish
In the clear, sunlit waters of a coral reef,
colours send signals about species identities,
as well as whether individuals are old enough
to reproduce. Blue predominates in many
species, including this humphead wrasse
(*Chelinus undulatus*), as blue light travels
further in water.

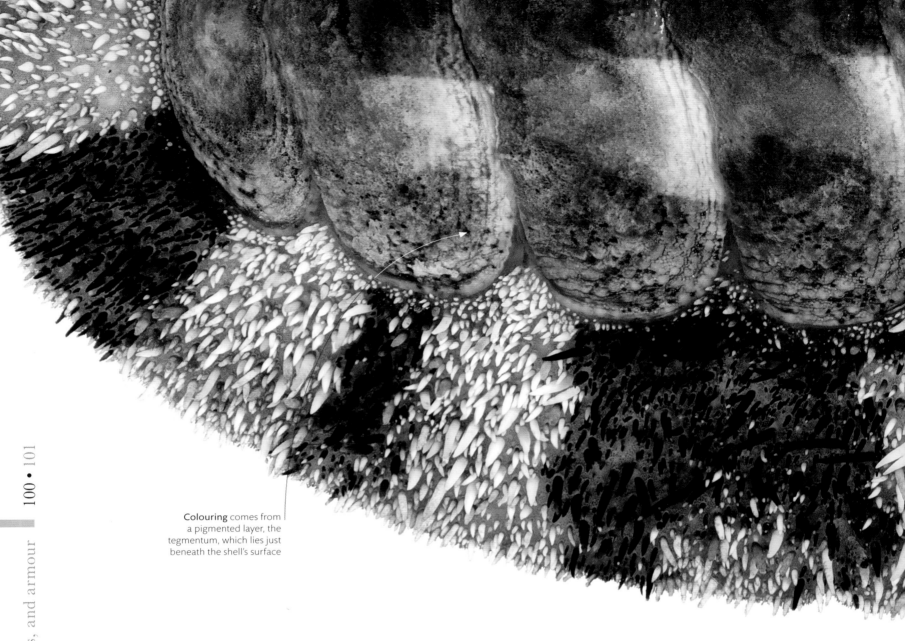

Colouring comes from a pigmented layer, the tegmentum, which lies just beneath the shell's surface

forming a shell

A shell is a characteristic feature of snails and other molluscs. While some molluscs, such as slugs and octopuses, get by without it, for many, a shell is vital protection for the soft body lying underneath. The shell is formed on a sheet of skin and muscle called the mantle, which stretches over the animal's back. The mantle encloses a cavity that contains gills and openings for excretory and reproductive systems. On its upper surface, however, the mantle releases substances that harden into a shell, which can be as simple as a limpet's cone or as complex as a twisted conch.

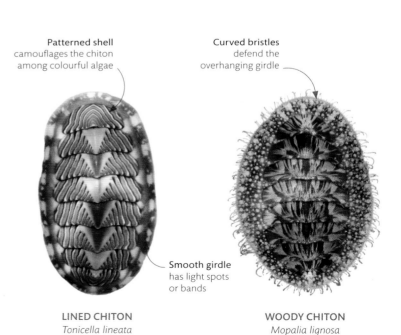

Patterned shell camouflages the chiton among colourful algae

Curved bristles defend the overhanging girdle

Smooth girdle has light spots or bands

LINED CHITON
Tonicella lineata

WOODY CHITON
Mopalia lignosa

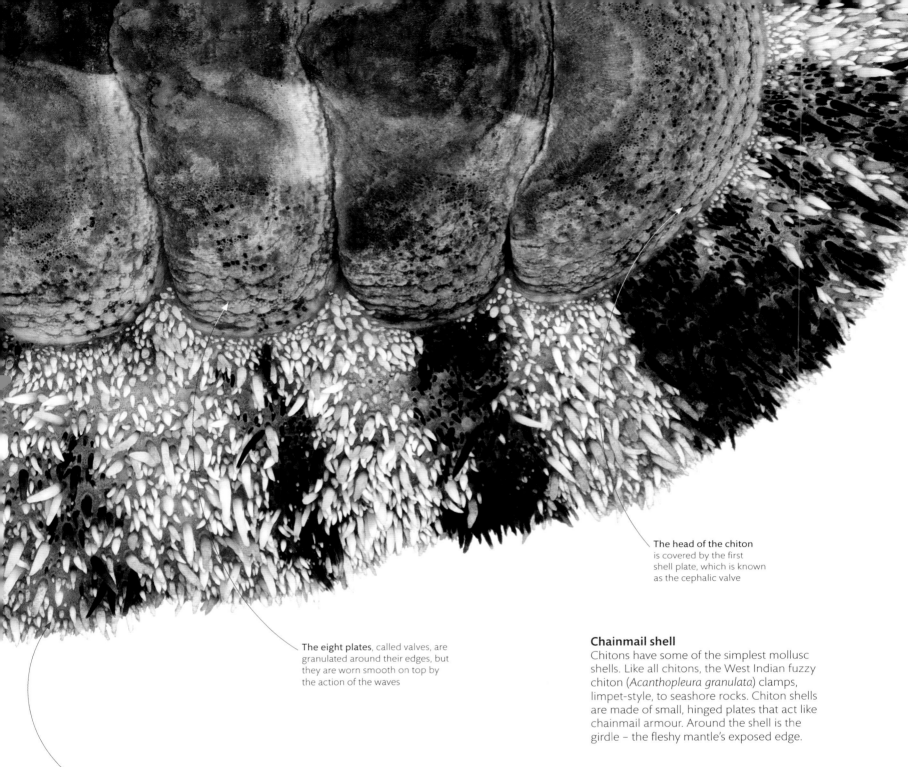

The head of the chiton is covered by the first shell plate, which is known as the cephalic valve

The eight plates, called valves, are granulated around their edges, but they are worn smooth on top by the action of the waves

The girdle – the overhanging edge of the mantle – shields the gills beneath; in this species, the mantle is protected on top by sharp calcareous spines

Chainmail shell

Chitons have some of the simplest mollusc shells. Like all chitons, the West Indian fuzzy chiton (*Acanthopleura granulata*) clamps, limpet-style, to seashore rocks. Chiton shells are made of small, hinged plates that act like chainmail armour. Around the shell is the girdle – the fleshy mantle's exposed edge.

HOW SHELLS ARE MADE

The mantle of a mollusc, such as a chiton, is not only highly muscular, helping the animal to move, it also has shell-building glands in its epidermis. These secrete a firm protein called conchin, which is impregnated with chalky minerals like those in the skeletons of stony corals and sea urchins.

Mantle

Shell plate (valve)

Girdle of mantle

CHITON CROSS-SECTION

Periostracum: shell's thin organic surface layer of conchin

Calcareous layer of shell: conchin with calcium for hardness

Shell

Epithelium of mantle secretes substances for making the shell

Mantle

Mantle muscles

SHELL-FORMING MANTLE

mollusc shells

Most molluscs belong to one of two major classes – gastropods (snails and slugs) or bivalves (which includes oysters and clams). Gastropods have a single shell, often twisted into a spiral, whereas bivalves are formed of two parts (called valves) hinged together. Each class can be further divided into groups based on the shape of the shell.

gastropod shells

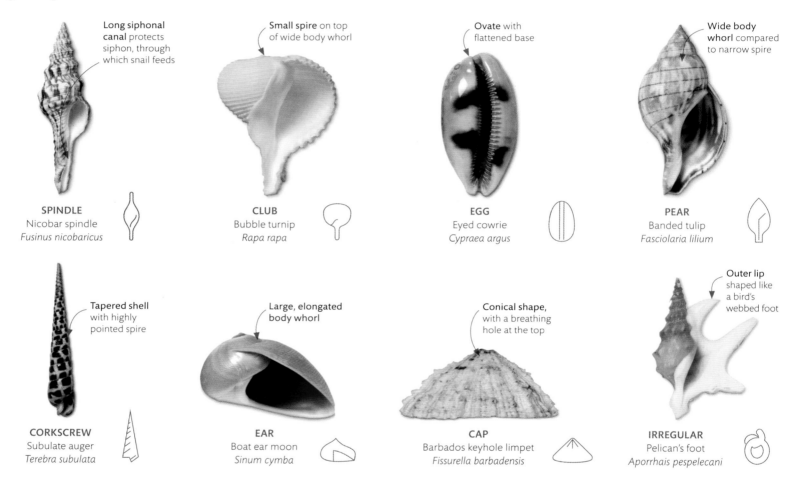

Long siphonal canal protects siphon, through which snail feeds

SPINDLE
Nicobar spindle
Fusinus nicobaricus

Small spire on top of wide body whorl

CLUB
Bubble turnip
Rapa rapa

Ovate with flattened base

EGG
Eyed cowrie
Cypraea argus

Wide body whorl compared to narrow spire

PEAR
Banded tulip
Fasciolaria lilium

Tapered shell with highly pointed spire

CORKSCREW
Subulate auger
Terebra subulata

Large, elongated body whorl

EAR
Boat ear moon
Sinum cymba

Conical shape, with a breathing hole at the top

CAP
Barbados keyhole limpet
Fissurella barbadensis

Outer lip shaped like a bird's webbed foot

IRREGULAR
Pelican's foot
Aporrhais pespelecani

bivalve shells

Hinge where the two valves connect

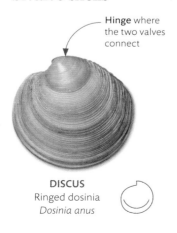

DISCUS
Ringed dosinia
Dosinia anus

Auricles (earlike flaps) present in scallops are asymmetrical

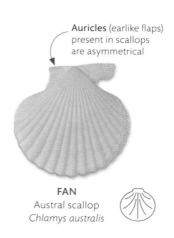

FAN
Austral scallop
Chlamys australis

Wide base forms triangular outline

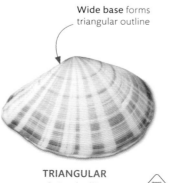

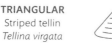

TRIANGULAR
Striped tellin
Tellina virgata

Thin, oblong-shaped valve

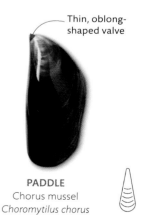

PADDLE
Chorus mussel
Choromytilus chorus

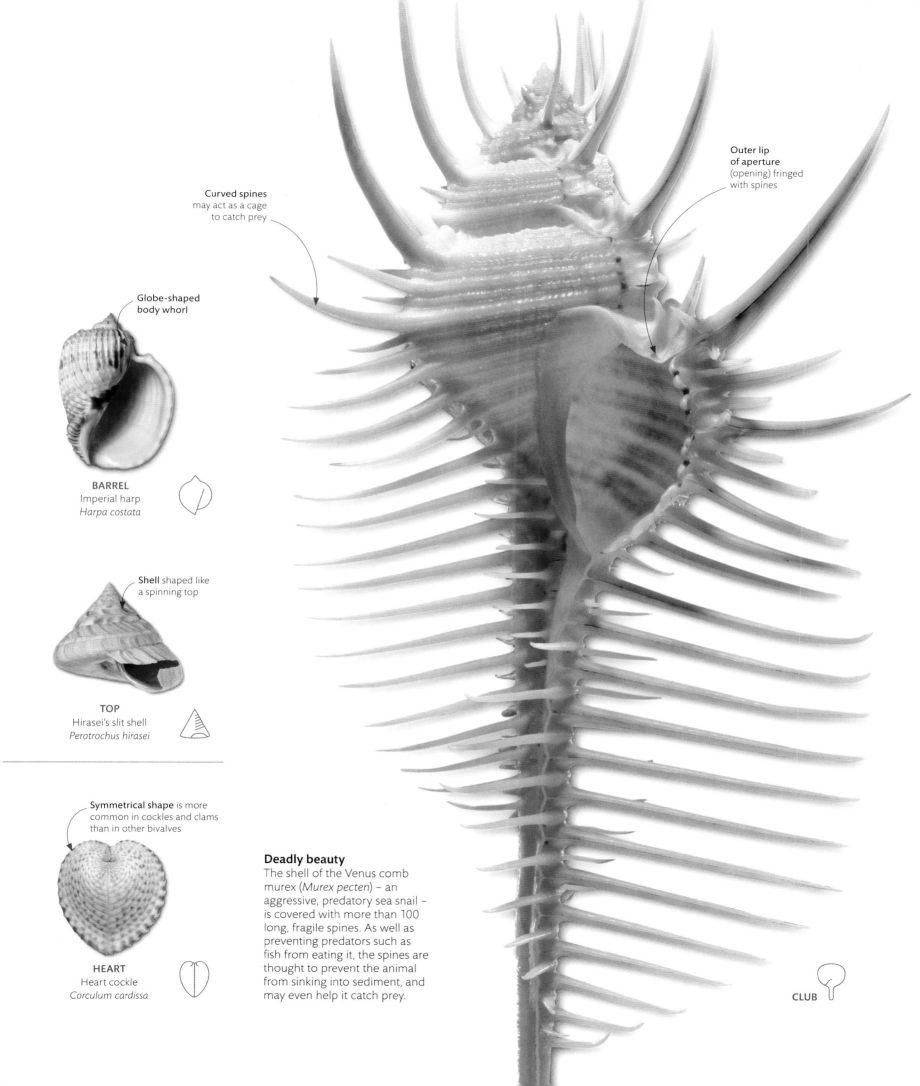

Curved spines
may act as a cage
to catch prey

Outer lip
of aperture
(opening) fringed
with spines

Globe-shaped
body whorl

BARREL
Imperial harp
Harpa costata

Shell shaped like
a spinning top

TOP
Hirasei's slit shell
Perotrochus hirasei

Symmetrical shape is more
common in cockles and clams
than in other bivalves

HEART
Heart cockle
Corculum cardissa

Deadly beauty

The shell of the Venus comb
murex (*Murex pecten*) – an
aggressive, predatory sea snail –
is covered with more than 100
long, fragile spines. As well as
preventing predators such as
fish from eating it, the spines are
thought to prevent the animal
from sinking into sediment, and
may even help it catch prey.

CLUB

vertebrate scales

Formed from folds of skin, a coating of scales provides flexible armour.
Fish scales have a bony core that grows from the deeper dermis of the skin.
Reptile scales are restricted to the surface epidermis and usually bone-free.
They contain hard keratin, plus oils that stop the skin below from drying out.

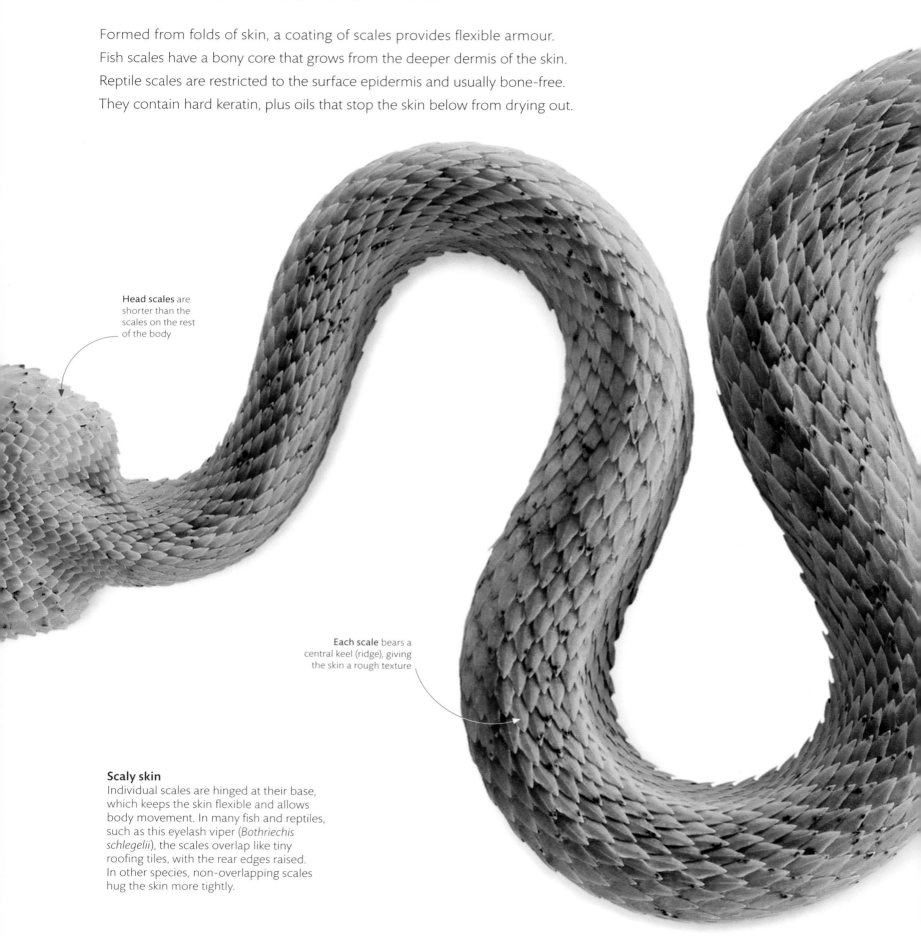

Head scales are shorter than the scales on the rest of the body

Each scale bears a central keel (ridge), giving the skin a rough texture

Scaly skin
Individual scales are hinged at their base, which keeps the skin flexible and allows body movement. In many fish and reptiles, such as this eyelash viper (*Bothriechis schlegelii*), the scales overlap like tiny roofing tiles, with the rear edges raised. In other species, non-overlapping scales hug the skin more tightly.

Scale diversity

Fish scales evolved as toothlike structures: placoid and ganoid fish scales still retain layers of enamel and dentine. Thinner bone makes up cycloid and ctenoid scales in most living fish species. The boneless scales of reptiles likely evolved independently of fish scales.

fish scales

Tiny placoid scales point backward, giving skin the texture of sandpaper

Platelike ganoid scales are hard and shiny, due to their enamel content

NURSEHOUND SHARK
Scyliorhinus stellaris

SPOTTED GAR
Lepisosteus oculatus

Cycloid scales contain concentric rings of thin bone

Ctenoid scales have comblike fringes, which reduce turbulence

ASIAN AROWANA
Scleropages formosus

RUSTY PARROTFISH
Scarus ferrugineus

reptile scales

Beadlike scales do not overlap

Wide, overlapping belly scales help grip branches

CRESTED GECKO
Correlophus ciliatus

WHITE-LIPPED PIT-VIPER
Trimeresurus albolabris

Golden-scaled phase is one of several colour variants of this species; others are pink, green, and brown, sometimes with darker markings

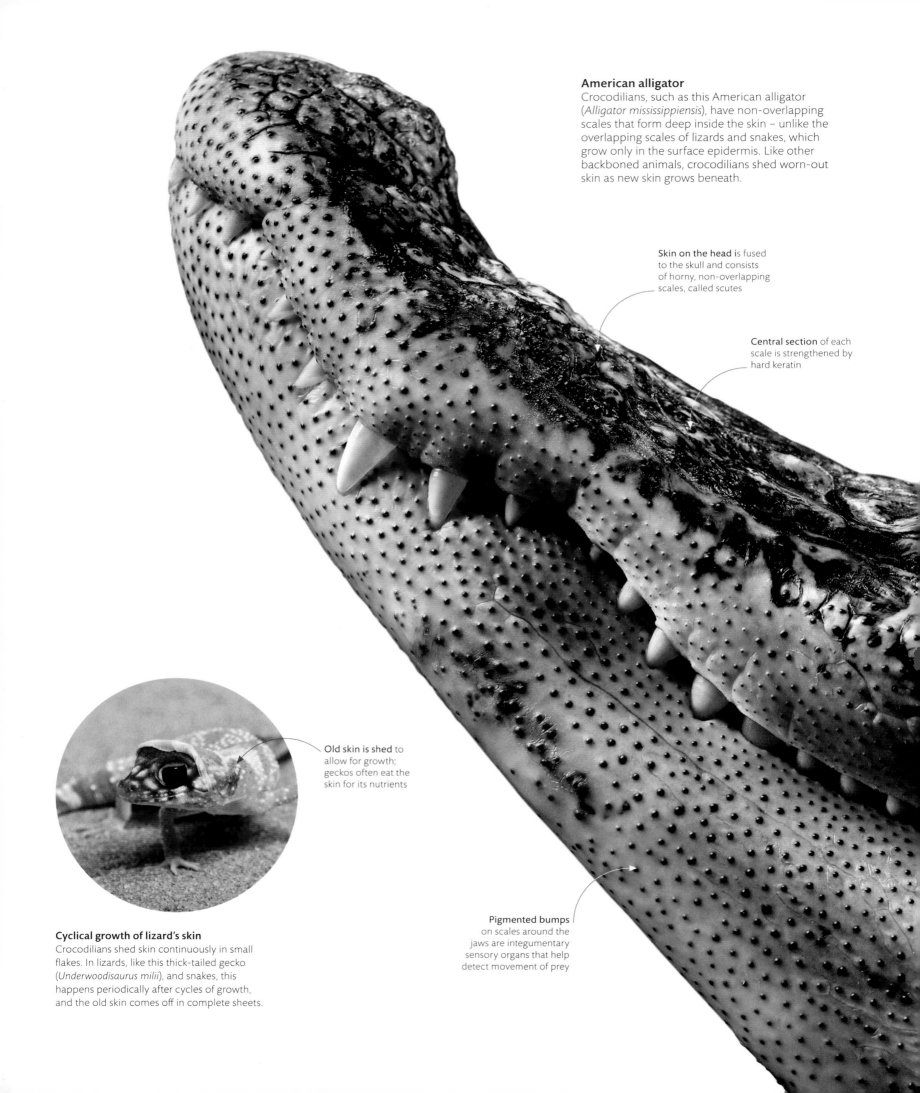

American alligator
Crocodilians, such as this American alligator (*Alligator mississippiensis*), have non-overlapping scales that form deep inside the skin – unlike the overlapping scales of lizards and snakes, which grow only in the surface epidermis. Like other backboned animals, crocodilians shed worn-out skin as new skin grows beneath.

Skin on the head is fused to the skull and consists of horny, non-overlapping scales, called scutes

Central section of each scale is strengthened by hard keratin

Old skin is shed to allow for growth; geckos often eat the skin for its nutrients

Pigmented bumps on scales around the jaws are integumentary sensory organs that help detect movement of prey

Cyclical growth of lizard's skin
Crocodilians shed skin continuously in small flakes. In lizards, like this thick-tailed gecko (*Underwoodisaurus milii*), and snakes, this happens periodically after cycles of growth, and the old skin comes off in complete sheets.

reptile skin

The scaly skin of reptiles has a tough, dry surface that keeps moisture locked in so they are better adapted for life on land than their scale-less amphibian ancestors. Reptile skin contains two types of the horny material keratin: one is hard and brittle, the other soft and pliable. The combination provides a flexible barrier that shields the body from abrasion and prevents the skin drying out.

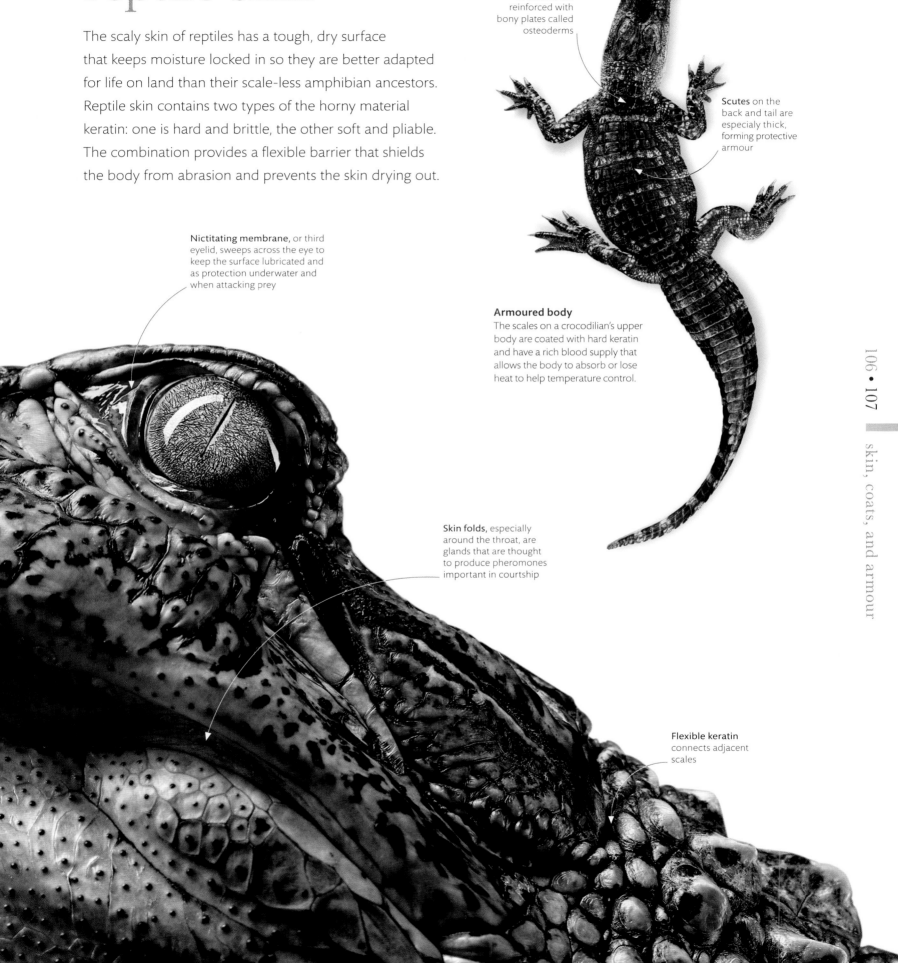

Upper body scales, or scutes, are reinforced with bony plates called osteoderms

Scutes on the back and tail are especially thick, forming protective armour

Nictitating membrane, or third eyelid, sweeps across the eye to keep the surface lubricated and as protection underwater and when attacking prey

Armoured body
The scales on a crocodilian's upper body are coated with hard keratin and have a rich blood supply that allows the body to absorb or lose heat to help temperature control.

Skin folds, especially around the throat, are glands that are thought to produce pheromones important in courtship

Flexible keratin connects adjacent scales

CHANGING COLOUR

Many animals change colour by expanding sacs of dark pigment in their skin (see pp.98–99), but others, such as chameleons, do it with crystals. Like many reptiles and amphibians, they have skin cells called iridophores, containing colour-reflecting crystals, but chameleons can move these crystals using neural control – changing their colour-reflecting properties within seconds.

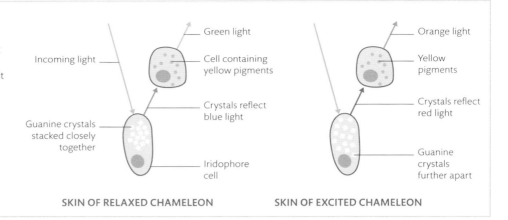

SKIN OF RELAXED CHAMELEON

Incoming light

Green light

Cell containing yellow pigments

Crystals reflect blue light

Guanine crystals stacked closely together

Iridophore cell

SKIN OF EXCITED CHAMELEON

Orange light

Yellow pigments

Crystals reflect red light

Guanine crystals further apart

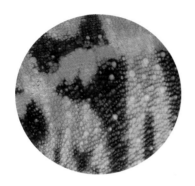

GREEN SKIN

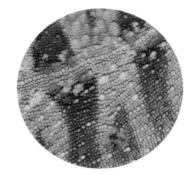

ORANGE SKIN

Showing off
In most panther chameleons, colour change involves a shift in background colour from green (when the animal is relaxed) to orange or red (when it is excited). In its rested state, the lizard is better camouflaged among the leaves of its treetop habitat.

advertisement colours

Colour sends a powerful signal in animals with the vision to detect it and, as they grow, many acquire strong colours that indicate sexual maturity or warn off rivals. Other animals can change colour at will by neural or hormonal control. Flashes of colour can send social signals – such as a change in disposition, a show of aggression in the face of competitors, or a willingness to mate – all without the danger of permanently attracting predators.

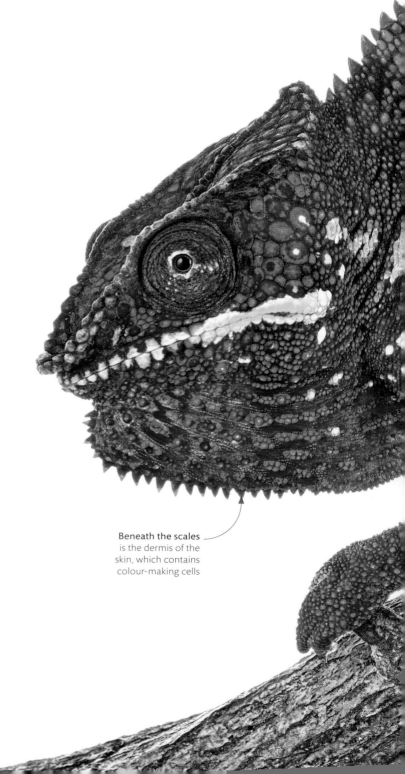

Beneath the scales is the dermis of the skin, which contains colour-making cells

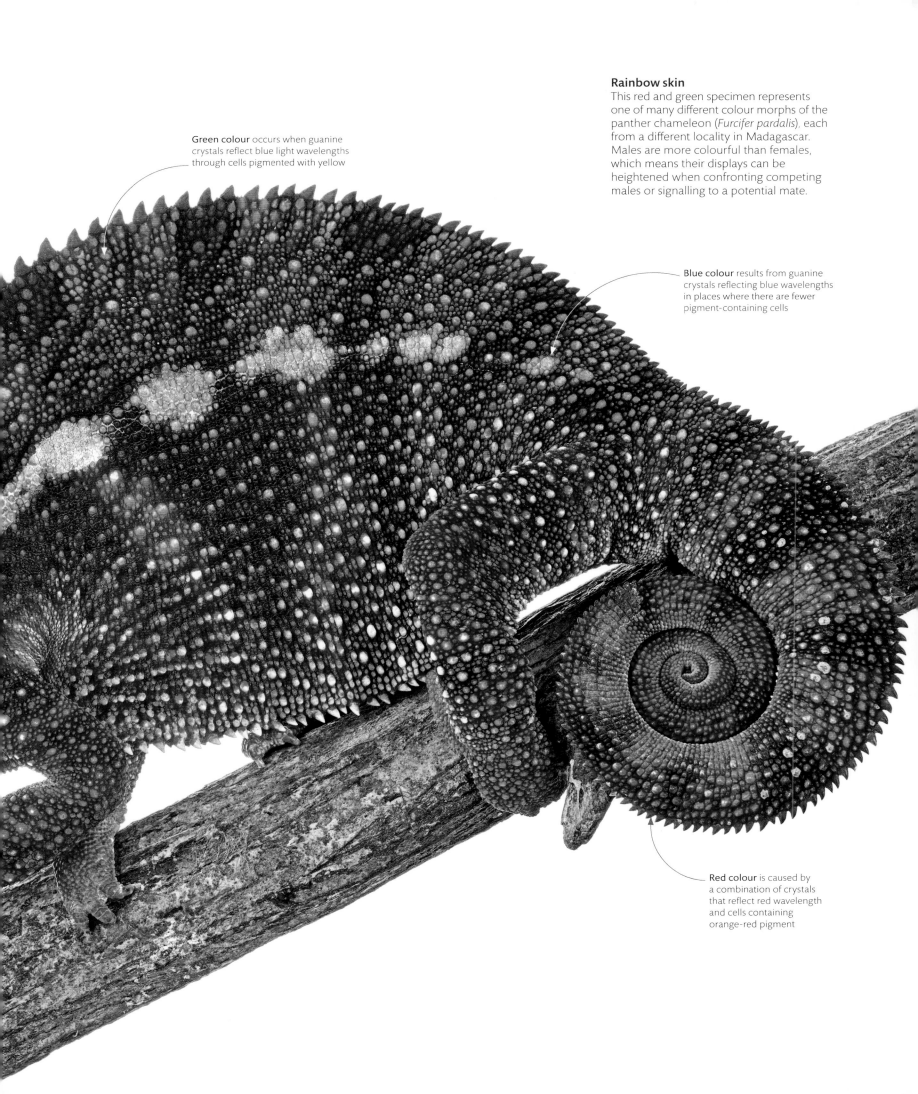

Green colour occurs when guanine crystals reflect blue light wavelengths through cells pigmented with yellow

Rainbow skin
This red and green specimen represents one of many different colour morphs of the panther chameleon (*Furcifer pardalis*), each from a different locality in Madagascar. Males are more colourful than females, which means their displays can be heightened when confronting competing males or signalling to a potential mate.

Blue colour results from guanine crystals reflecting blue wavelengths in places where there are fewer pigment-containing cells

Red colour is caused by a combination of crystals that reflect red wavelength and cells containing orange-red pigment

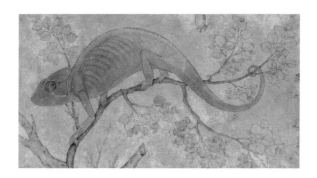

Chameleon (c.1612)
Larger than life on a branch amid hovering insects, Ustad Mansur's chameleon is a study of the creature's physiology and habitat.

Squirrels in a Plane Tree (c.1610)
A miniature of playful squirrels is attributed to Mughal court favourite Abu al-Hasan. They appear to be European red squirrels absent from Emperor Jahangir's realm – so these were most likely observed in his private zoo. This meticulous representation of animals and birds suggests contribututions from fellow artist Ustad Mansur.

animals in art

in the mughal courts

The wealth and power of the Mughals, who ruled India and much of south Asia from the 16th to the 18th century, was matched by their love of aesthetics. Jewel-like miniatures of legends, battles, portraits, and hunting scenes were prized in the royal courts. The fourth emperor, Jahangir, also indulged a passion for natural history by commissioning accurate paintings of flora and fauna that are now regarded as exquisite works of art.

During the successive reigns of Humayun, Akbar, and Jahangir, leading artists from Persia (present-day Iran), central Asia, and Afghanistan were lured by the wealth and prestige of the Mughal courts. Miniatures were often a collaboration between calligraphers, designers, and artists: the drawings first covered with a wash of white before colour was applied in thin layers with fine hair brushes. The paintings were then burnished to an enamel finish with an agate stone.

Jahangir's favourites, Abu al-Hasan and Ustad Mansur, travelled with him across the empire and were given the title "wonder of the age" for their work. Jahangir's love of animals is revealed in memoirs that record the specimens that were brought to him: a rare falcon from the Shah of Persia; a zebra from Abyssinia; a turkey cock and Himalayan cheer pheasant from an agent in Goa, India. Two dodos in Jahangir's menagerie in Surat were probably gifts from traders.

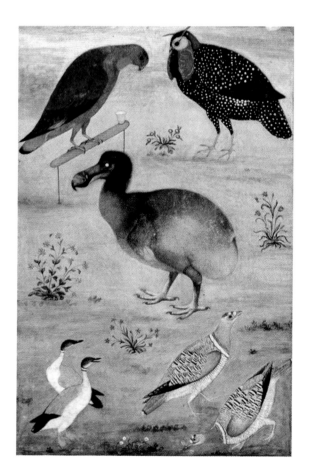

The Dodo in colour (c.1627)
Ustad Mansur, a leading artist in the 17th-century court of the Emperor Jahangir, worked under the direct orders of his patron to record rare specimens of birds, animals, and plants. Mansur is thought to have painted this rare picture of a dodo in colour.

66 He had brought several strange and unusual animals… I ordered the artists to draw their likenesses. 99

THE JAHANGIRNAMA: MEMOIRS OF JAHANGIR, EMPEROR OF INDIA, 1627

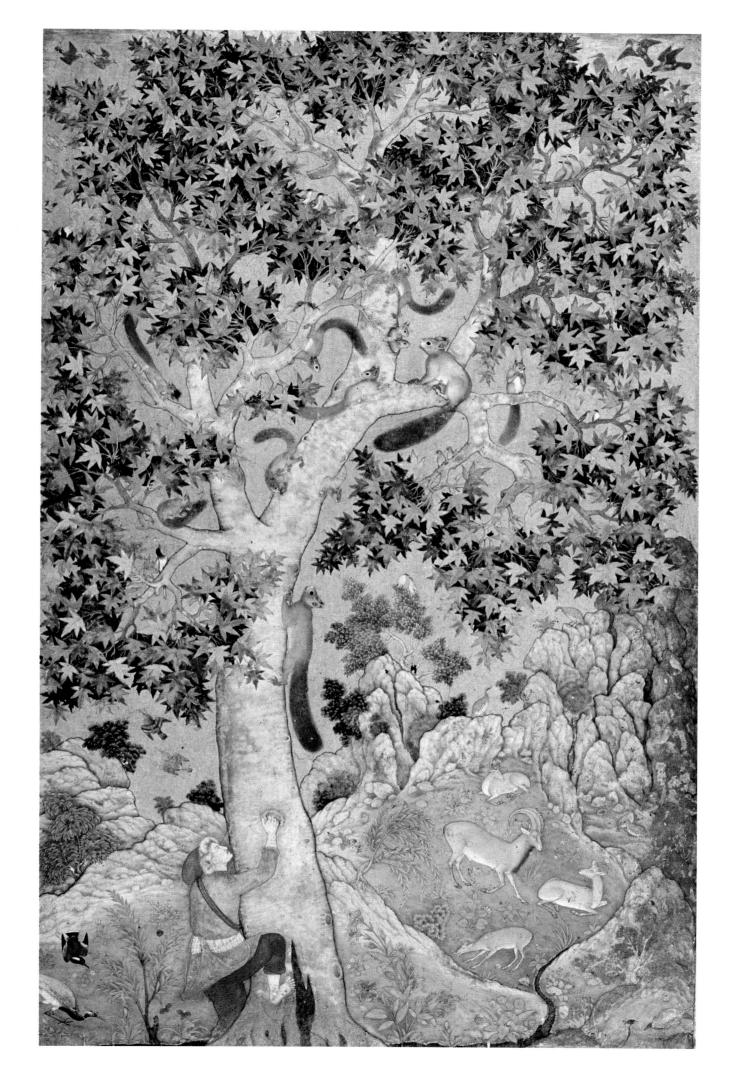

frills and dewlaps

Some of the best kinds of ornaments used for display are those that can be flaunted only when needed. Displays that are permanently on show risk attracting the attention of predators, and those that can be flashed in a moment can be even more eye-catching. Some lizards use a moveable bony apparatus located in the bottom of their throat to spreads flaps of skin. This can act as a social signal to others of the same species, or can make the lizard appear more threatening to predators.

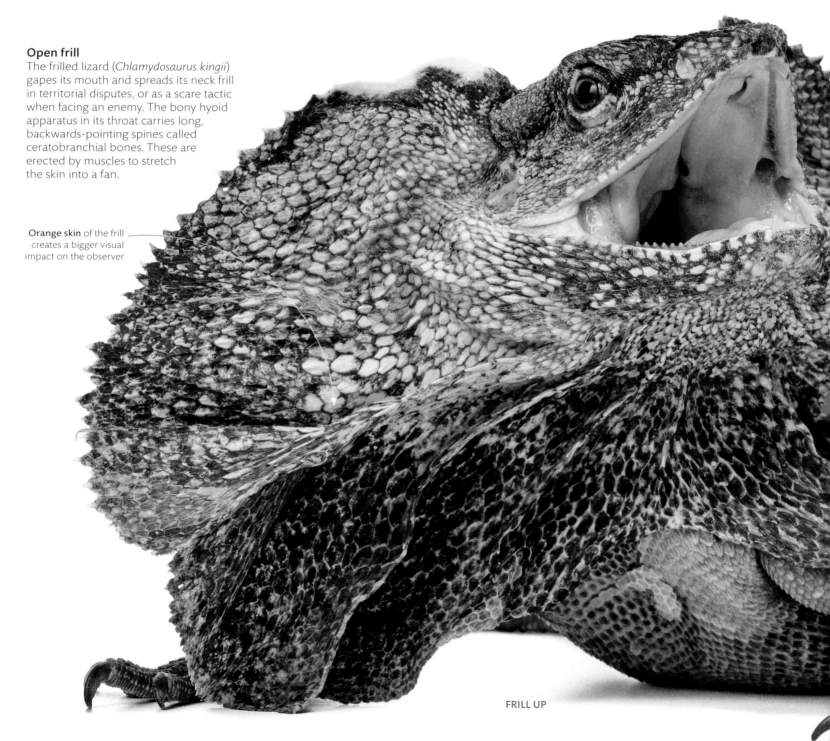

Open frill
The frilled lizard (*Chlamydosaurus kingii*) gapes its mouth and spreads its neck frill in territorial disputes, or as a scare tactic when facing an enemy. The bony hyoid apparatus in its throat carries long, backwards-pointing spines called ceratobranchial bones. These are erected by muscles to stretch the skin into a fan.

Orange skin of the frill creates a bigger visual impact on the observer

FRILL UP

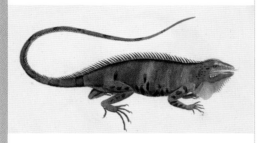
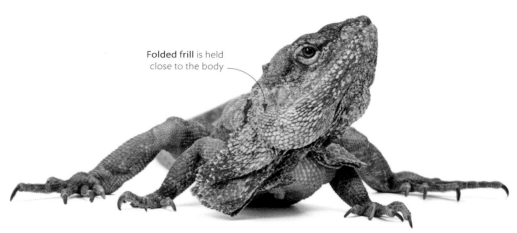

Folded frill is held close to the body

FRILL DOWN

"**Spokes**" of the neck frill are made from ceratobranchial bones extending from the lower throat

Strong hind legs allow the lizard to escape on two legs if its display fails

FRILL HALFWAY DOWN

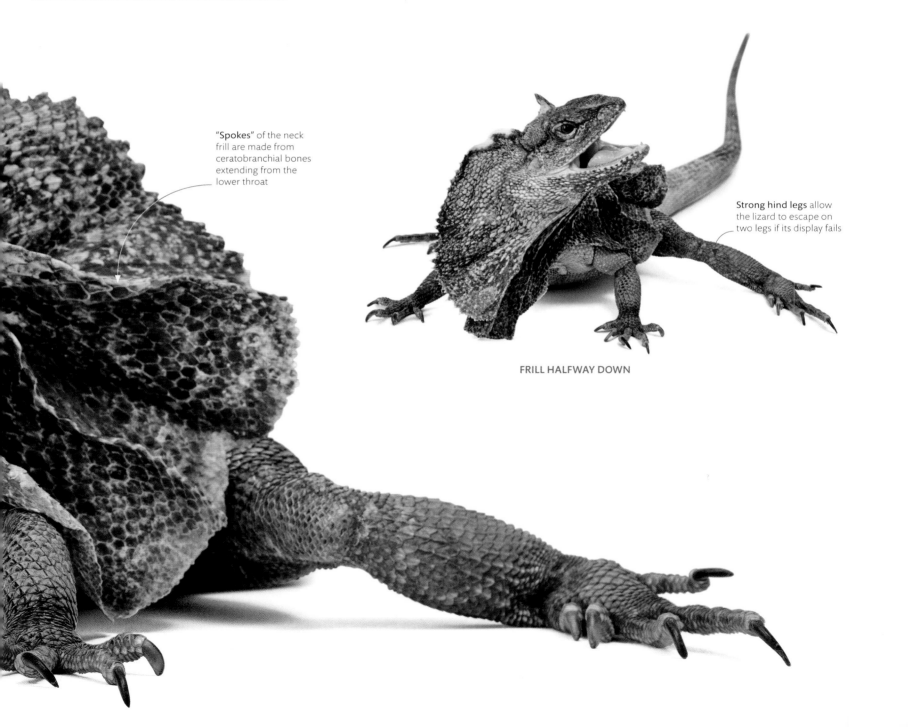

weapons and fighting

Physical conflict can be dangerous, so most animals avoid it even if equipped with weaponry – but sometimes the prize is worth the risk. The jaws of male stag beetles grow so large that they cannot even close them to bite. But they can wield them in jaw-to-jaw pushing matches over sources of food or a potential mate, and like their mammalian namesakes (see pp.88–89), it is brute strength that determines the victor.

Fighting male stag beetles
These bronze elkhorn stag beetles (*Lamprima adolphinae*) are using their antlerlike jaws, or mandibles, in combat. Those with bigger jaws have an advantage, and can lift their opponent and throw him to the ground.

Serrated edge can hook onto opponent's exoskeleton

Enlarged jaw curves upwards and is useless for feeding

Mouthparts lap sweet liquid food such as tree sap or juice, but cannot bite

Forked-tip jaw can lift adversary off the ground

Antenna projects outwards to prevent damage during conflict

GROWING BODIES

Body parts that are used for display or as weapons grow more rapidly than the rest of the body, so end up disproportionately larger. The example here shows the growth rate of the male fiddler crab claw relative to its body size. These crabs use their single enlarged claws in combat when competing with males, and they wave them to attract females.

COMPARING BODY AND WEAPON GROWTH

Adult claw proportionately larger than in young crab

Body and legs stay in proportion throughout development

Accelerated (geometric) claw growth rate

Crab's body growth rate (arithmetic growth)

Weapons for both sexes

Both the male and female Amazonian purple scarab (*Coprophanaeus lancifer*) are equipped with hornlike weapons. Males employ theirs in combat over a mate, while females use them to defend carcasses from other females, so they can bury the carcass as food for their larvae.

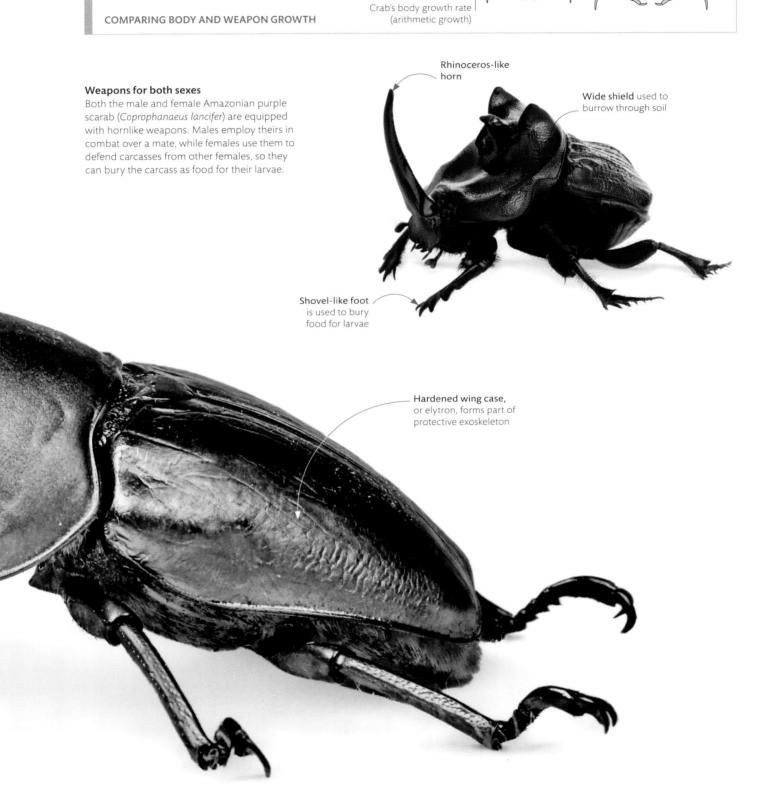

Rhinoceros-like horn

Wide shield used to burrow through soil

Shovel-like foot is used to bury food for larvae

Hardened wing case, or elytron, forms part of protective exoskeleton

Camouflaged flatfish

The peacock flounder (*Bothus lunatus*) starts life swimming upright but changes shape dramatically as it grows. The eye on the right side of the fish moves over to the left side. The right side of the flounder becomes its underside, while the left side – now with both eyes – acquires the mottled pigmentation that conceals it so successfully on the stony sea bed.

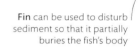

Long, continuous dorsal fin extends along the length of the flounder's body

Fin can be used to disturb sediment so that it partially buries the fish's body

blending in

A successful disguise can benefit animals in different ways. Vulnerable animals that otherwise lack defensive weapons can avoid the attention of hunters, while predators themselves can use camouflage to ambush prey. For most animals, blending in is a matter of using their inherited shape and colour to hide in an appropriate habitat. However, other animals are more versatile, and – under the influence of their nervous system – are able to change colour or pattern to match their background.

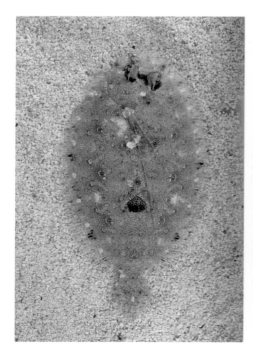

FLOUNDER BLENDING IN TO FINE SEDIMENT

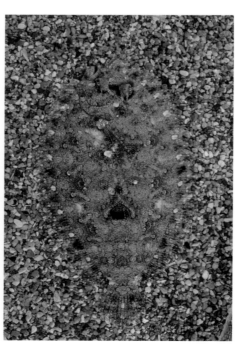

FLOUNDER BLENDING IN TO COARSE SEDIMENT

Matching patterns

Peacock flounders have the ability to match the colour and pattern of their background due to specialized colour-changing skin cells (see p.98). These cells contain microscopic granules of pigment that can bunch together or disperse to lighten or darken the skin. According to what environment the peacock flounder detects with its vision, hormones are released to change the distribution of pigment in the skin within seconds.

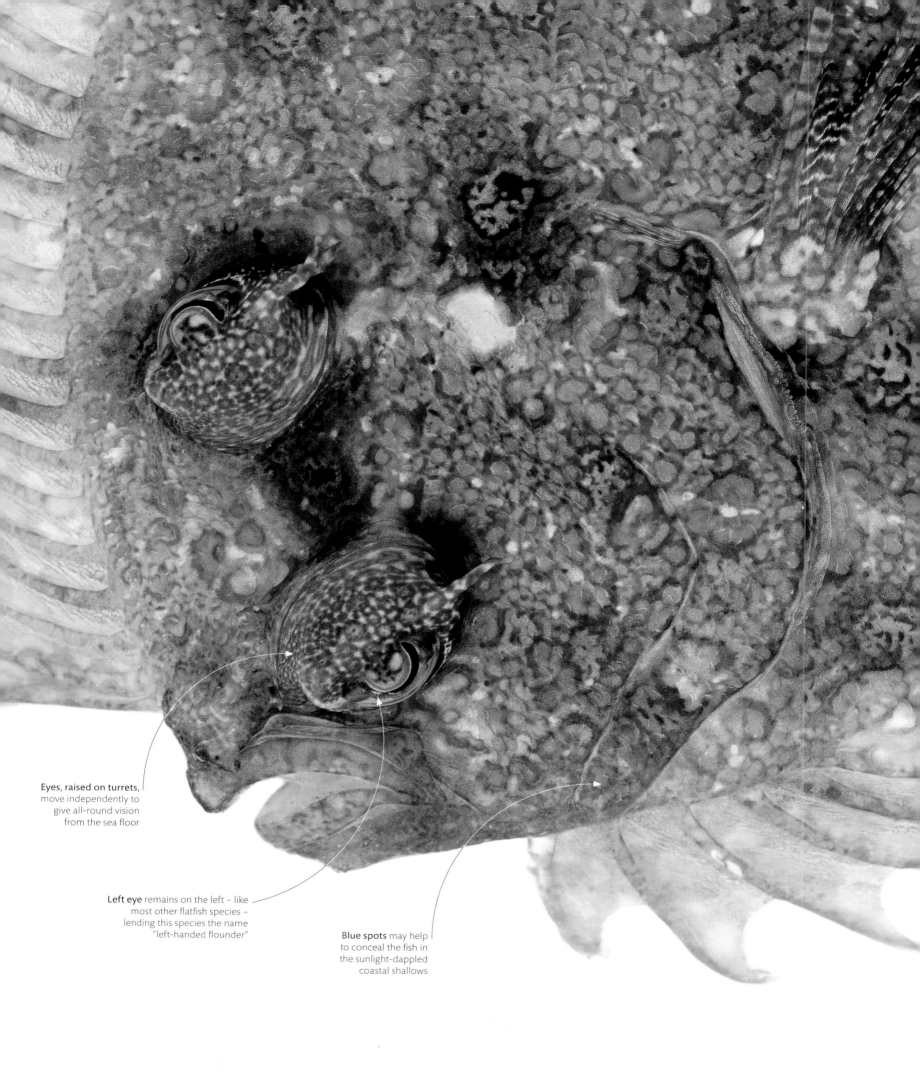

Eyes, raised on turrets, move independently to give all-round vision from the sea floor

Left eye remains on the left – like most other flatfish species – lending this species the name "left-handed flounder"

Blue spots may help to conceal the fish in the sunlight-dappled coastal shallows

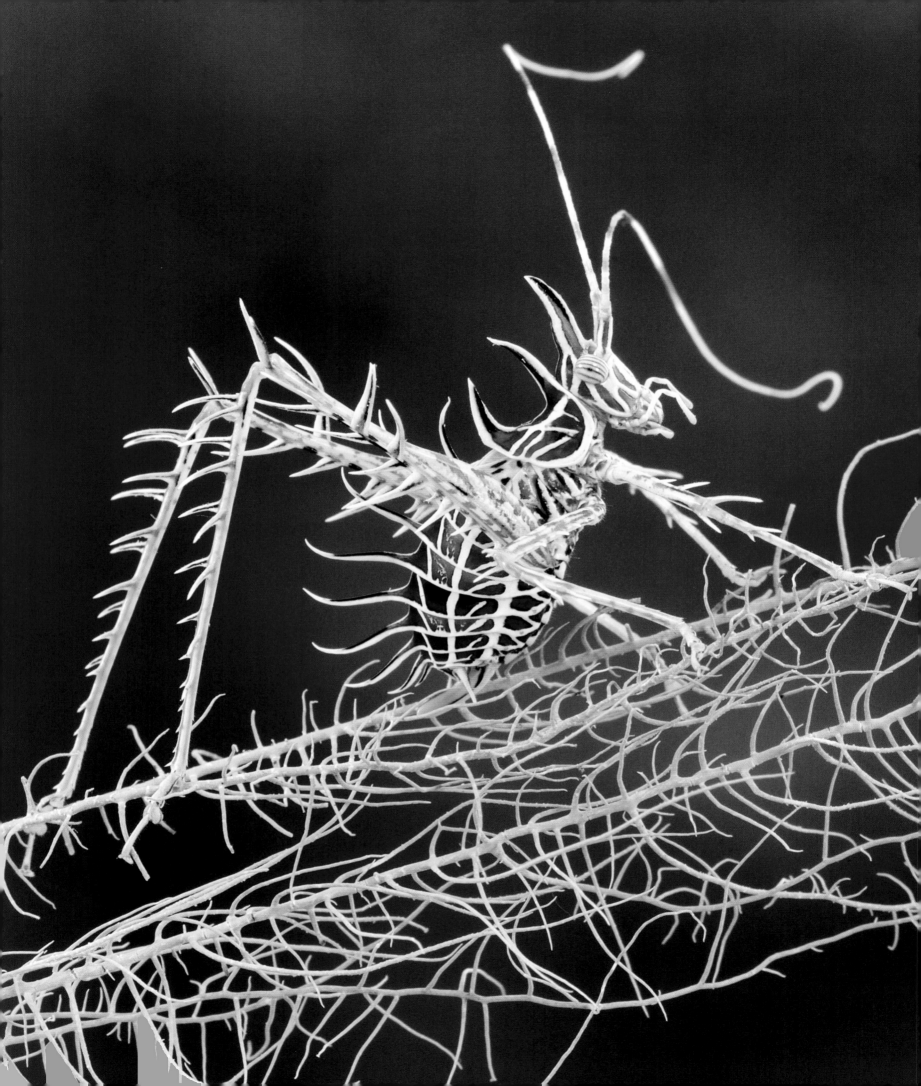

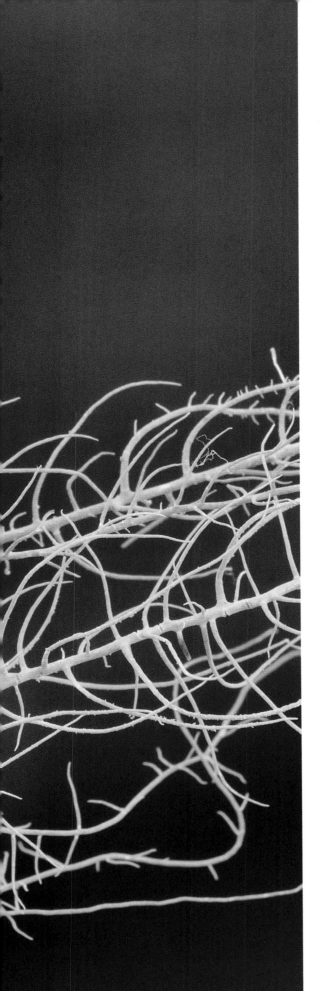

lichen katydid

As a prey item for birds, reptiles, bats, ground- and tree-dwelling mammals, other insects, and spiders, the lichen katydid (*Markia hystrix*) needs to conceal itself in its rainforest habitat. What it lacks in chemical and physical defences, it makes up for with exceptional camouflage and mimicry.

Native to Central and South America, the lichen katydid is a nocturnal and tree-dwelling insect; these traits help it to avoid predators that hunt in daylight or mainly on the ground. Its chances of survival are enhanced by adaptations to its colour, body shape, and movements that make it virtually indistinguishable from the beard lichen (*Usnea* sp.) on which it feeds.

As well as blending in with the lichen's green-and-white colouring, the katydid's body and legs have spines that mimic the latticelike form of the beard lichen. This combination makes the katydid especially difficult for predators to spot, particularly since the wings of the adults are also marked with lichenlike patterns. To further evade detection, the katydid typically moves with very slow, deliberate steps – unless it feels threatened, in which case it takes flight to escape danger.

The variety of mimicry in katydids is astonishing: there are also species that resemble leaves, moss, bark, and even rock. Leaf mimicry includes colour-matching, leaf-shaped bodies with veinlike patterns, and sometimes patches resembling decay or holes. As a result, individuals in the same species can look very different, making the predator's task of sorting leaf from insect even harder.

Perfect disguise
This juvenile lichen katydid has not yet grown its wings, but it still perfectly matches the colour of the lichen – even down to the darker patches that mimic the spaces between the thin branches.

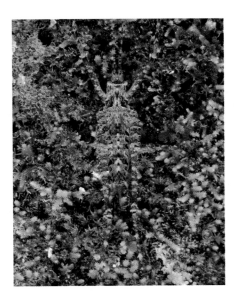

Moss mimic
The plantlike ornamentations of the moss katydid (*Championica montana*) of Costa Rica and Panama are actually protective spines.

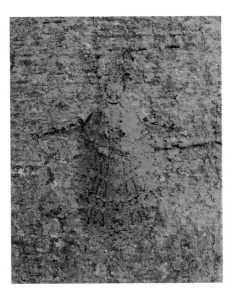

Fake scar tissue
Like unrelated scale insects, some katydids' appearance resembles the ridges produced when bark has been torn or pierced.

An array of colours
Bird's feathers come in many colours. The black and browns result from melanin pigment granules made in skin cells; yellows and reds are intensified by diet. Blues, violets, and greens occur because of the way feather structure bends or scatters light. (See Cells for colour, p.98.)

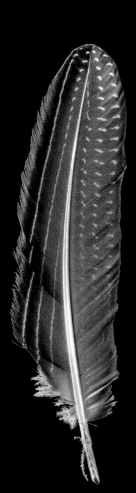

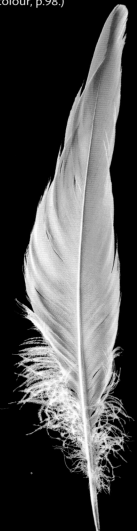

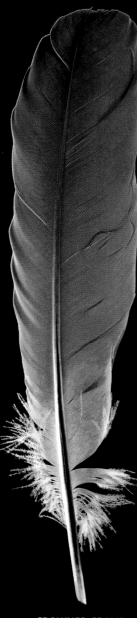

LILAC BREASTED ROLLER
Coracias caudatus

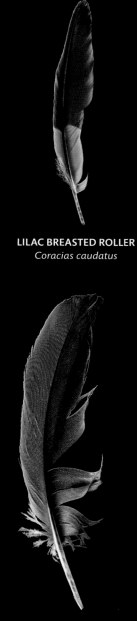

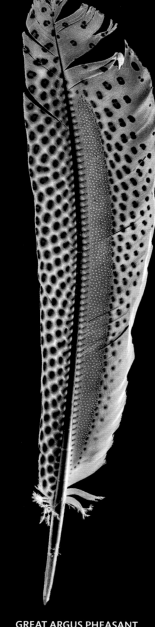

VULTURINE GUINEAFOWL
Acryllium vulturinum

AMERICAN FLAMINGO
Phoenicopterus ruber

CROWNED CRANE
Balearica regulorum

MALE ECLECTUS PARROT
Eclectus roratus

GREAT ARGUS PHEASANT
Argusianus argus

feathers

Birds are the only living animals that are feathered. Feathers probably evolved from modified scales in their dinosaur ancestors, but whether these "protofeathers" initially helped with trapping body heat or were aerodynamic aids for gliding or flight is uncertain. In modern birds, feathers serve both purposes – an undercoat of downy feathers provides thermal insulation, while stiff-bladed feathers make the body streamlined and help provide the necessary lift for carrying the bird through the air.

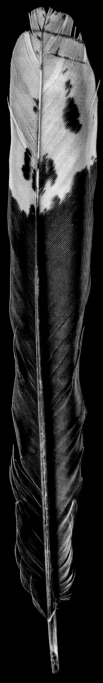

MALAYAN BLACK HORNBILL
Anthracoceros malayanus

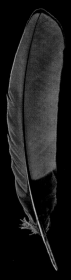

VIOLET TURACO
Musophaga violacea

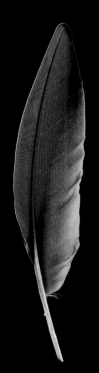

RED-FRONTED MACAW
Ara rubrogenys

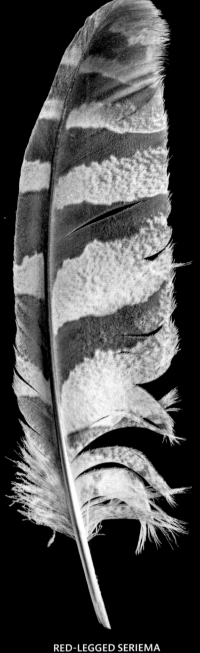

RED-LEGGED SERIEMA
Cariama cristata

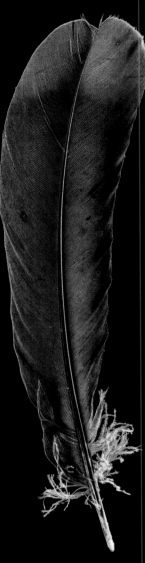

VICTORIA CROWNED PIGEON
Goura victoria

TYPES OF FEATHER

Feathers grow from follicles, each forming a shaft and wide vane. The vane begins as a sheet of keratin-making cells, then a complex pattern of slits opens to separate the barbs. Flight and contour feathers have self-sealing vanes held together by microscopic hooks (see p.123). Fluffy down feathers insulate from the cold, while filoplumes are sensory.

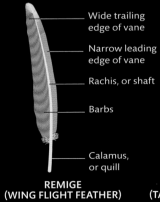

- Wide trailing edge of vane
- Narrow leading edge of vane
- Rachis, or shaft
- Barbs
- Calamus, or quill

**REMIGE
(WING FLIGHT FEATHER)**

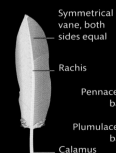

- Symmetrical vane, both sides equal
- Rachis
- Pennaceous barbs
- Plumulaceous barbs
- Calamus

**RECTRIX
(TAIL FLIGHT FEATHER)**

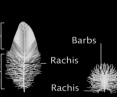

- Barbs
- Rachis
- Rachis
- Barbs
- Rachis
- Calamus

**CONTOUR
FEATHER**

**DOWN
FEATHER**

FILOPLUME

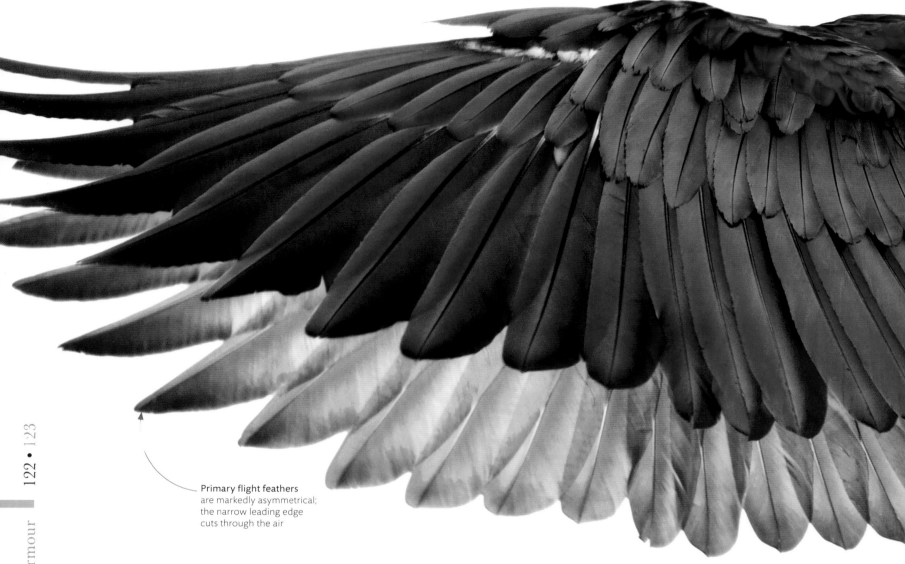

Primary flight feathers are markedly asymmetrical; the narrow leading edge cuts through the air

flight feathers

It takes a special kind of feather to enable a bird to fly. The biggest, stiffest ones of the wings and tail – the flight feathers – are attached directly to the skeleton and account for most of the surface of the wings and the tail. These feathers have self-sealing vanes that depend on a complex system of microscopic hooks (see box opposite) to maintain the bladelike shape so important for lifting a bird into the air.

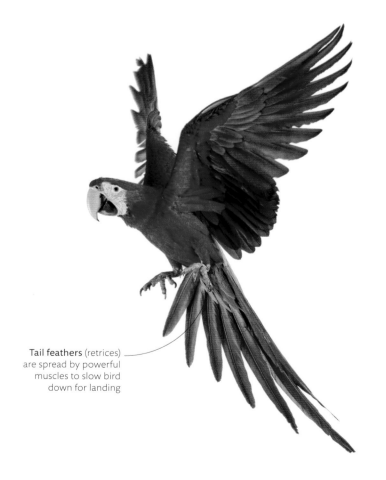

Feathery brake
The flight feathers in both the wings and the tail are also used as brakes for landing. In this green-winged macaw, more than half of the wing is made up of flight feathers, and its distinctive tail feathers are as long as the head and body combined.

Tail feathers (retrices) are spread by powerful muscles to slow bird down for landing

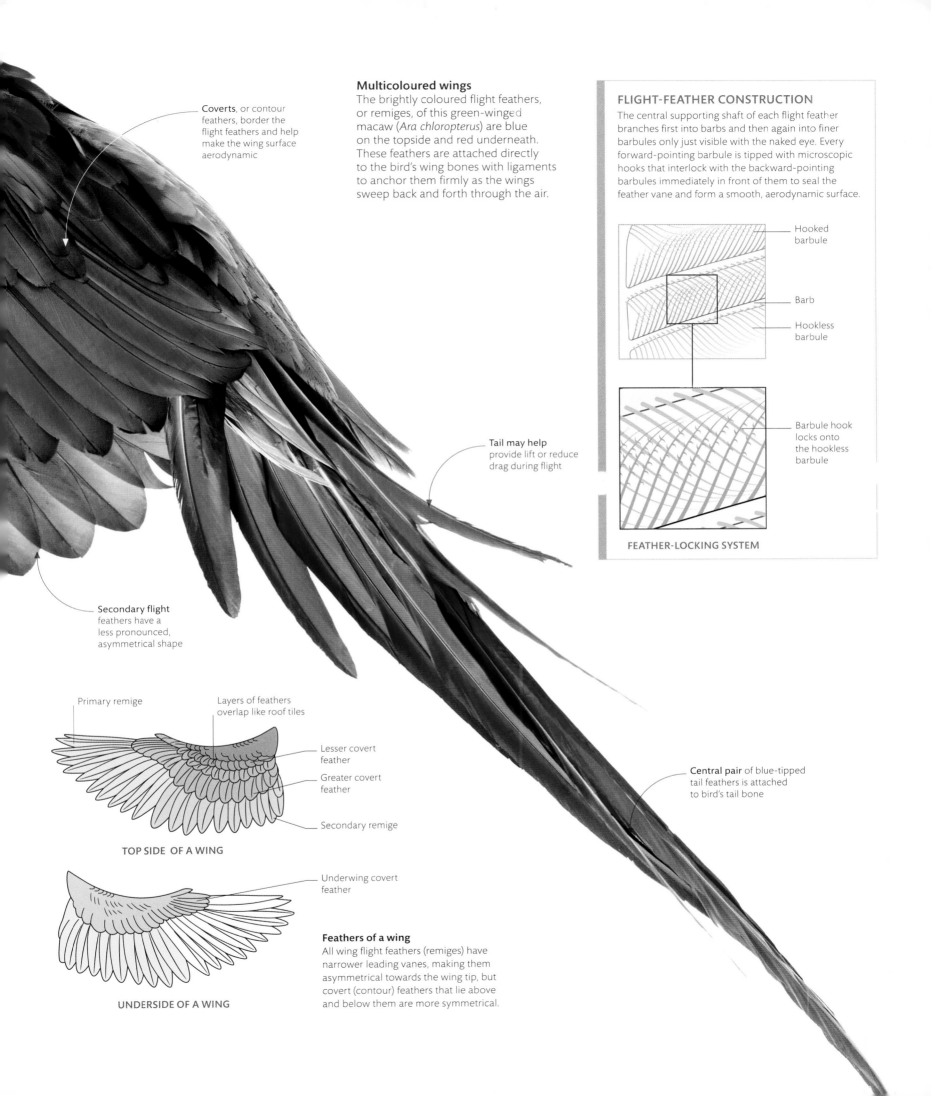

Coverts, or contour feathers, border the flight feathers and help make the wing surface aerodynamic

Multicoloured wings
The brightly coloured flight feathers, or remiges, of this green-winged macaw (*Ara chloropterus*) are blue on the topside and red underneath. These feathers are attached directly to the bird's wing bones with ligaments to anchor them firmly as the wings sweep back and forth through the air.

FLIGHT-FEATHER CONSTRUCTION
The central supporting shaft of each flight feather branches first into barbs and then again into finer barbules only just visible with the naked eye. Every forward-pointing barbule is tipped with microscopic hooks that interlock with the backward-pointing barbules immediately in front of them to seal the feather vane and form a smooth, aerodynamic surface.

Hooked barbule

Barb

Hookless barbule

Barbule hook locks onto the hookless barbule

FEATHER-LOCKING SYSTEM

Tail may help provide lift or reduce drag during flight

Secondary flight feathers have a less pronounced, asymmetrical shape

Primary remige

Layers of feathers overlap like roof tiles

Lesser covert feather

Greater covert feather

Secondary remige

TOP SIDE OF A WING

Underwing covert feather

UNDERSIDE OF A WING

Central pair of blue-tipped tail feathers is attached to bird's tail bone

Feathers of a wing
All wing flight feathers (remiges) have narrower leading vanes, making them asymmetrical towards the wing tip, but covert (contour) feathers that lie above and below them are more symmetrical.

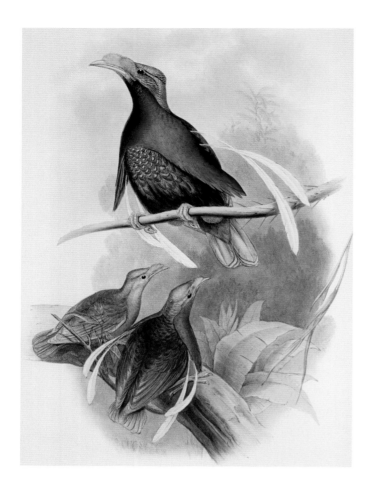

Clandestine show-off
Some of the most extravagant plumages belong to birds that hide deep in thick forests, such as the southern crowned pigeon (*Goura scheepmakeri*) of New Guinea. This bird is seen only by those that really matter – potential mates that are attracted to its visual cues. Males and females are the same colour and both sport identical crests.

Competing for attention
Males of many bird species, such as these standardwing birds-of-paradise (*Semioptera wallacii*), display in groups called leks, and the females gather to choose a mate. The male's elaborate plumage reflects his primary need to attract a mate. Conversely, the female is plain so she is well camouflaged as she raises her brood alone.

Lacy crest feathers have hookless barbs and lack solid shafts

display plumage

Feathers are strikingly versatile instruments of visual display. Structurally, their vanes can be long and stiff such as those in a pheasant's tail, or wispy like ostrich plumes, and colours can be bold and brash for showing off or cryptic for hiding. Birds flaunt extravagant ornamentation more than many mammals, perhaps because they can fly away from danger, but also because their better colour vision makes this kind of social signalling a highly effective way both to attract mates and to intimidate competitors.

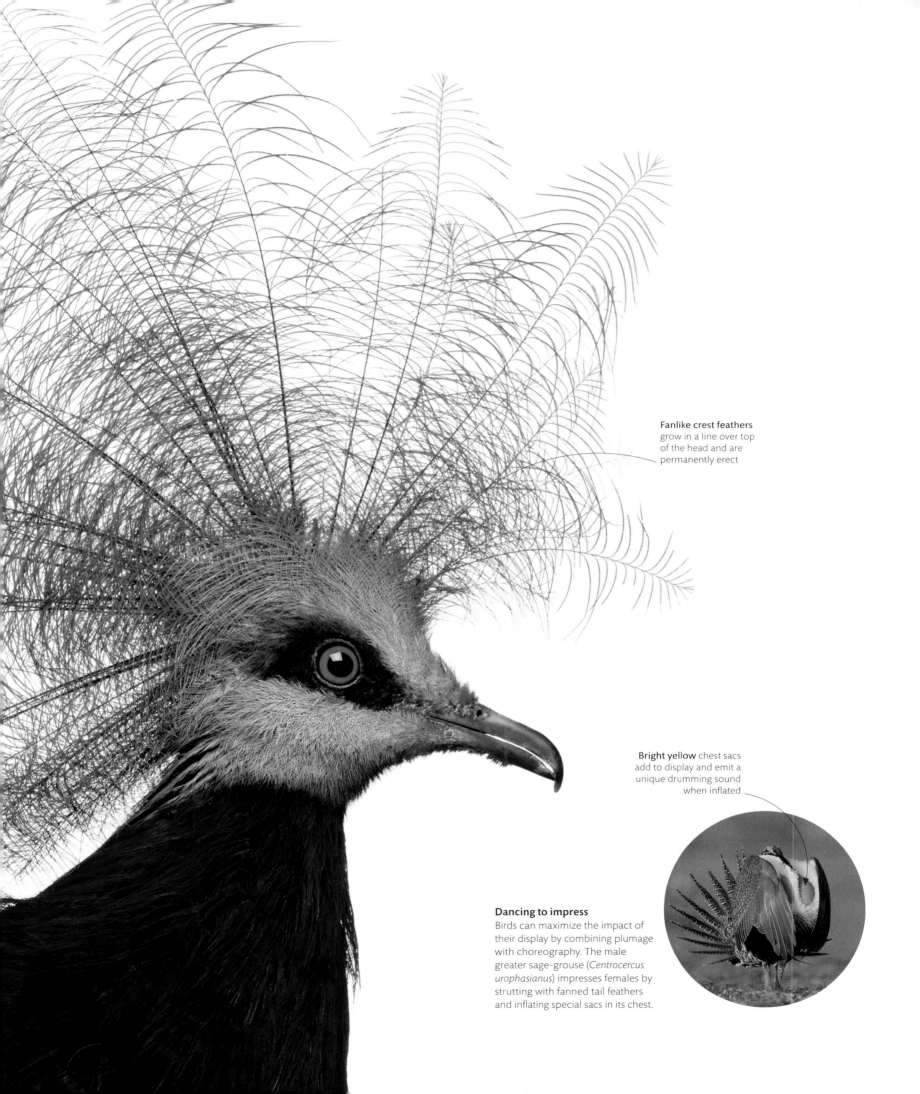

Fanlike crest feathers grow in a line over top of the head and are permanently erect

Bright yellow chest sacs add to display and emit a unique drumming sound when inflated

Dancing to impress
Birds can maximize the impact of their display by combining plumage with choreography. The male greater sage-grouse (*Centrocercus urophasianus*) impresses females by strutting with fanned tail feathers and inflating special sacs in its chest.

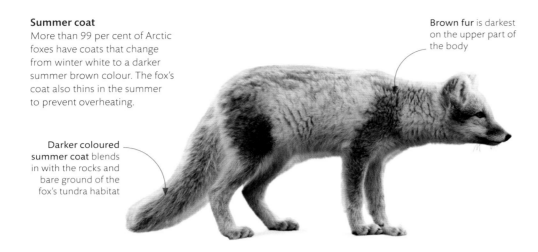

Summer coat
More than 99 per cent of Arctic foxes have coats that change from winter white to a darker summer brown colour. The fox's coat also thins in the summer to prevent overheating.

Darker coloured summer coat blends in with the rocks and bare ground of the fox's tundra habitat

Brown fur is darkest on the upper part of the body

seasonal protection

A furry skin is a quintessential part of being a mammal. Hair is made from keratin – the same tough protein that hardens the surface skin of all vertebrates that live on land. The ability to grow especially dense layers of hair in the coldest places and the harshest conditions means that mammals can trap enough vital body heat near the skin to survive and stay active, even when temperatures drop to freezing point and below.

DOUBLE FUR LAYERS

All hairs grow from specialized pockets of the skin's surface epidermis known as follicles, and there are two types: guard hairs and smaller secondary, or accessory, hairs that make up the underfur. The latter are effective at trapping air close to the skin surface, reducing heat loss. In addition, minute muscles attached to larger guard hairs can pull them straight, erecting the fur for extra insulation in very cold conditions.

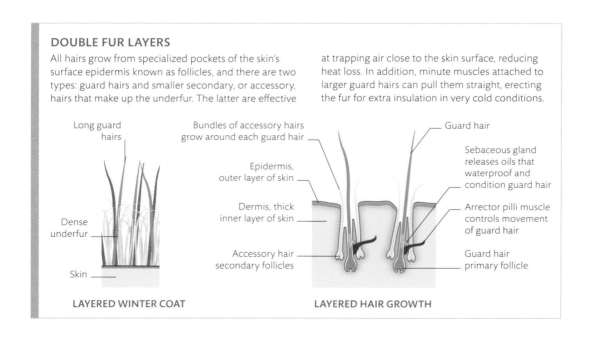

Long guard hairs

Dense underfur

Skin

Bundles of accessory hairs grow around each guard hair

Epidermis, outer layer of skin

Dermis, thick inner layer of skin

Accessory hair secondary follicles

Guard hair

Sebaceous gland releases oils that waterproof and condition guard hair

Arrector pilli muscle controls movement of guard hair

Guard hair primary follicle

LAYERED WINTER COAT

LAYERED HAIR GROWTH

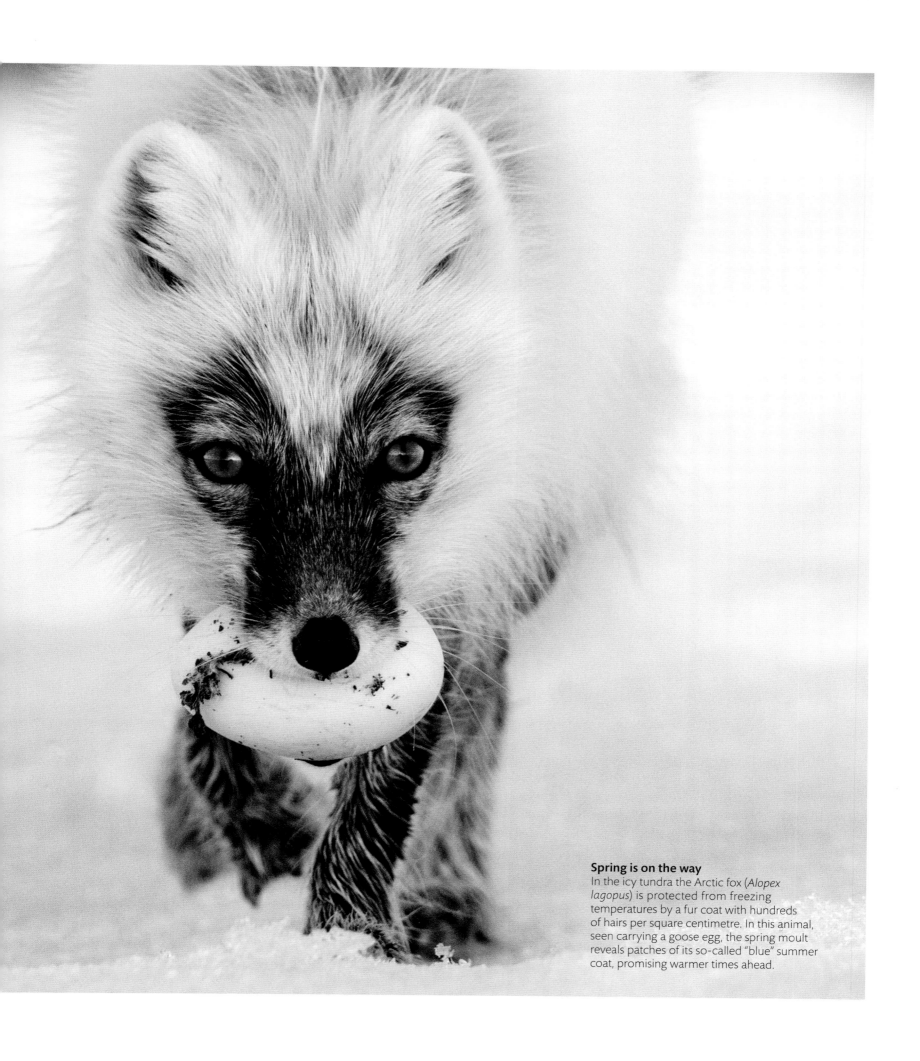

Spring is on the way
In the icy tundra the Arctic fox (*Alopex lagopus*) is protected from freezing temperatures by a fur coat with hundreds of hairs per square centimetre. In this animal, seen carrying a goose egg, the spring moult reveals patches of its so-called "blue" summer coat, promising warmer times ahead.

Colour

Fur colours and the patterns they form, including countershading (having a light belly to contrast a darker back) and dappling, often hide animals in their habitats. As well as acting as camouflage, some patterns, such as a giraffe's spots and a zebra's stripes, are thought to help control body temperature and, in the zebra's case, deter flies. Black-and-white coloration is sometimes a warning that the animal can exude a toxic substance or retaliate fiercely if threatened.

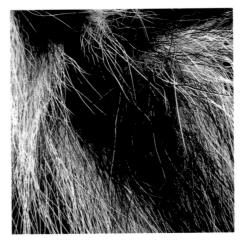

WARNING
Striped skunk
Mephitis mephitis

COUNTERSHADING
Blackbuck
Antilope cervicapra

Form

Fur can be thin or thick, smooth or coarse. Animals in temperate and hot climates are typically covered in shorter, uniform hair, but those facing extreme cold have ultra-thick or double coats, consisting of soft, insulating underfur topped with coarse, water-resistant guard hair. Moles usually have velvety fur that can lie in any direction, minimizing friction with the soil. Even the spines, quills, and scales seen on some mammals are modified hairs.

SINGLE LENGTH, SHORT COAT
Lion
Panthera leo

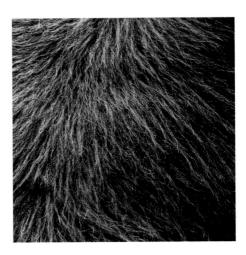

DOUBLE COAT
Musk ox
Ovibos moschatus

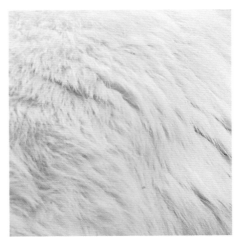

WOOL
Dall sheep
Ovis dalli

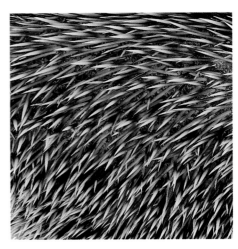

SPINES
Lesser hedgehog tenrec
Echinops telfairi

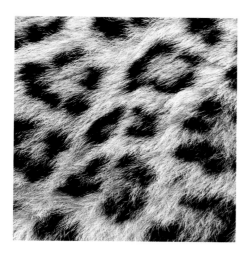

DAPPLED
North Chinese leopard
Panthera pardus japonensis

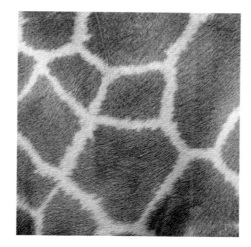

SPOTTED
Reticulated giraffe
Giraffa camelopardalis reticulata

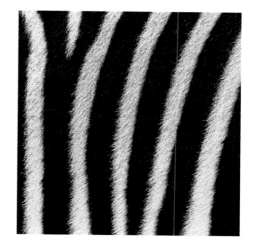

STRIPED
Plains zebra
Equus quagga

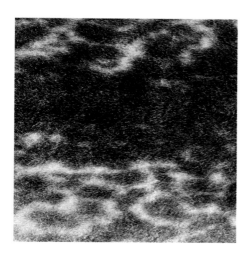

WATERPROOF
Common seal
Phoca vitulina

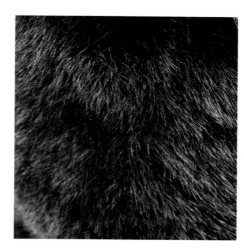

LOW FRICTION
European mole
Talpa europaea

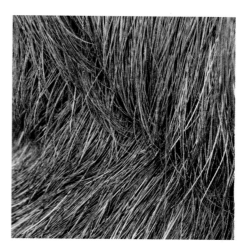

COARSE
Brown-throated three-toed sloth
Bradypus variegatus

mammal fur

Whether called fur, hair, or whiskers, all mammalian skin coverings are made of the same material: a protein called keratin. While providing insulation and protection, fur also camouflages predators and prey, reduces friction, and signals sexual maturity, and certain colours and patterns may have additional functions.

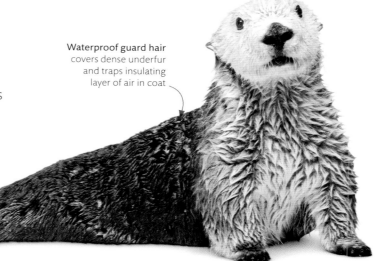

Waterproof guard hair
covers dense underfur and traps insulating layer of air in coat

Densest fur on Earth
With up to 155,000 hairs per sq cm (1 million per sq in) the southern sea otter (*Enhydra lutris*) maintains a comfortable body temperature in water as cold as 1°C (34°F), despite a lack of thick, insulating blubber.

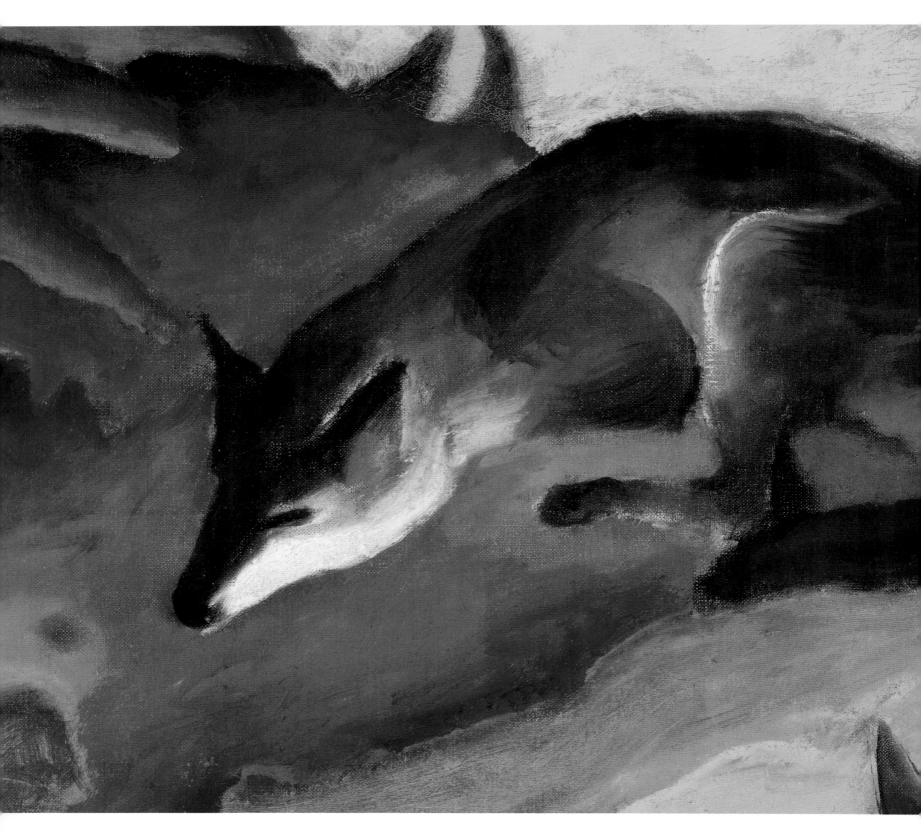

Blue Fox (1911)
Animals were the leitmotif of Franz Marc's
work, returned to continually throughout
his brief life. The German painter and
printmaker used simplified contours and
shapes, and he invested his colours with
spiritual associations to "penetrate
into the soul" of his subjects.

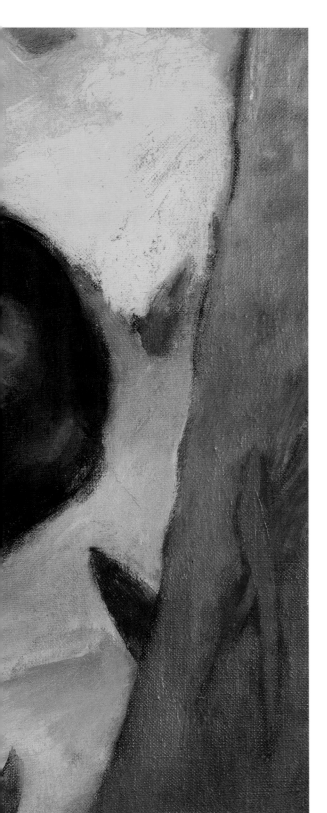

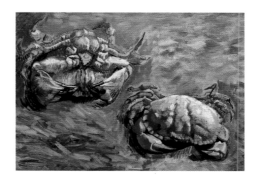

Two Crabs (1889)
Vincent van Gogh's still life painting of vibrant red crabs on a sea-green background pairs complementary colours to dazzling effect. The Dutch post-impressionist may have been prompted by a woodcut of crabs by the Japanese master Hokusai.

animals in art

expressionist nature

At the turn of the 20th century, artists sought revolutionary ways to reflect the pace and complexity of modern life. In a startling new approach, Expressionist painters elevated line, shape, and, most importantly, colour above fidelity to the natural world. Reviewers labelled French painters such as Henri Matisse and Georges Rouault, "Fauves" (wild beasts) because of the raw emotion on display in their vivid pigments and the visceral response of their audience.

While the late 19th-century Impressionists had been preoccupied with capturing fleeting changes in light in their landscapes, flowers, and portraits, post-Impressionists took a new direction. The obsession with colour and form in the works of Vincent van Gogh, for example, were a step on the way to abstract painting, and an inspiration to the Expressionist German artist Franz Marc. Like the Fauves, Marc was concerned with man's empathy with the natural world. Animals were key: in Munich he studied animal anatomy; in Berlin he spent countless hours sketching the shapes of animals and birds and observing their behaviour at Berlin Zoo. "How does a horse see the world, or an eagle, a deer, or a dog? How poor and how soulless is our convention of placing animals in a landscape which belongs to our eyes, instead of penetrating into the soul of the animal in order to imagine his perception?" he wrote in a treatise in 1915.

In 1911 Marc co-founded *Der Blaue Reiter* (*Blue Rider*) magazine and art movement with Vassily Kandinsky, who shared the belief that their abstract art was a counterbalance to a toxic world. In any painting, colour was a separate entity with transcendent associations: blue was male, stern and spiritual; yellow, female, gentle and happy; red, brutal and heavy – the juxtaposition and mixing of colours added insight and balance to works.

Marc's paintings of the *Blue Fox* and *The Little Blue Horses* suggest innocence, while his *The Yellow Cow* has a boundless joy. In an apocalyptic painting of wildlife, *Fate of the Animals* (1913), his coloured beasts, trapped in a forest of burning red, foreshadow the approaching World War in which he lost his life.

66 The natural feeling for life possessed by animals set in vibration everything good in me. 99

FRANZ MARC, *LETTER FROM THE WESTERN FRONT*, APRIL 1915

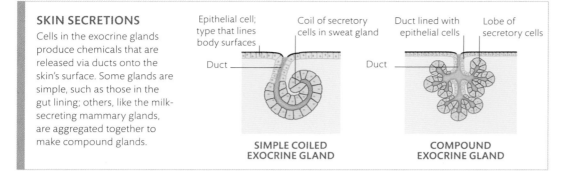

SKIN SECRETIONS

Cells in the exocrine glands produce chemicals that are released via ducts onto the skin's surface. Some glands are simple, such as those in the gut lining; others, like the milk-secreting mammary glands, are aggregated together to make compound glands.

Epithelial cell; type that lines body surfaces

Coil of secretory cells in sweat gland

Duct lined with epithelial cells

Lobe of secretory cells

Duct

Duct

SIMPLE COILED EXOCRINE GLAND

COMPOUND EXOCRINE GLAND

skin glands

Glands are organs that secrete useful substances. Endocrine glands release body chemicals called hormones into the bloodstream, while exocrine glands discharge their contents through ducts onto a surface epithelium, for example, digestive juices into the gut from its walls, or as secretions over the skin. Mammals have many skin glands; some produce watery sweat that cools it or oil for waterproofing, while others generate chemical scents to mark territory, identify individuals, or stimulate courtship.

Facial scent glands
Many hoofed mammals have scent glands on the face. The preorbital glands of the Kirk's dik-dik exude a dark tarlike secretion that it smears on twigs and branches near dung heaps and regular pathways through its territory.

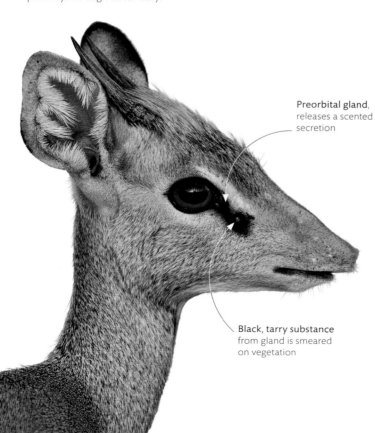

Preorbital gland, releases a scented secretion

Black, tarry substance from gland is smeared on vegetation

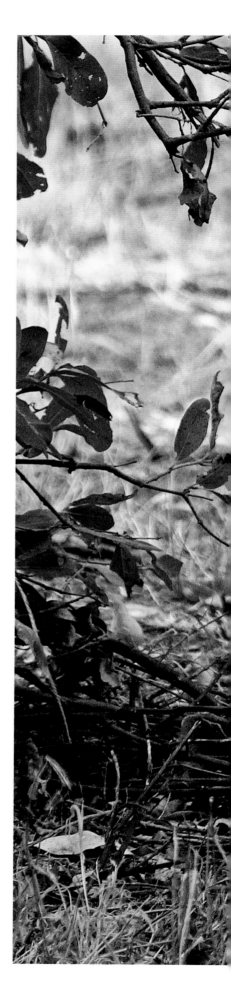

Social signalling
Group organization for this African antelope, the Kirk's dik-dik (*Madoqua kirkii*), depends on scent. Horned males (centre) accompany their female and recent offspring and both adults (males more than the females) deter enemies by marking their territory with facial secretions.

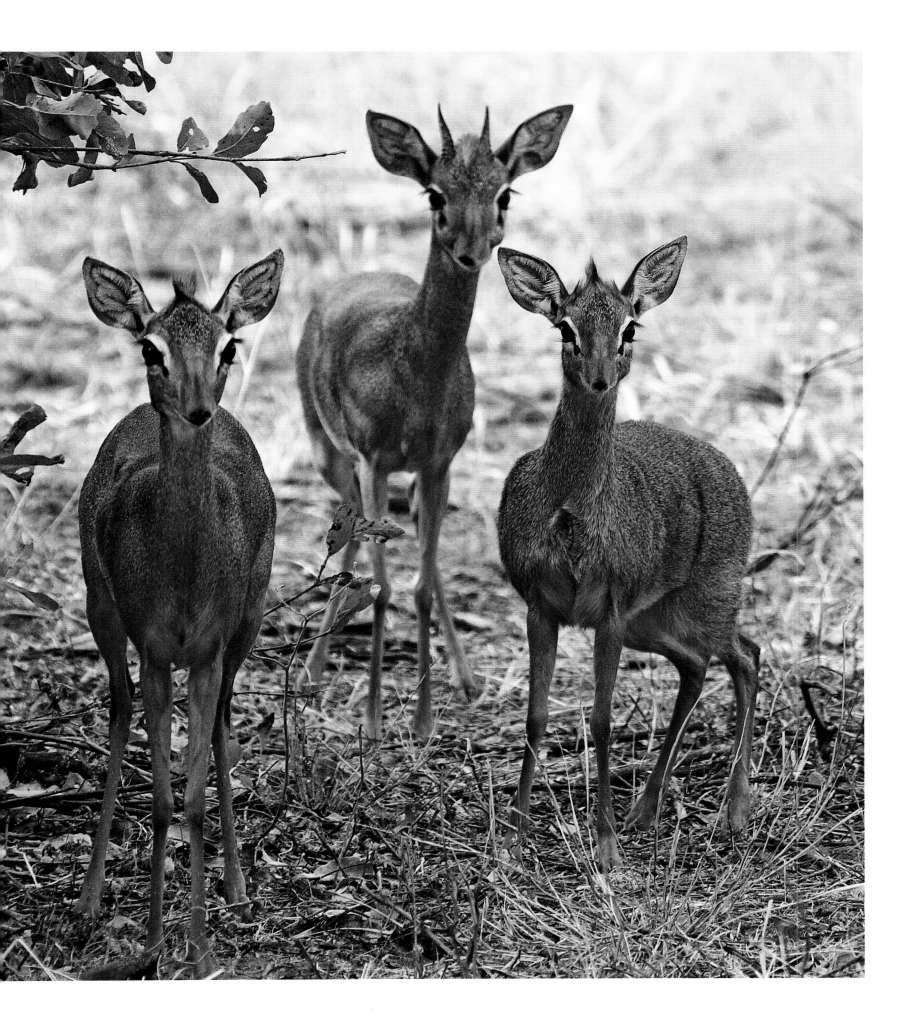

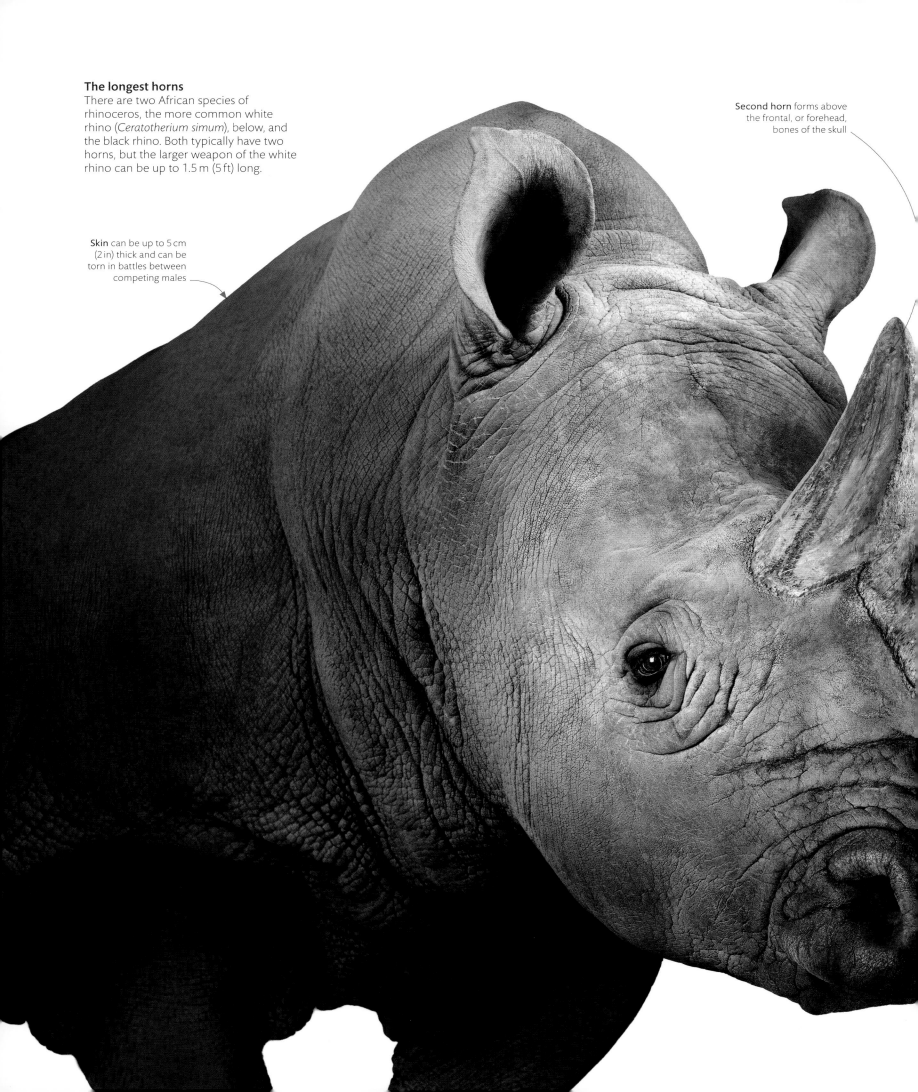

The longest horns
There are two African species of rhinoceros, the more common white rhino (*Ceratotherium simum*), below, and the black rhino. Both typically have two horns, but the larger weapon of the white rhino can be up to 1.5 m (5 ft) long.

Second horn forms above the frontal, or forehead, bones of the skull

Skin can be up to 5 cm (2 in) thick and can be torn in battles between competing males

One-horned rhinoceros
There are three rhinoceros species in Asia. The Indian Rhinoceros (*Rhinoceros unicornis*), right, has a single horn, while females of the one-horned Javan rhinoceros (*R. sondaicus*) are hornless.

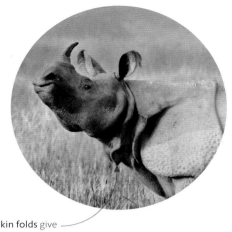

Front horn forms above nasal bones and averages 90 cm (3 ft) in length

Distinctive skin folds give the Indian rhinoceros a more armoured look than its African counterpart

horns from skin

No other animal has horns like those of a rhinoceros, whose name comes from the Greek for "nose-horn". The horns are unique not only in that they sit on the skull, but also in how they are formed. Other animal's horns are bone covered with a sheath of hardened skin (see p.87), while those of a rhinoceros are made from the protein keratin, which also makes claws and hair, compacted to form a defensive weapon that can be wielded just as effectively.

Platform of skin fastens horn to a roughened region of bone

STRUCTURE OF A RHINOCEROS HORN
Scans of rhinoceros horns reveal that, although lacking a bony core, the centre is reinforced with calcium and a pigment called melanin, which protects it from the sun's rays. The outer layers are softer and can become worn. But there is no scientific evidence for the purported therapeutic value that is driving rhinos to extinction.

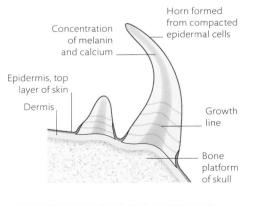

Horn formed from compacted epidermal cells

Concentration of melanin and calcium

Epidermis, top layer of skin

Dermis

Growth line

Bone platform of skull

CROSS-SECTION OF KERATIN-BASED HORN

armoured skin

Keratin is a horny protein that strengthens skin. It is found in its purest form in hairs, claws, and feathers, and in some animals it is used as armour. Pangolins are protected by hard, keratinized, nail-like scales that are highly sensitive to touch – only the underside of the body is unprotected. To make and maintain the scales, pangolins need a high-protein diet – something they satisfy by eating countless ants and termites.

Muscles control the orientation of the scales – raising them when a pangolin rolls into a ball

Adult's scales retain their slightly pointed tips, even after years of abrasion

Protective armour accounts for up to one-third of a pangolin's body weight

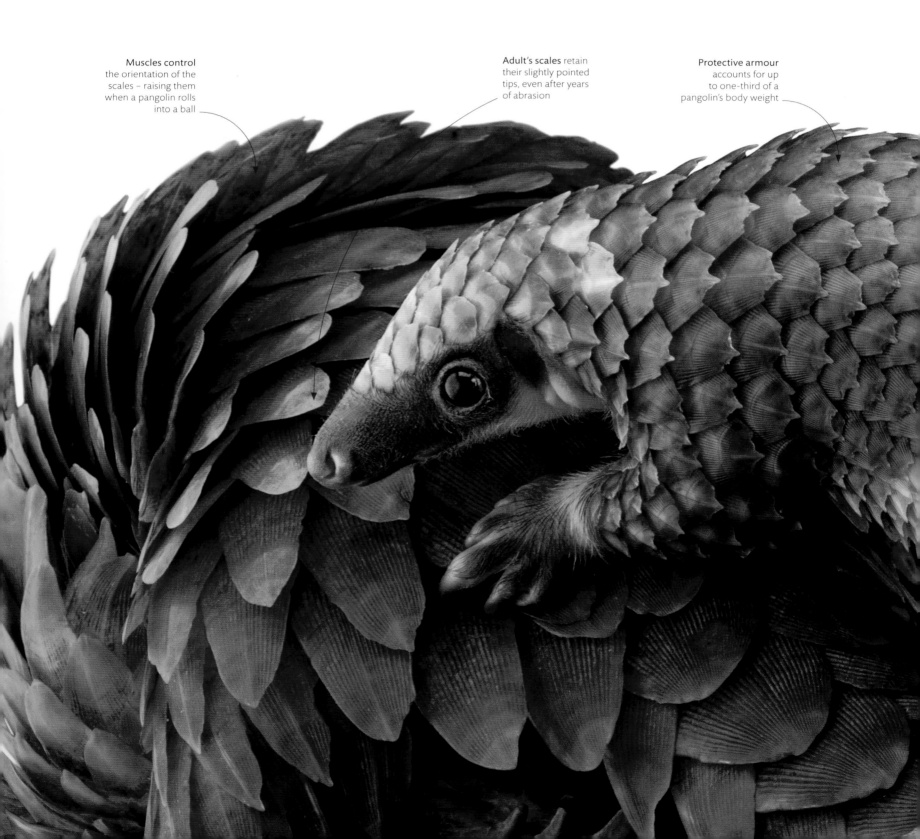

Armadillo protection

Like pangolins, armadillos have hard scales, but they are fused together to form a continuous shield supported on bony plates. The pink fairy armadillo (*Chlamyphorus truncatus*), the smallest species, uses its rump plate to compact sand in its burrows, or possibly to plug the entrance in self-defence.

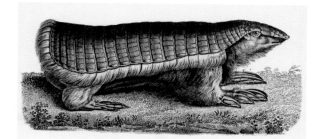

PANGOLIN SCALES

The scales of pangolins are produced by cells in the skin that become filled with hard keratin (or keratinized) as they form this non-living, hornlike material – a process known as cornification. The scales most resemble the nails of primates. As their exposed edges are worn down, they are "repaired" with new keratin made at the base of the scale from cornifying cells in the epidermis.

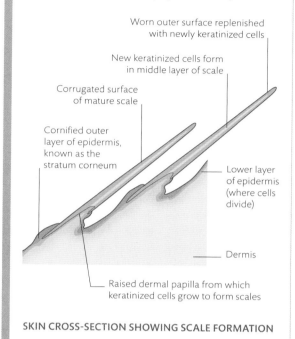

Worn outer surface replenished with newly keratinized cells

New keratinized cells form in middle layer of scale

Corrugated surface of mature scale

Cornified outer layer of epidermis, known as the stratum corneum

Lower layer of epidermis (where cells divide)

Dermis

Raised dermal papilla from which keratinized cells grow to form scales

SKIN CROSS-SECTION SHOWING SCALE FORMATION

Scales of an infant

have a characteristic three-pointed shape, which becomes smoother with age

Overlapping scales

protect against bites from larger predators but provide little defence against stinging insects

Protected from birth

A newborn common African pangolin (*Manis tricuspis*) comes into the world covered with soft scales, which rapidly harden. Shortly after birth the infant learns to cling to safety on its mother's back and – like her – rolls into a ball if it is in danger.

senses

sense. a faculty – such as sight, hearing, smell, taste, or touch – by which an animal receives information about the external world.

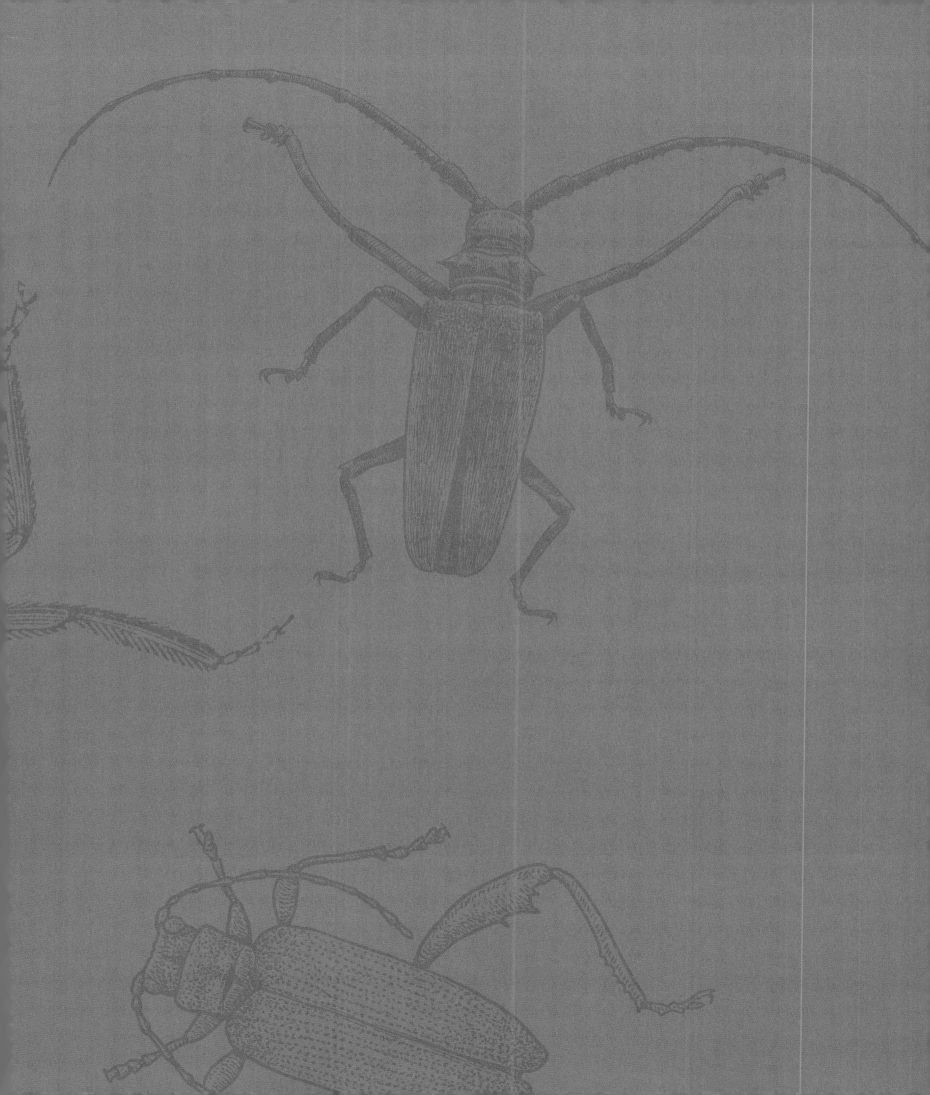

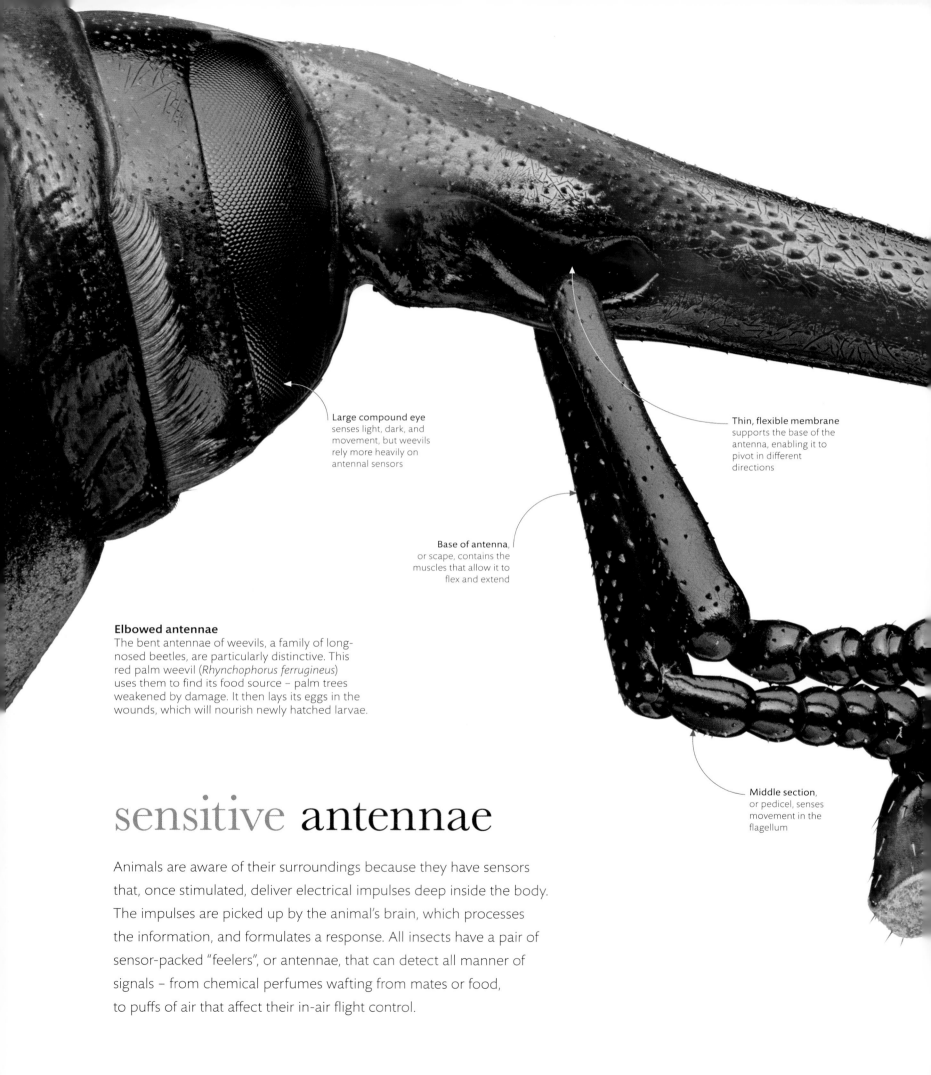

Large compound eye
senses light, dark, and movement, but weevils rely more heavily on antennal sensors

Thin, flexible membrane
supports the base of the antenna, enabling it to pivot in different directions

Base of antenna,
or scape, contains the muscles that allow it to flex and extend

Elbowed antennae
The bent antennae of weevils, a family of long-nosed beetles, are particularly distinctive. This red palm weevil (*Rhynchophorus ferrugineus*) uses them to find its food source – palm trees weakened by damage. It then lays its eggs in the wounds, which will nourish newly hatched larvae.

Middle section,
or pedicel, senses movement in the flagellum

sensitive antennae

Animals are aware of their surroundings because they have sensors that, once stimulated, deliver electrical impulses deep inside the body. The impulses are picked up by the animal's brain, which processes the information, and formulates a response. All insects have a pair of sensor-packed "feelers", or antennae, that can detect all manner of signals – from chemical perfumes wafting from mates or food, to puffs of air that affect their in-air flight control.

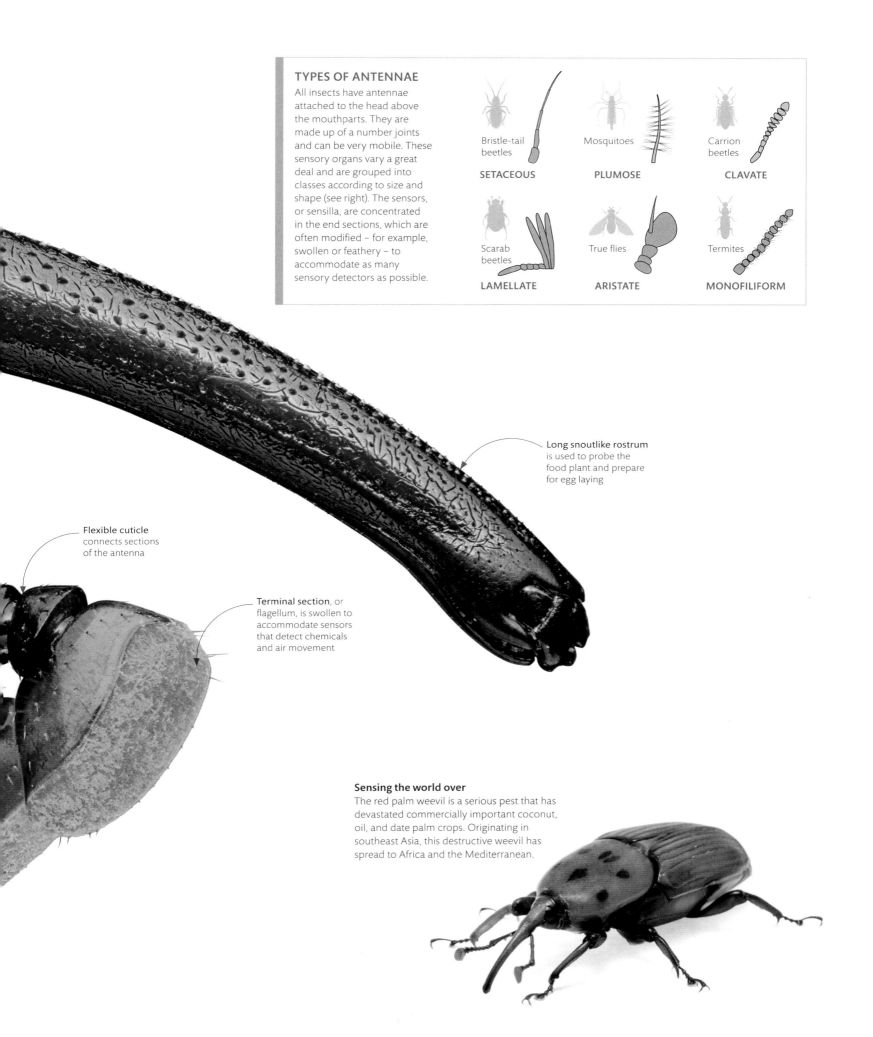

TYPES OF ANTENNAE

All insects have antennae attached to the head above the mouthparts. They are made up of a number joints and can be very mobile. These sensory organs vary a great deal and are grouped into classes according to size and shape (see right). The sensors, or sensilla, are concentrated in the end sections, which are often modified – for example, swollen or feathery – to accommodate as many sensory detectors as possible.

Bristle-tail beetles
SETACEOUS

Mosquitoes
PLUMOSE

Carrion beetles
CLAVATE

Scarab beetles
LAMELLATE

True flies
ARISTATE

Termites
MONOFILIFORM

Long snoutlike rostrum
is used to probe the food plant and prepare for egg laying

Flexible cuticle
connects sections of the antenna

Terminal section, or flagellum, is swollen to accommodate sensors that detect chemicals and air movement

Sensing the world over

The red palm weevil is a serious pest that has devastated commercially important coconut, oil, and date palm crops. Originating in southeast Asia, this destructive weevil has spread to Africa and the Mediterranean.

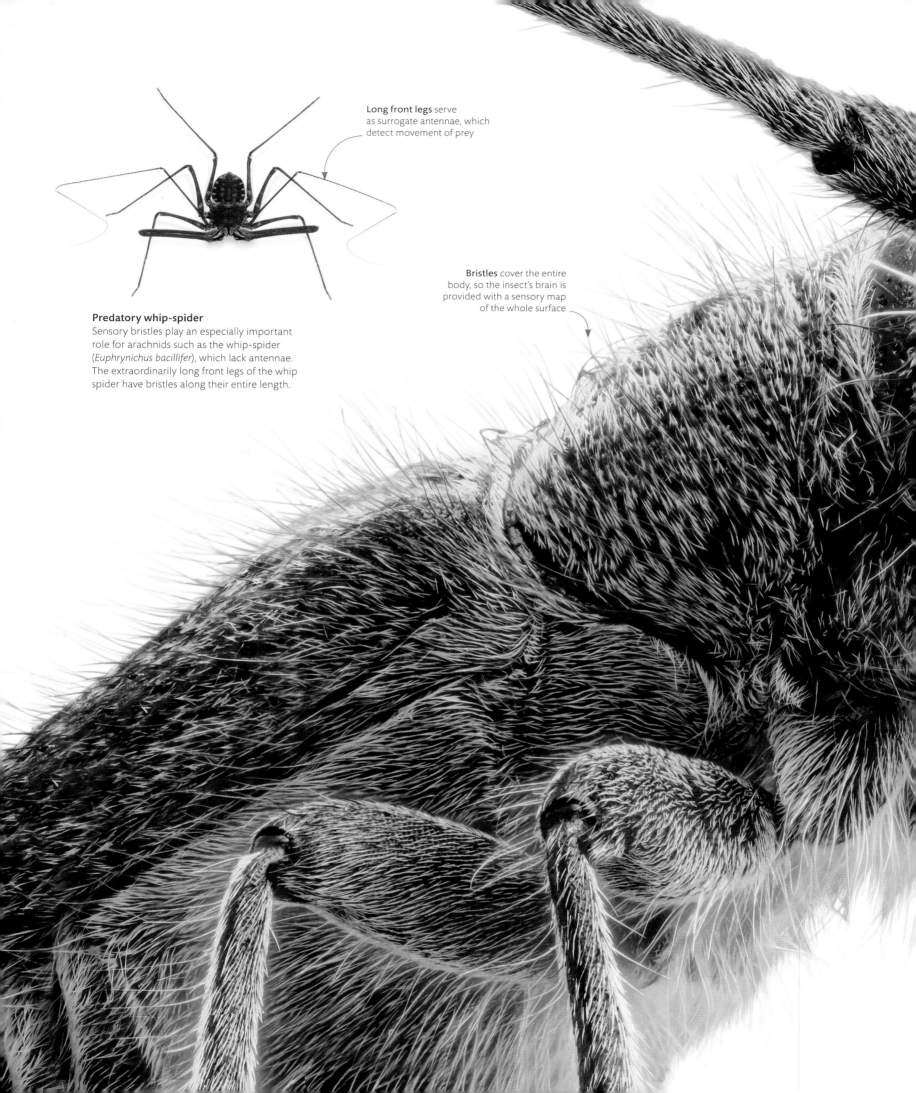

Long front legs serve
as surrogate antennae, which
detect movement of prey

Bristles cover the entire
body, so the insect's brain is
provided with a sensory map
of the whole surface

Predatory whip-spider
Sensory bristles play an especially important
role for arachnids such as the whip-spider
(*Euphrynichus bacillifer*), which lack antennae.
The extraordinarily long front legs of the whip
spider have bristles along their entire length.

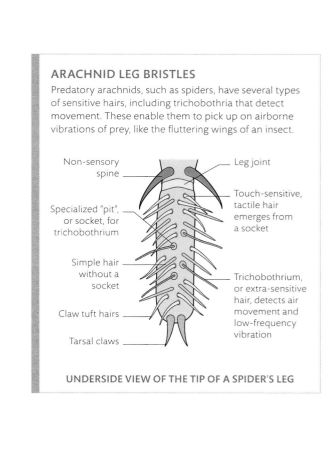

Densely coated beetle
When viewed under high magnification, the extraordinary coat of sensory bristles (or setae) on this bluish longhorn beetle (*Opsilia coerulescens*) shows just how much insects rely on tactile stimulation. Each hair – when deflected – tells the insect where it is, as well as what is happening around it.

Every touch-sensitive hair (seta) emerges from a socket connected to sensory nerve endings

Concentrations of hairs at the joints – hair plates – sense movement of body parts and provide information about their relative positions

sensory bristles

The skin of any animal is at the frontline of its surroundings, so is filled with sensory nerve endings that pick up signals from the environment and carry them to the brain. For animals such as insects and arachnids with a hard exoskeleton outside their skin, cells in the epidermis produce an array of specialized, tactile bristles that poke through the tough outer layer to improve the body's sensitivity to touch and movement.

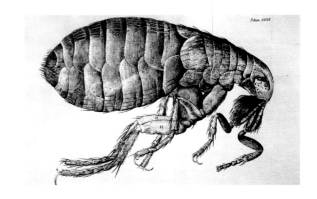

The flea (1665)
In his book *Micrographia*, natural philosopher Robert Hooke reveals the unexplored complexity of a flea made visible by his revolutionary adjustable microscope.

animals in art

small world

After 200 years of exploration had brought the world's wildlife to European shores, the 17th century saw a new age of discovery, which revealed the mysteries of a smaller world. The new science of entomology was fuelled by advances in microscopes, by the artistic skills of scientists, and by artists who used precise methods to portray insect life in acclaimed works of art.

Illustration skills became an imperative for the new entomologists of the 17th century. English inventor and scientist Robert Hooke was a consummate artist. Using a new compound microscope with illumination, he revealed the unimagined anatomy of insects, and identified plant cells. When he published his astounding drawings in his 1665 work *Micrographia*, the insects were so alien, some refused to believe they were real.

In the Netherlands, textile merchant Antonie van Leeuwenhoek created tiny microscopes using 1-mm lenses. Their manufacture was a closely guarded secret, but they were probably forged from the glass pearls that were used to examine cloth. Their unparalleled magnification revealed the structure of single-celled organisms, bacteria, and spermatozoa.

In the same period, notable artists immersed themselves in acute observation of natural subjects. Jan van Kessel II, from the illustrious Brueghel family, worked with live insect specimens and pored over illustrated scientific texts to bring complete accuracy to art.

In her teenage years, Frankfurt-born Maria Sibylla Merian (1647–1717) was fascinated by the metamorphosis of caterpillars into moths and butterflies. She went on to become one of the first artists to draw insects and butterflies on their host plants. After moving to Amsterdam, Merian was given, at age 52, a rare government sponsorship to record insect life in the Dutch colony of Surinam. Carl Linnaeus later used the magnificent drawings from her book *Metamorphosis of the Insects of Surinam* to classify new species.

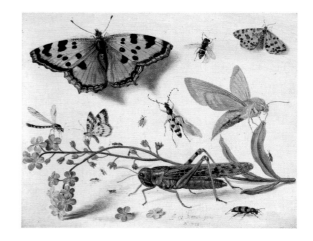

Insects and forget-me-nots (1653)
Jan van Kessel II used scientific texts as sources for delicate watercolours on parchment of beetles, moths, butterflies, and grasshoppers. He sometimes created pictures of his name spelled out with caterpillars.

Caterpillars, butterflies, and flowers (1705)
Two saturniid moths (*Arsenura armida*), widespread in Mexico and South America, frame the flowering branch of a towering tropical tree in a watercolour from Maria Sibylla Merian's remarkable book on the insects of Surinam. The larvae shown are not the early stage of this moth, as Merian supposed, but from an unknown species.

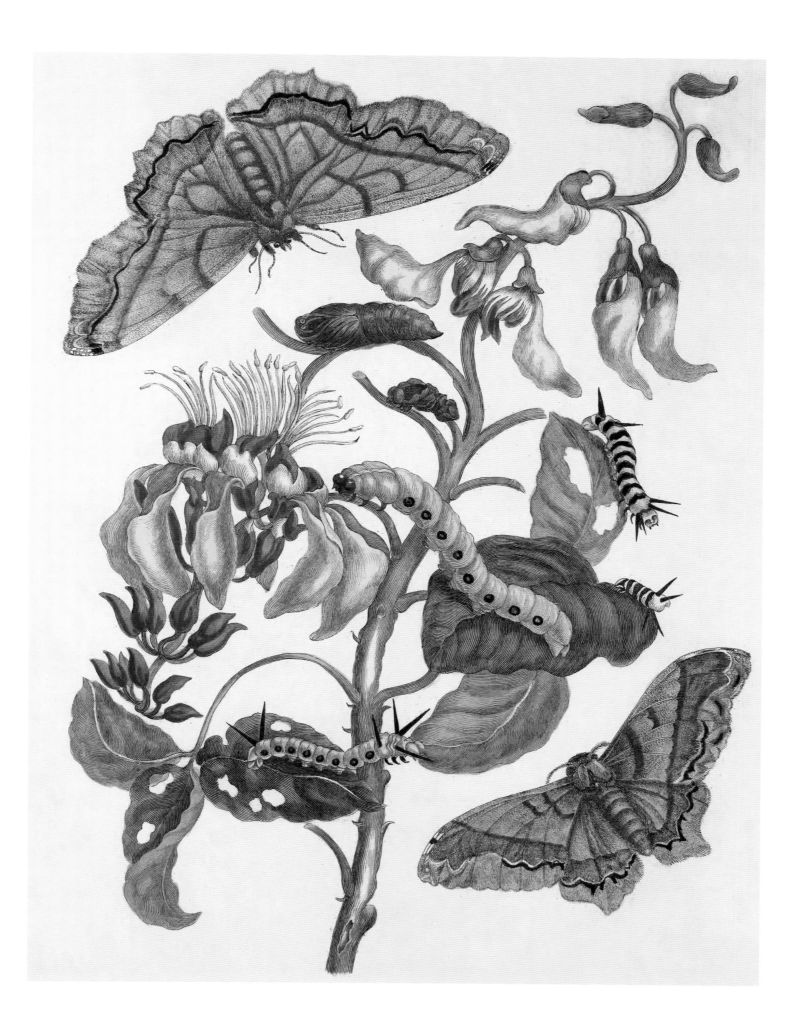

SENSITIVE HAIRS

Each whisker – technically called a vibrissa – is linked to a few sensory nerve endings near the skin's surface, which mainly wrap around the whisker. However, 80 per cent of the nerve endings associated with the whisker are deeper down and lie parallel with the root of the whisker. When the whisker bends away from its follicle base, the nerve endings are stimulated to fire electrical impulses to the brain.

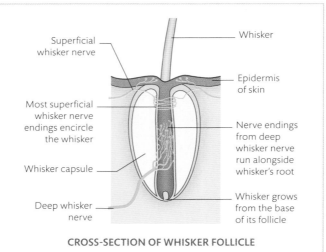

Superficial whisker nerve

Whisker

Epidermis of skin

Most superficial whisker nerve endings encircle the whisker

Nerve endings from deep whisker nerve run alongside whisker's root

Whisker capsule

Deep whisker nerve

Whisker grows from the base of its follicle

CROSS-SECTION OF WHISKER FOLLICLE

Feeling for fish

Whiskers enable animals to "feel" their surroundings by sensing minute vibrations in the water or air. The whiskers of a California sea lion (*Zalophus californianus*) can detect the tiny wake caused by its fish prey against the background movement of the currents, and even determine a target's size, shape, and texture by touch. They allow the sea lion to find food in the limited visibility of turbid coastal waters.

sensory whiskers

The hairs of a mammal are rooted to nerves and muscles, enabling the animal to detect their movement. Hairs on the face – notably in carnivores and rodents – have evolved to become especially sensitive, being attached to complex bundles of nerve fibres that fire impulses in response to the slightest touch. It is possible that, millions of years ago, the very first "protohairs" in the reptilian ancestors of mammals evolved with this tactile function, as nocturnal or burrowing pioneers struggled to find their way in darkness.

Whiskered bird

Some birds have stiff, modified feathers, called rictal bristles, that project from the base of the bill and may work like whiskers. Rictal bristles are conspicuous in nightjars and flycatchers, which perhaps use them to detect contact with insects when hunting on the wing. They are most developed in nocturnal species: kiwi use their rictal bristles to feel for invertebrates on the ground, aided by a sense of smell that is unusually strong for a bird.

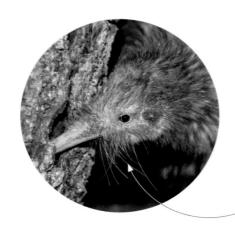

Each rictal bristle is a modified feather with a stiffened shaft that lacks barbs

Coat consists of guard hairs overlying shorter fine underhairs; oil secreted by glands under the skin keeps the coat waterproofed

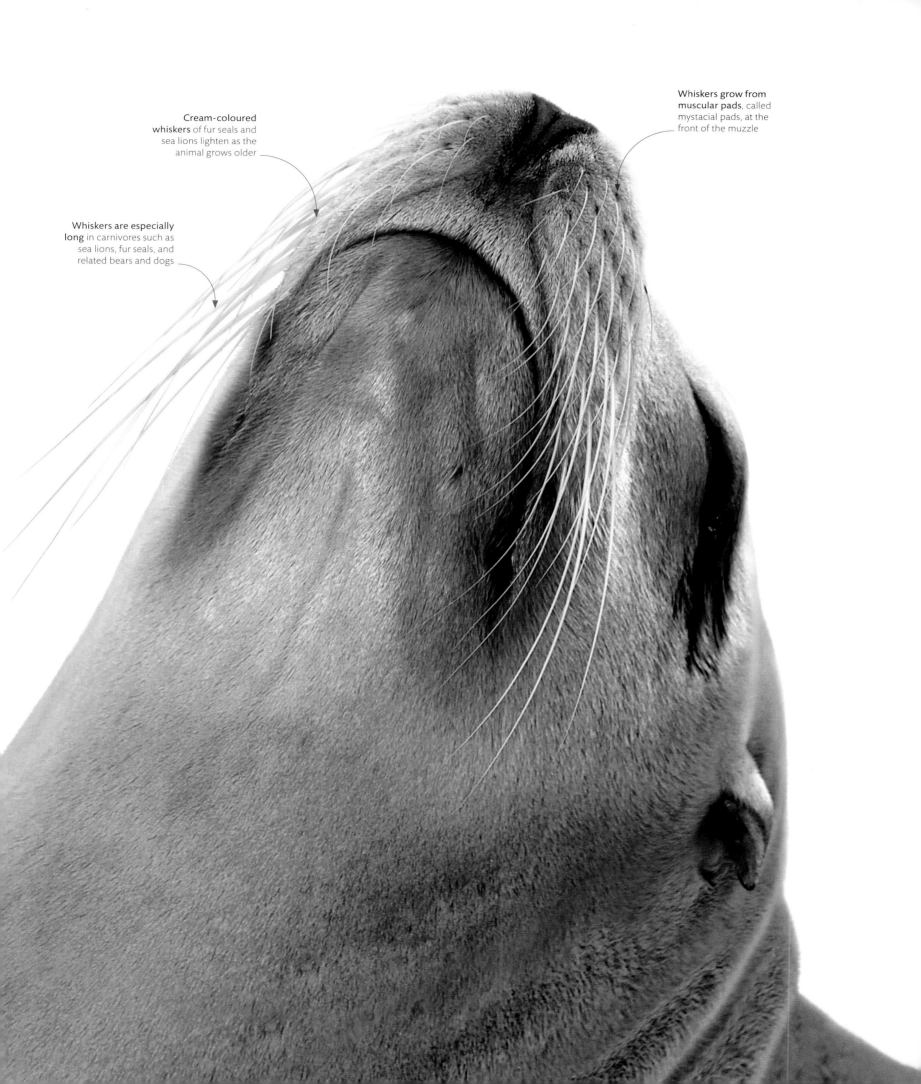

Cream-coloured **whiskers** of fur seals and sea lions lighten as the animal grows older

Whiskers grow from muscular pads, called mystacial pads, at the front of the muzzle

Whiskers are especially long in carnivores such as sea lions, fur seals, and related bears and dogs

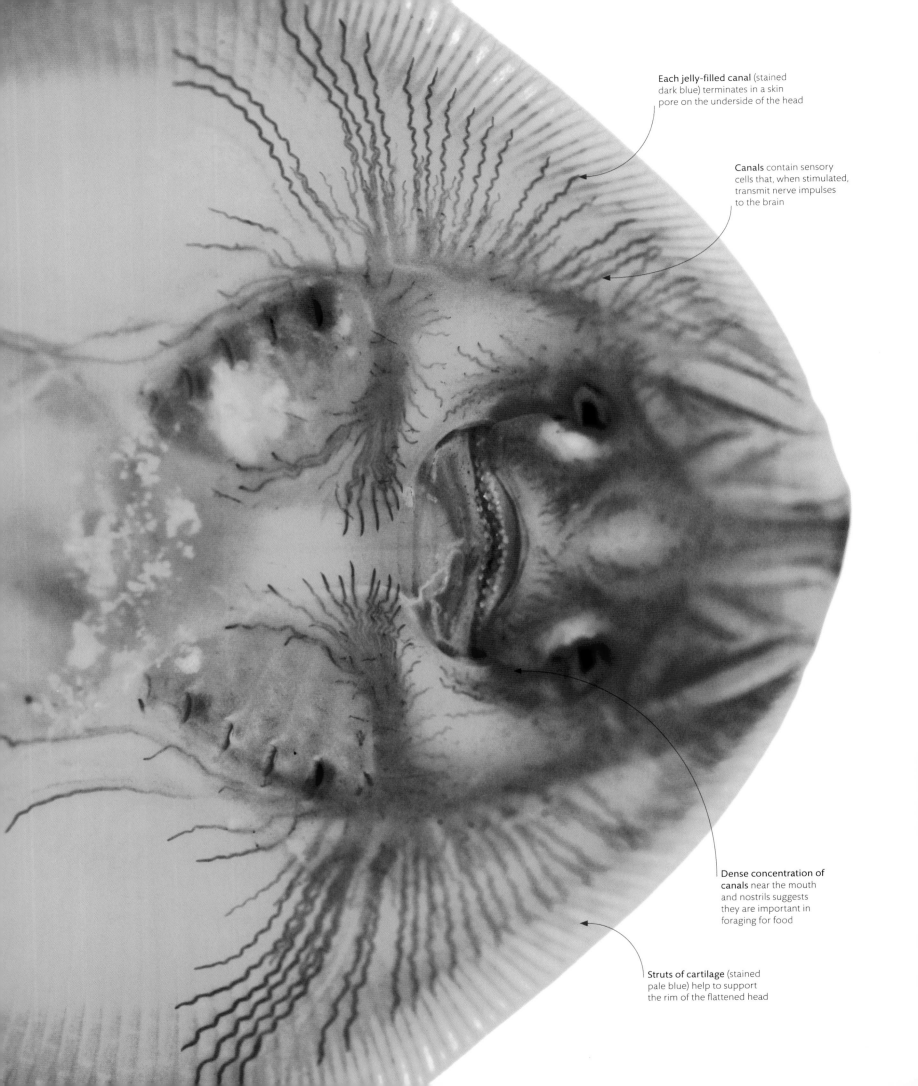

Each jelly-filled canal (stained dark blue) terminates in a skin pore on the underside of the head

Canals contain sensory cells that, when stimulated, transmit nerve impulses to the brain

Dense concentration of canals near the mouth and nostrils suggests they are important in foraging for food

Struts of cartilage (stained pale blue) help to support the rim of the flattened head

Sensory canals

This underside view of the flattened head of a little skate (*Leucoraja erinacea*) – here stained blue – reveals dark, radiating, jelly-filled canals called ampullary organs. The canals contain sensory cells that detect electrical fields produced by the muscles of prey animals. Once triggered, the cells fire off nerve impulses to the skate's brain, guiding the fish's underslung mouth towards invertebrates buried, and unseen, in sea-floor sediment.

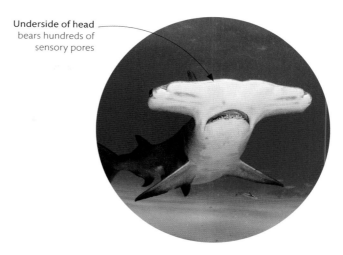

Underside of head
bears hundreds of
sensory pores

Targeting prey

An arrangement of sensors along the wide front of its modified head helps a great hammerhead shark (*Sphyrna mokarran*) triangulate chemical and electrical signals to pinpoint prey.

sensing under water

Underwater animals live in a medium that is denser than air – a place where sound travels better, but light is diminished by turbidity or depth, and odours disperse more slowly. Aquatic animals have evolved sensory systems adapted to these conditions: tactile receptors that detect the slightest ripples and remarkable sensors that pick up chemical traces or even faint electric signals from prey.

DETECTING MOVEMENT

Fish sense water movement via their lateral line system: a series of canals that run along the body, under the skin. The canals channel water from the surroundings to bend neuromasts – jelly-topped bundles of sensory cells – that send nerve impulses to the brain as they move back and forth.

Tube opening to
surrounding sea water

Scale

Current bends
jelly cone

Bending
cone triggers
nerve signal

Neuromast

Sensory nerve links
to neuromasts

Sensory hairs
embedded in
jellylike cone

Sensory
hair cell

CROSS-SECTION OF SHARK BODY SURFACE

NEUROMAST

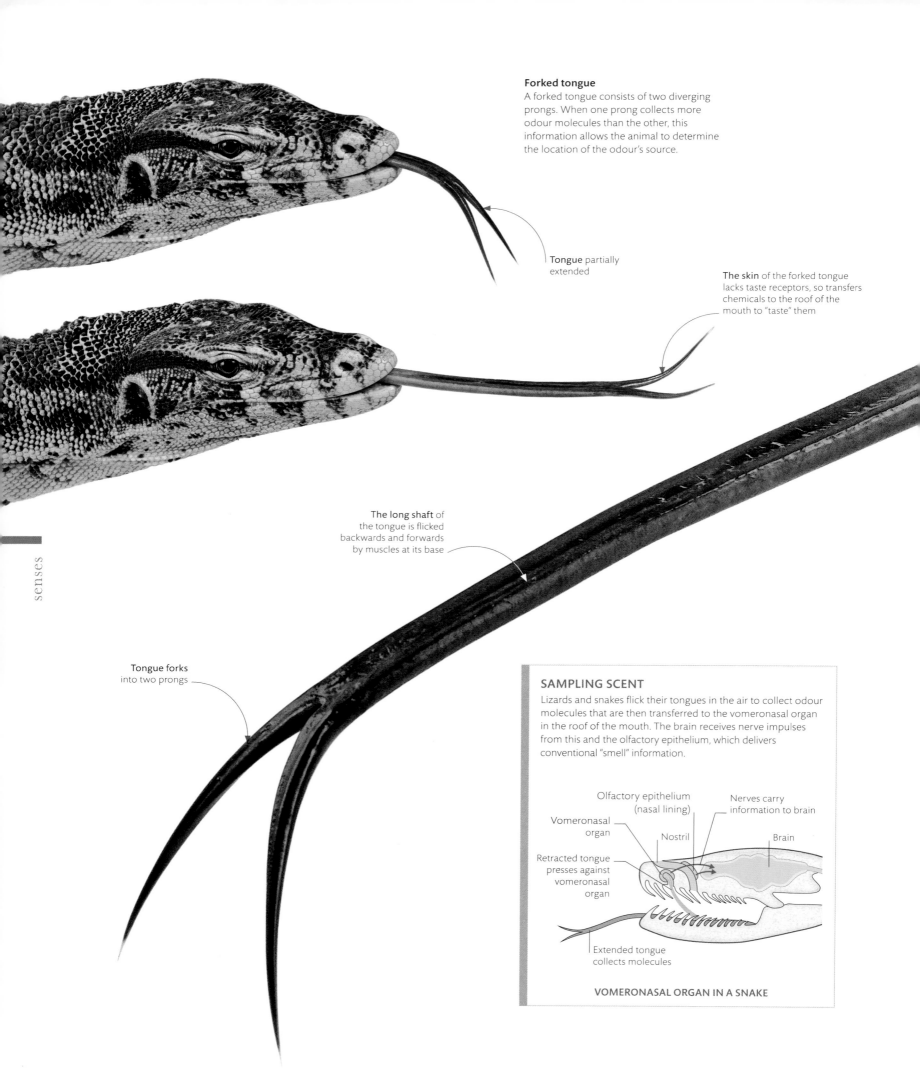

Forked tongue
A forked tongue consists of two diverging prongs. When one prong collects more odour molecules than the other, this information allows the animal to determine the location of the odour's source.

Tongue partially extended

The skin of the forked tongue lacks taste receptors, so transfers chemicals to the roof of the mouth to "taste" them

The long shaft of the tongue is flicked backwards and forwards by muscles at its base

Tongue forks into two prongs

SAMPLING SCENT

Lizards and snakes flick their tongues in the air to collect odour molecules that are then transferred to the vomeronasal organ in the roof of the mouth. The brain receives nerve impulses from this and the olfactory epithelium, which delivers conventional "smell" information.

Olfactory epithelium (nasal lining)

Nerves carry information to brain

Vomeronasal organ

Nostril

Brain

Retracted tongue presses against vomeronasal organ

Extended tongue collects molecules

VOMERONASAL ORGAN IN A SNAKE

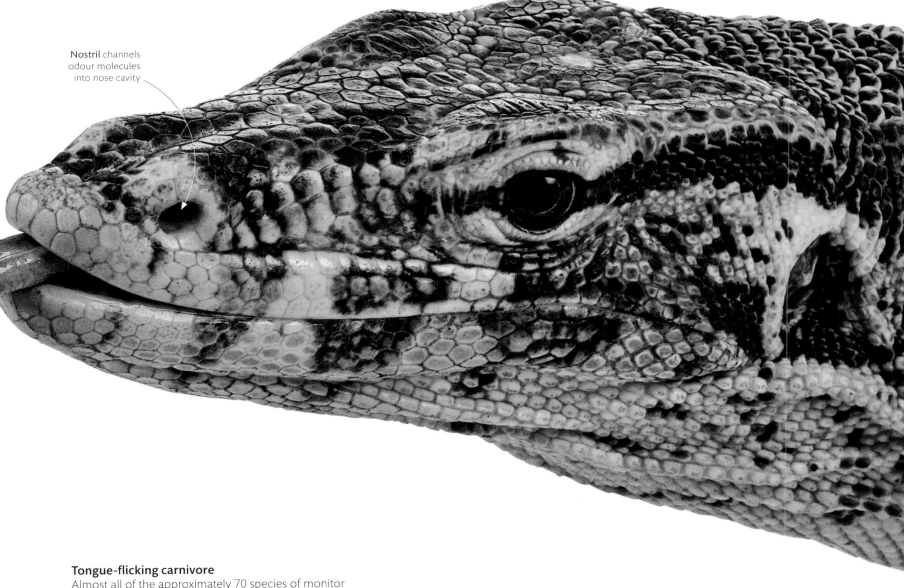

Nostril channels
odour molecules
into nose cavity

Tongue-flicking carnivore
Almost all of the approximately 70 species of monitor
lizard have a taste for meat and can sense live prey or
carrion by flicking their tongue. The Asian water monitor
(*Varanus salvator*) is one of the largest when full-grown,
attacking animals up to the size of young crocodiles.

tasting the air

Chemoreception, or sensing chemicals, can take the form of smelling odours
(olfaction) or tasting flavours in the mouth – but sometimes the distinction is
blurred. The vomeronasal (or Jacobson's) organ is a sense organ that supplements
the main olfactory system in many amphibians, reptiles, and mammals. In lizards
and snakes, it senses odour molecules collected from the air by a forked tongue,
helping the animal to "taste" the air for prey, predators, or even potential mates.

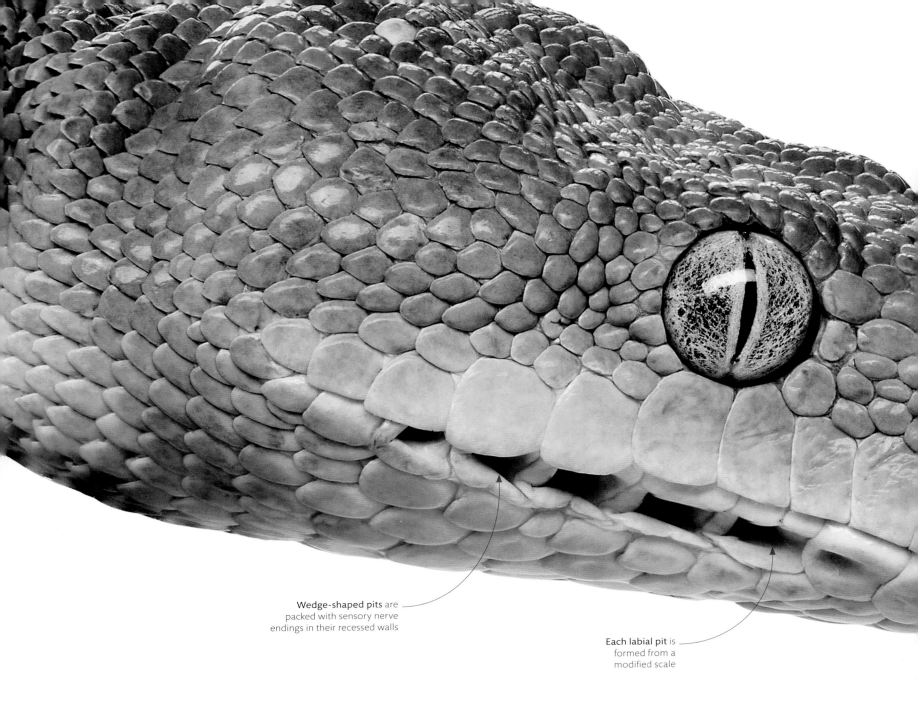

Wedge-shaped pits are packed with sensory nerve endings in their recessed walls

Each labial pit is formed from a modified scale

sensing heat

Many animals sense a source of heat not by using temperature sensors, but with cells that detect infrared radiation. Infrared is a form of electromagnetic radiation – with a wavelength slightly longer than that of visible red light – emitted by warm objects. Animals with infrared sensors can pick up these radiation signals at great distances, helping predators, such as certain kinds of snakes, track down warm-blooded prey.

Heat sensors

The infrared sensors of pythons, such as this green tree python (*Morelia viridis*), are carried in pits along the upper and lower lip and on the snout. These pits help the snake detect its prey at night. The python reaches out at lightning-fast speed to grab prey with its jaws, which it then kills by constriction.

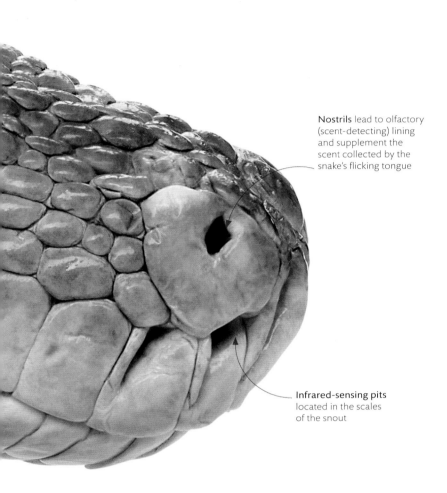

Nostrils lead to olfactory (scent-detecting) lining and supplement the scent collected by the snake's flicking tongue

Infrared-sensing pits located in the scales of the snout

Labial scales along the upper and lower lips of boas support sensory nerve endings

INFRARED RECEPTORS

Sensory nerve endings act as infrared receptors and are triggered when surrounding tissue is warmed by infrared. In boas, nerve endings are embedded in surface scales, and in pythons at the base of pits. In both, some heat dissipates into the skin. Pit-vipers have nerve endings that are suspended in membranes, which warm more quickly, making them more sensitive.

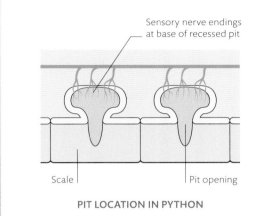

Sensory nerve endings at base of recessed pit

Scale   Pit opening

PIT LOCATION IN PYTHON

Parallel lives

Despite evolving in a different part of the world, the emerald tree boa (*Corallus caninus*), from South America, shares a number of characteristics with the green tree python, from New Guinea. Both are nocturnal hunters that live in the low branches of the rainforest, and both use infrared detectors to hunt for warm-blooded prey in the dark.

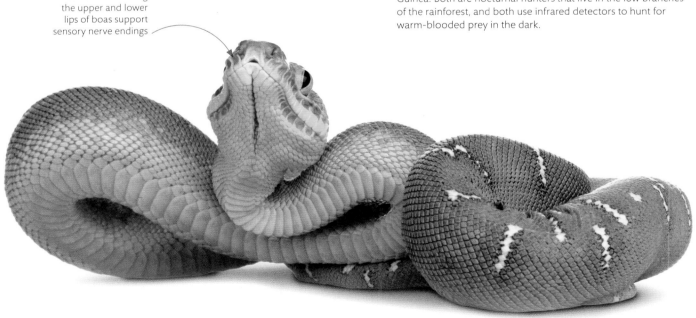

electrical sense

Swimming in murky water can make it difficult to move around safely and find food. Some animals take advantage of the fact that minerals in water cause them to conduct electricity, which can be detected with special sensors. Most fish and a group of egg-laying mammals – the monotremes – use their sensors to pick up electrical signals that indicate the presence of prey or predators.

senses

Eyes and ears remain closed when in water, so animal is dependent on sensory bill

Frontal shield, made of sensor-packed skin, extends the area for detecting electrical signals

Flared nostrils close during diving

Rubbery skin carrying sensors covers beak-shaped facial bones

Horny pads instead of teeth grind the exoskeletons of invertebrate prey

---

ELECTROSENSING PREY

There are two kinds of egg-laying mammals. Spiny echidnas have some electrical sensors in their pointed noses, enabling them to probe soil for worms. But the aquatic platypus's duck bill is packed with them. Its prey gives off electrical signals that are picked up almost instantaneously by the bill. Other sensors detect movement in the water slightly later. The animal's brain uses this time difference to determine the position of the prey in the water.

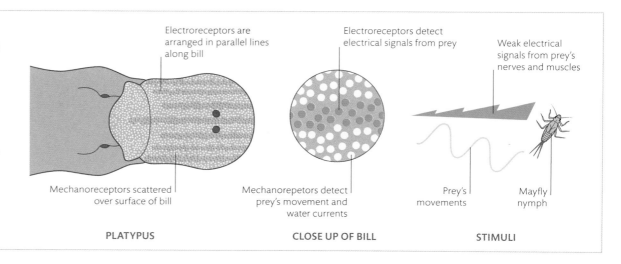

Electroreceptors are arranged in parallel lines along bill

Mechanoreceptors scattered over surface of bill

Electroreceptors detect electrical signals from prey

Mechanorepetors detect prey's movement and water currents

Weak electrical signals from prey's nerves and muscles

Prey's movements

Mayfly nymph

PLATYPUS

CLOSE UP OF BILL

STIMULI

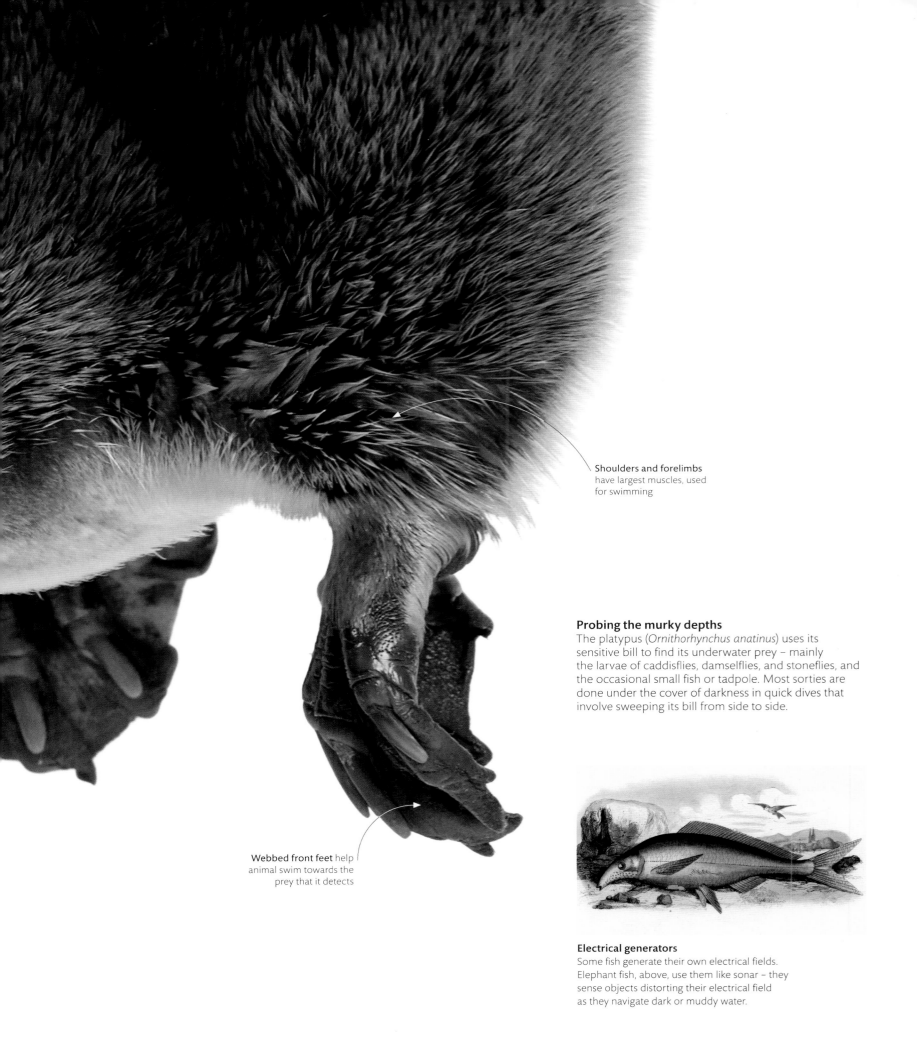

Shoulders and forelimbs have largest muscles, used for swimming

Webbed front feet help animal swim towards the prey that it detects

Probing the murky depths

The platypus (*Ornithorhynchus anatinus*) uses its sensitive bill to find its underwater prey – mainly the larvae of caddisflies, damselflies, and stoneflies, and the occasional small fish or tadpole. Most sorties are done under the cover of darkness in quick dives that involve sweeping its bill from side to side.

Electrical generators

Some fish generate their own electrical fields. Elephant fish, above, use them like sonar – they sense objects distorting their electrical field as they navigate dark or muddy water.

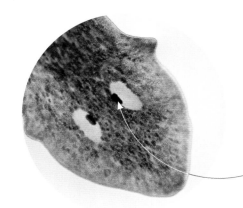

Eyespots
This planarian flatworm (*Dugesia* sp.) has a dark spot on the inner surface of each eye. The spots screen out light so that each eye senses light coming from a different direction.

Each eye is a "cup" of nerve fibres that detects light on only one side of the dark eye-spot

detecting light

Sensing light requires a pigment that is chemically altered when illuminated. Even bacteria and plants have this, but only animals generate information from light in a way that provides genuine vision. Light stimulates pigment-containing cells in their eyes to send impulses to the brain for processing. The simplest animal eyes – flatworm eyes – sense light and its direction, but complex eyes, like spider eyes, use lenses to focus and create images.

Eyes of a hunter
Most types of spider have eight eyes, each with a lens. Web-building spiders rely more on tactile cues than vision, but pursuit-hunters, such as this jumping spider, use their large, forward-facing eyes to judge both detail and depth when ambushing prey.

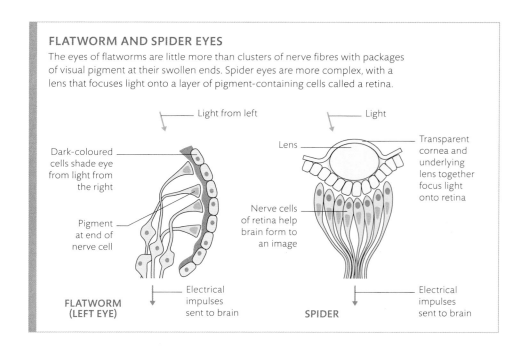

FLATWORM AND SPIDER EYES
The eyes of flatworms are little more than clusters of nerve fibres with packages of visual pigment at their swollen ends. Spider eyes are more complex, with a lens that focuses light onto a layer of pigment-containing cells called a retina.

Light from left

Light

Lens

Dark-coloured cells shade eye from light from the right

Transparent cornea and underlying lens together focus light onto retina

Pigment at end of nerve cell

Nerve cells of retina help brain form to an image

FLATWORM (LEFT EYE)

Electrical impulses sent to brain

SPIDER

Electrical impulses sent to brain

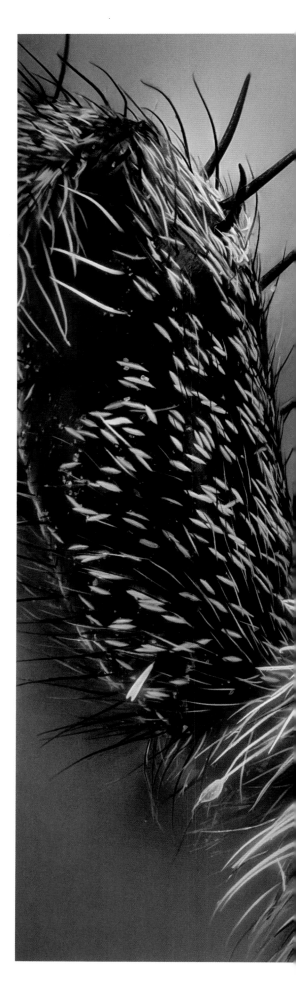

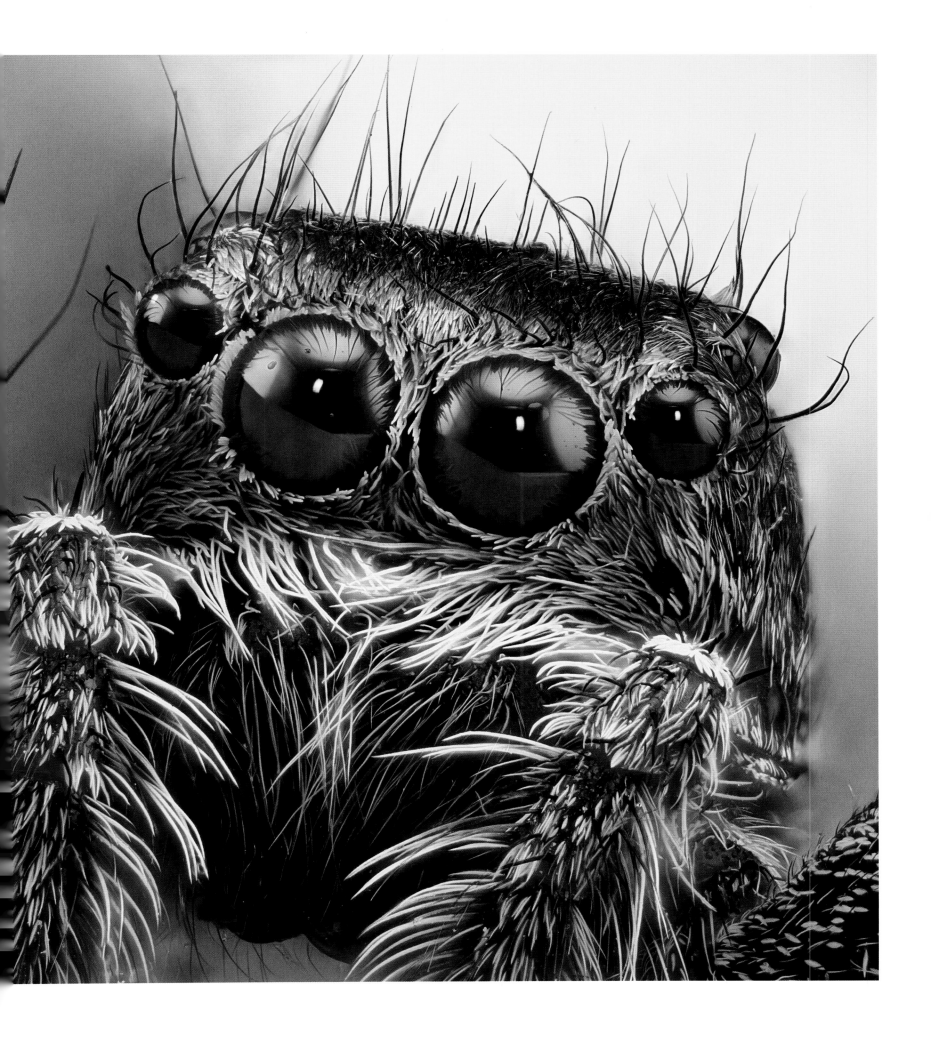

Horizontal pupils

Apart from exceptions such as the predatory mongoose, most mammals with horizontal pupils are grazing herbivores, such as deer and gazelles. Grazing animals spend much of their time with their heads lowered. Horizontal pupils allow them not only to keep the ground in sharp focus, they also help provide a panoramic-type view – vital for predator detection. The eyes constantly rotate to align the horizontal pupil with the ground's focal plane.

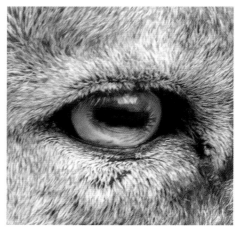

ALPINE IBEX
Capra ibex

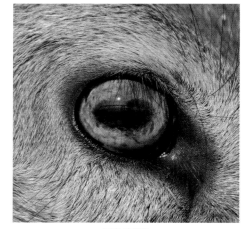

RED DEER
Cervus elaphus

Round pupils

Generally, the greater the distance from the ground, the rounder the pupil – as in taller mammals such as great apes or elephants. Active hunters that rely on strength or speed to catch their prey – such as big cats or wolves – also have circular pupils, as do birds of prey such as owls and eagles, which need to pinpoint prey positions from great heights.

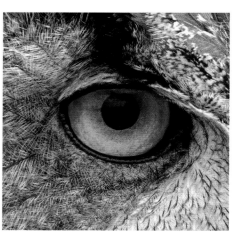

EAGLE OWL
Bubo bubo

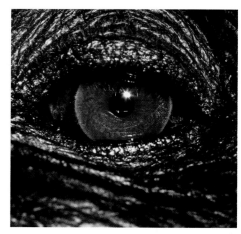

WESTERN GORILLA
Gorilla gorilla

Vertical pupils

Small ambush predators, including smaller cats such as bobcats, that use hide-and-strike techniques to catch their prey often have vertical pupils. This shape is optimal for judging distance without moving their heads – which could alert prey to their presence. Vertical pupils tend to be found in animals that hunt in changing light conditions, as the pupil shape reacts quickly; enlarging in dim light and rapidly constricting in bright conditions.

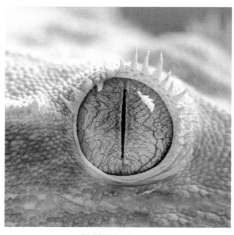

CRESTED GECKO
Correlophus ciliatus

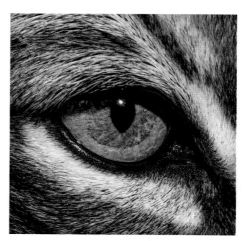

BOBCAT
Lynx rufus

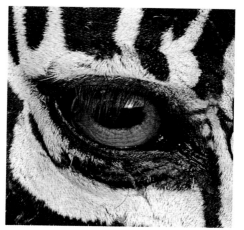

PLAINS ZEBRA
Equus quagga

pupil shapes

Pupil shape can be a strong indicator not only of where an animal fits into the food chain, but what types of hunting techniques it uses if it is a predator, or, if it is a prey species, how, what, and where it eats. While some animals have pupils shapes that cannot easily be categorized, most pupils fall into three basic types: horizontal, circular, or vertical.

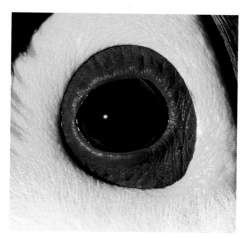

TOCO TOUCAN
Ramphastos toco

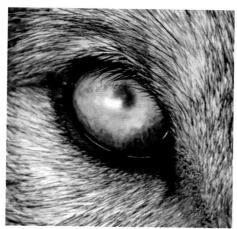

GREY WOLF
Canis lupus

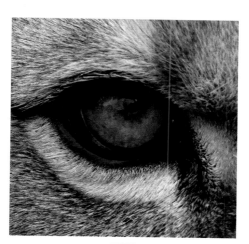

LION
Panthera leo

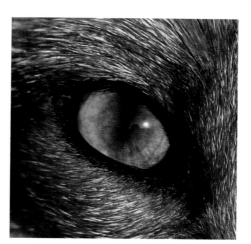

RED FOX
Vulpes vulpes

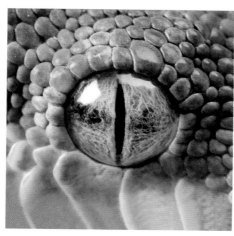

GREEN TREE PYTHON
Morelia viridis

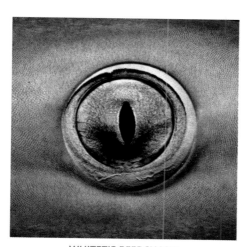

WHITETIP REEF SHARK
Triaenodon obesus

compound eyes

Insects and related animals see the world with the tiniest of lenses, clustered together in hundreds or thousands to form compound eyes. Each lens is part of a visual unit called an ommatidium, or facet, complete with light detectors and nerves that lead to the brain. Individual ommatidia cannot make a sharp image, but collectively they can detect the slightest movement – an object, such as a predatory bird, passing across the eye stimulates one ommatidium after another.

Sensitive eyes
Males of the tapered dronefly (*Eristalis pertinax*) have compound eyes with bigger facets than those of females, enabling them to collect more light for chasing mates. Bigger facets usually means poorer resolution, but the intricate neural wiring in the eyes of droneflies and other fast-moving true flies produces a visual system that is both ultra-sensitive and high in resolution.

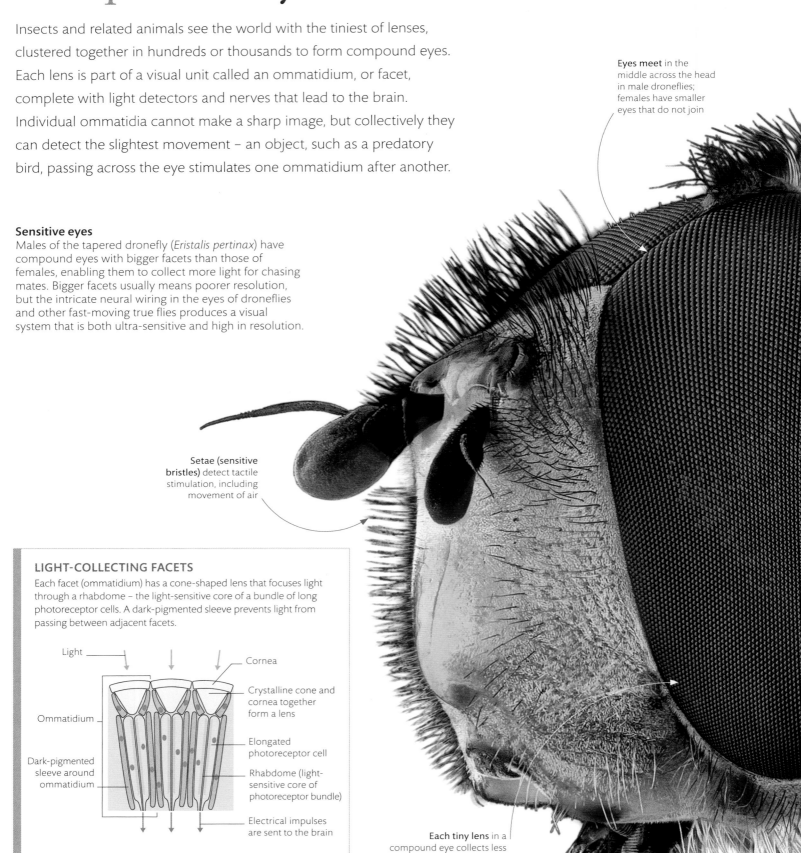

Eyes meet in the middle across the head in male droneflies; females have smaller eyes that do not join

Setae (sensitive bristles) detect tactile stimulation, including movement of air

LIGHT-COLLECTING FACETS
Each facet (ommatidium) has a cone-shaped lens that focuses light through a rhabdome – the light-sensitive core of a bundle of long photoreceptor cells. A dark-pigmented sleeve prevents light from passing between adjacent facets.

Light

Cornea

Crystalline cone and cornea together form a lens

Ommatidium

Elongated photoreceptor cell

Dark-pigmented sleeve around ommatidium

Rhabdome (light-sensitive core of photoreceptor bundle)

Electrical impulses are sent to the brain

ADJACENT FACETS IN A COMPOUND EYE

Each tiny lens in a compound eye collects less light than do the larger, single-lens eyes of vertebrates

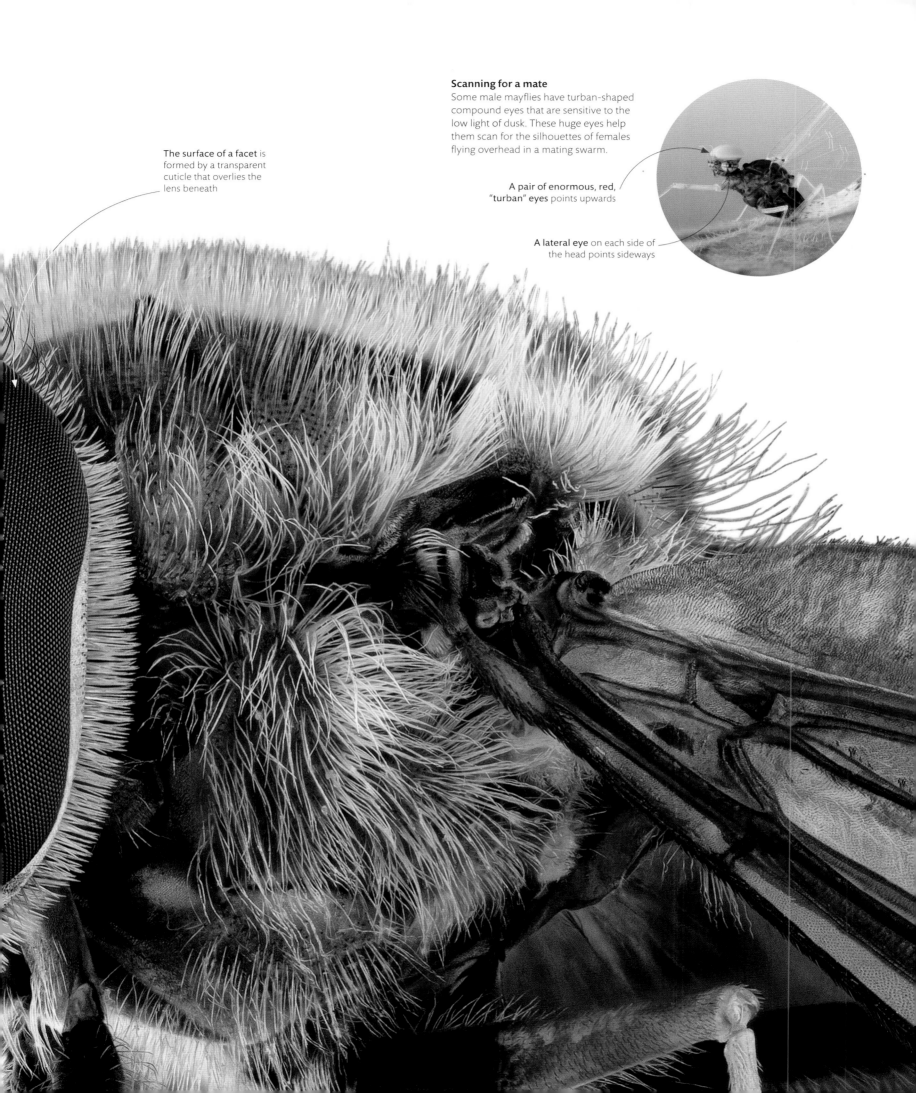

The surface of a facet is formed by a transparent cuticle that overlies the lens beneath

Scanning for a mate
Some male mayflies have turban-shaped compound eyes that are sensitive to the low light of dusk. These huge eyes help them scan for the silhouettes of females flying overhead in a mating swarm.

A pair of enormous, red, "turban" eyes points upwards

A lateral eye on each side of the head points sideways

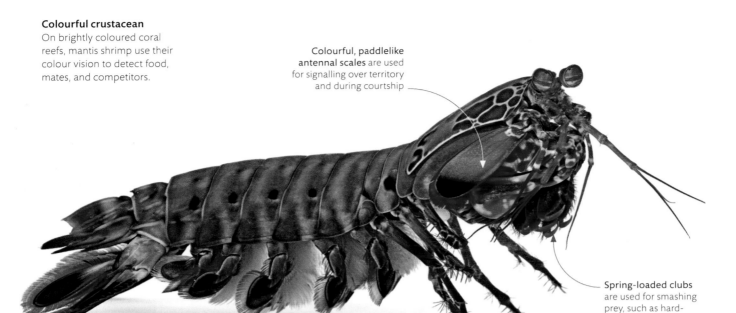

Colourful crustacean
On brightly coloured coral reefs, mantis shrimp use their colour vision to detect food, mates, and competitors.

Colourful, paddlelike antennal scales are used for signalling over territory and during courtship

Spring-loaded clubs are used for smashing prey, such as hard-shelled crabs

colour vision

The eyes of many animals go further than just detecting light – they can discriminate wavelengths of light, too. This means that they sense colour, from short-wave blue to long-wave red, and the spectrum in-between. They are able to do so because their eyes contain different kinds of visual pigments that absorb the different wavelengths. Living in a colourful world means animals can pick up on many more visual signals, such as enticements to mate during courtship, or warnings to stay away from danger.

Broadening the spectrum
Humans have three kinds of colour-sensitive visual pigments; the compound eyes of a peacock mantis shrimp (*Odontodactylus scyllarus*) contain 12. It can see wavelengths that are invisible to humans, such as ultraviolet and infrared.

Complex, multicoloured pattern is important in social signalling

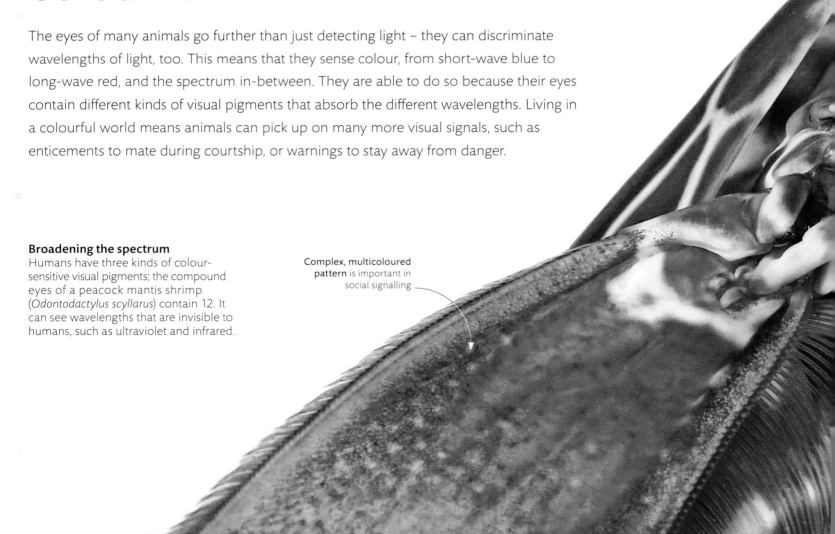

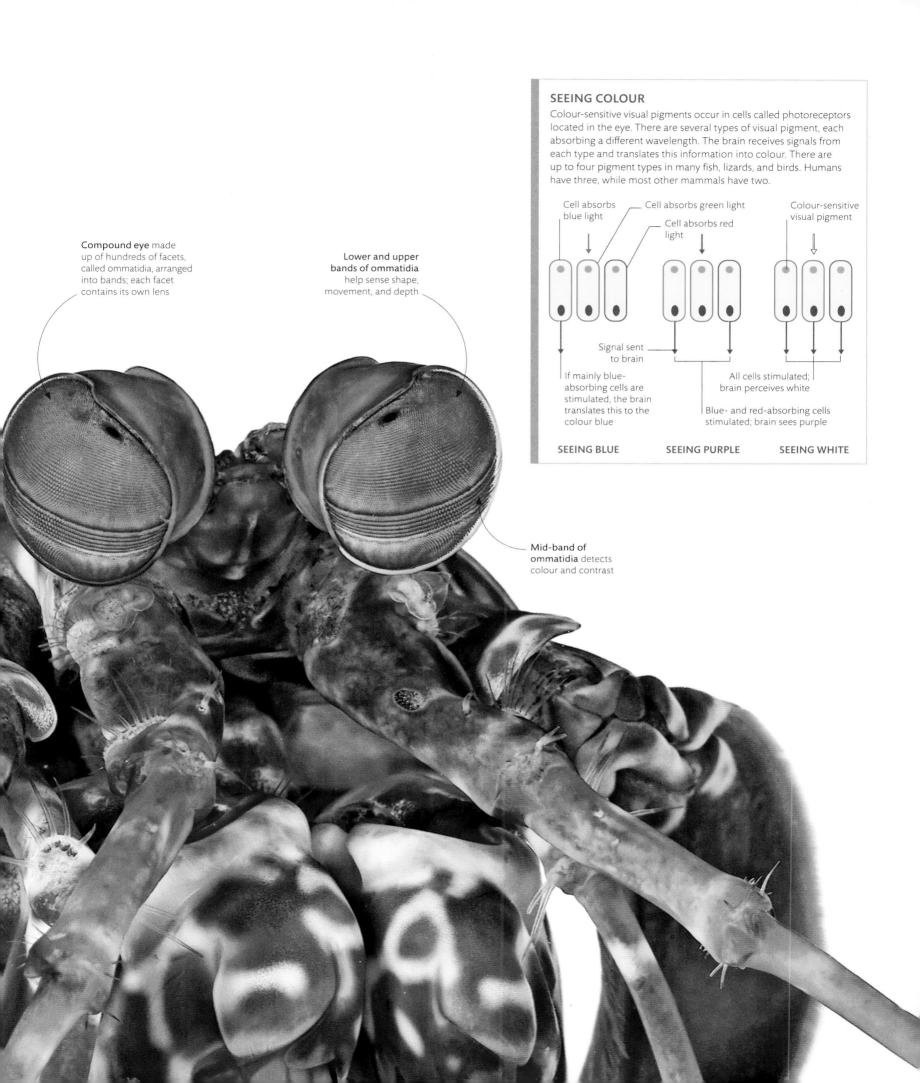

Compound eye made up of hundreds of facets, called ommatidia, arranged into bands; each facet contains its own lens

Lower and upper bands of ommatidia help sense shape, movement, and depth

Mid-band of ommatidia detects colour and contrast

SEEING COLOUR

Colour-sensitive visual pigments occur in cells called photoreceptors located in the eye. There are several types of visual pigment, each absorbing a different wavelength. The brain receives signals from each type and translates this information into colour. There are up to four pigment types in many fish, lizards, and birds. Humans have three, while most other mammals have two.

Cell absorbs blue light

Cell absorbs green light

Cell absorbs red light

Colour-sensitive visual pigment

Signal sent to brain

If mainly blue-absorbing cells are stimulated, the brain translates this to the colour blue

All cells stimulated; brain perceives white

Blue- and red-absorbing cells stimulated; brain sees purple

SEEING BLUE

SEEING PURPLE

SEEING WHITE

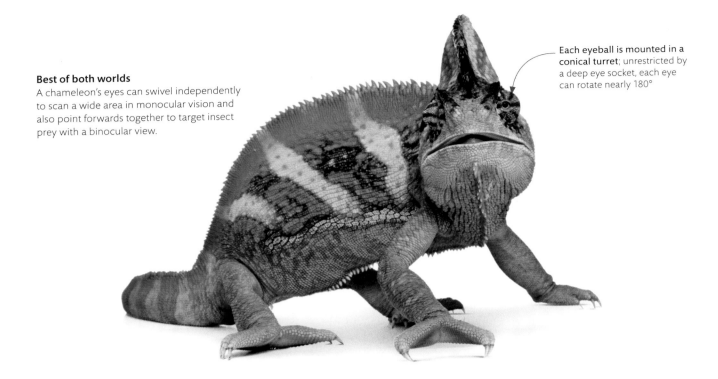

Each eyeball is mounted in a conical turret; unrestricted by a deep eye socket, each eye can rotate nearly 180°

Best of both worlds
A chameleon's eyes can swivel independently to scan a wide area in monocular vision and also point forwards together to target insect prey with a binocular view.

seeing depth

Unlike the simpler eyes of many invertebrates, the paired eyes of vertebrates have lenses that adjust their position or shape to focus at different distances. Their light-detecting cells, or photoreceptors, form a layer called the retina at the back of the eye. The cells' extraordinary sensitivity provides the brain with sufficient data for the animal to see a detailed, even three-dimensional, picture of the world around them.

JUDGING DISTANCE
Animals with their eyes on the sides of the head scan a wide area in separate monocular ("one-eye") fields of view. Those with forward-facing eyes have overlapping fields combined into a binocular ("two-eye") view. Although this narrows the total field of view, the brain combines the slightly different views coming from left and right within the region of overlap to create a sense of depth, allowing more accurate judging of distances.

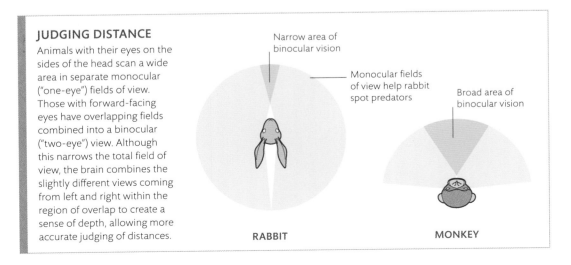

Narrow area of binocular vision

Monocular fields of view help rabbit spot predators

Broad area of binocular vision

RABBIT

MONKEY

Nocturnal tree-dweller
The lifestyle of the spectral tarsier (*Tarsius tarsier*) places great demands on its binocular vision. Not only must the tarsier judge distances accurately with its huge, forward-facing eyes when leaping through trees, it must also do so in the dark of a rainforest night.

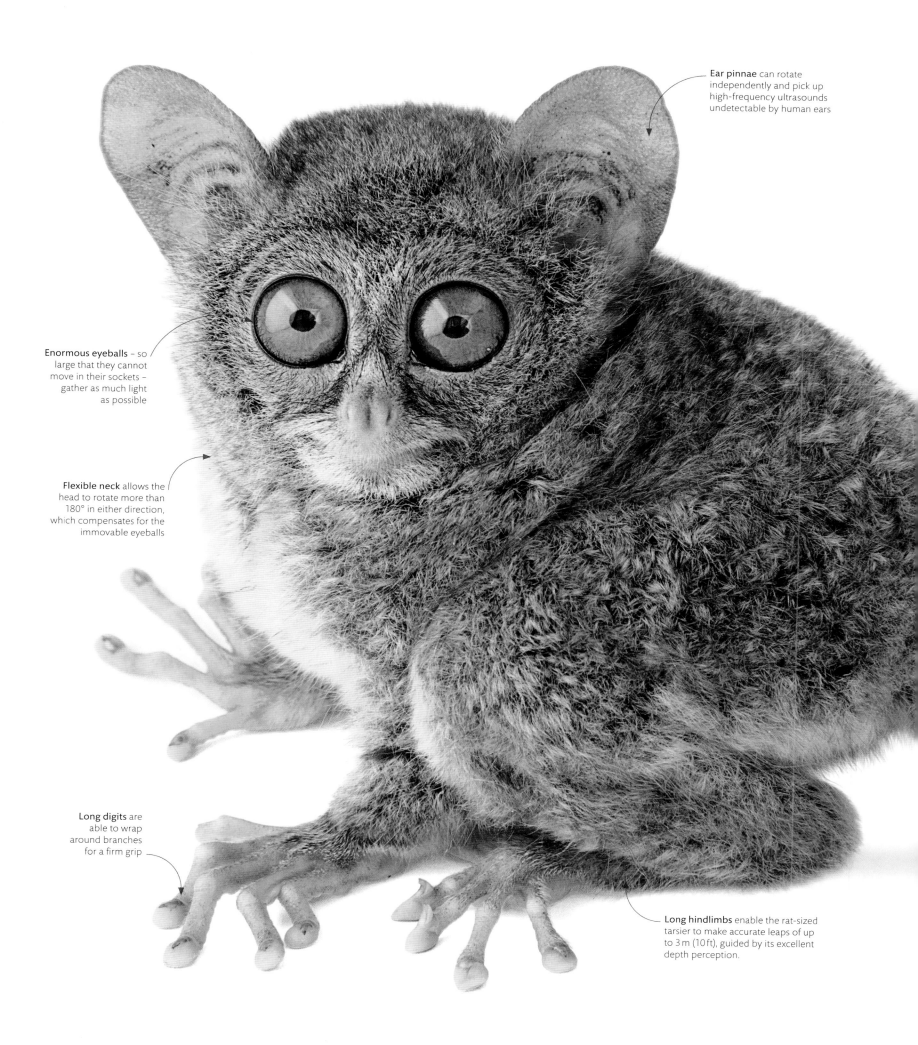

Ear pinnae can rotate independently and pick up high-frequency ultrasounds undetectable by human ears

Enormous eyeballs – so large that they cannot move in their sockets – gather as much light as possible

Flexible neck allows the head to rotate more than 180° in either direction, which compensates for the immovable eyeballs

Long digits are able to wrap around branches for a firm grip

Long hindlimbs enable the rat-sized tarsier to make accurate leaps of up to 3 m (10 ft), guided by its excellent depth perception.

common kingfisher

The distinctive, blue common kingfisher (*Alcedo atthis*) certainly lives up to its name. This bird cannot pursue prey under water like some seabirds: for example, penguins have denser bones and tight-fitting plumage that works like a swimsuit, so they can plunge the depths for food. Instead, the kingfisher must first target fish accurately from its perch above the water.

There are more than 100 species of kingfishers, but most – such as the Australasian kookaburras – hunt on land. It is likely that kingfishers evolved in tropical forests, where they are still at their most diverse, and used their daggerlike bills to snatch small animals from the ground. Only about 25 per cent of them – including the common kingfisher of Eurasia – specialize in diving for fish.

Diving kingfishers must be accurate and fast, and their hunting instinct is so strong that in winter they will smash thin ice to get a meal. Their hollow bones and waterproof plumage make them very buoyant and they cannot stay submerged, so kingfishers target their prey before they hit the water. From a favourite riverside spot, a bird selects a fish in the water beneath it. It adjusts its angle of attack to take account of the way light from its target bends at the water surface (see below). Then the dive begins. The bird plunges in towards its prey, pulling back its wings to form a streamlined shape that can cut through the water. An opaque eyelid – the nictitating membrane found in many vertebrates – pulls over its eyes to protect them. The bird grabs the fish with its bill, then buoyancy and a few wing flaps bring it back to the surface. Back on its perch, the bird holds the fish by its tail, smashes its head against the branch, then flips it and swallows it head-first, so the backward-pointing scales go down more easily. From start to finish, the entire episode lasts only a few seconds.

Flash of blue

The common kingfisher's diet is 60 per cent fish – the rest being aquatic invertebrates. This bird typically dives steeply from about 1–2 m (3–6 ft) above the water, plunging to a depth of 1 m (3 ft).

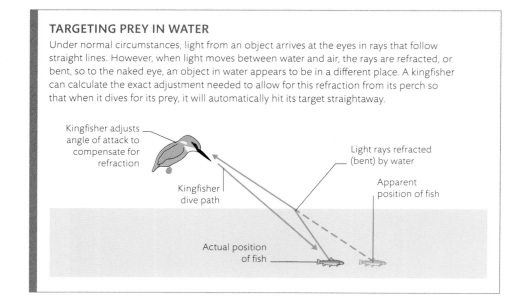

TARGETING PREY IN WATER

Under normal circumstances, light from an object arrives at the eyes in rays that follow straight lines. However, when light moves between water and air, the rays are refracted, or bent, so to the naked eye, an object in water appears to be in a different place. A kingfisher can calculate the exact adjustment needed to allow for this refraction from its perch so that when it dives for its prey, it will automatically hit its target straightaway.

Kingfisher adjusts angle of attack to compensate for refraction

Light rays refracted (bent) by water

Kingfisher dive path

Apparent position of fish

Actual position of fish

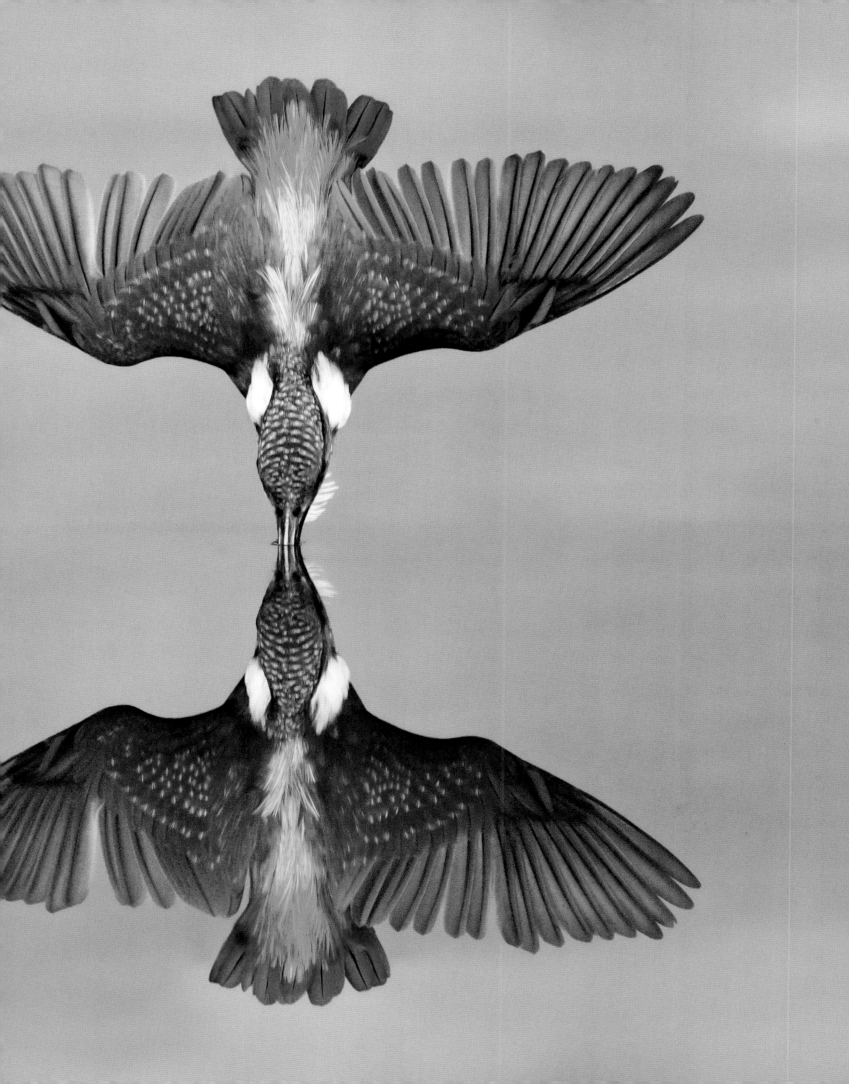

Scare tactics
The wing pattern of both the male (below) and female polyphemus moth serves to startle predators, such as frogs and birds, by flashing spots that resemble the eyes of a hunting owl.

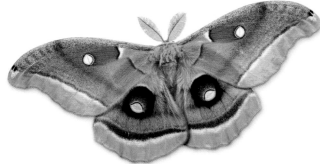

Longest part of the antennae, the flagellum, is highly modified for accommodating sensory receptors

scent detection

Smell or taste is often associated with food, but can also signal remarkably precise information about other animals nearby. Prey species, for instance, might sense a predator by detecting a substance that is unique to them. And many social signals, such advertisements to mate, take the form of chemical perfumes called pheromones that attract animals with the sensitivity to detect them – sometimes from a considerable distance.

Base of antenna is flexible and contains sensors that are triggered when flagellum is moved by air currents

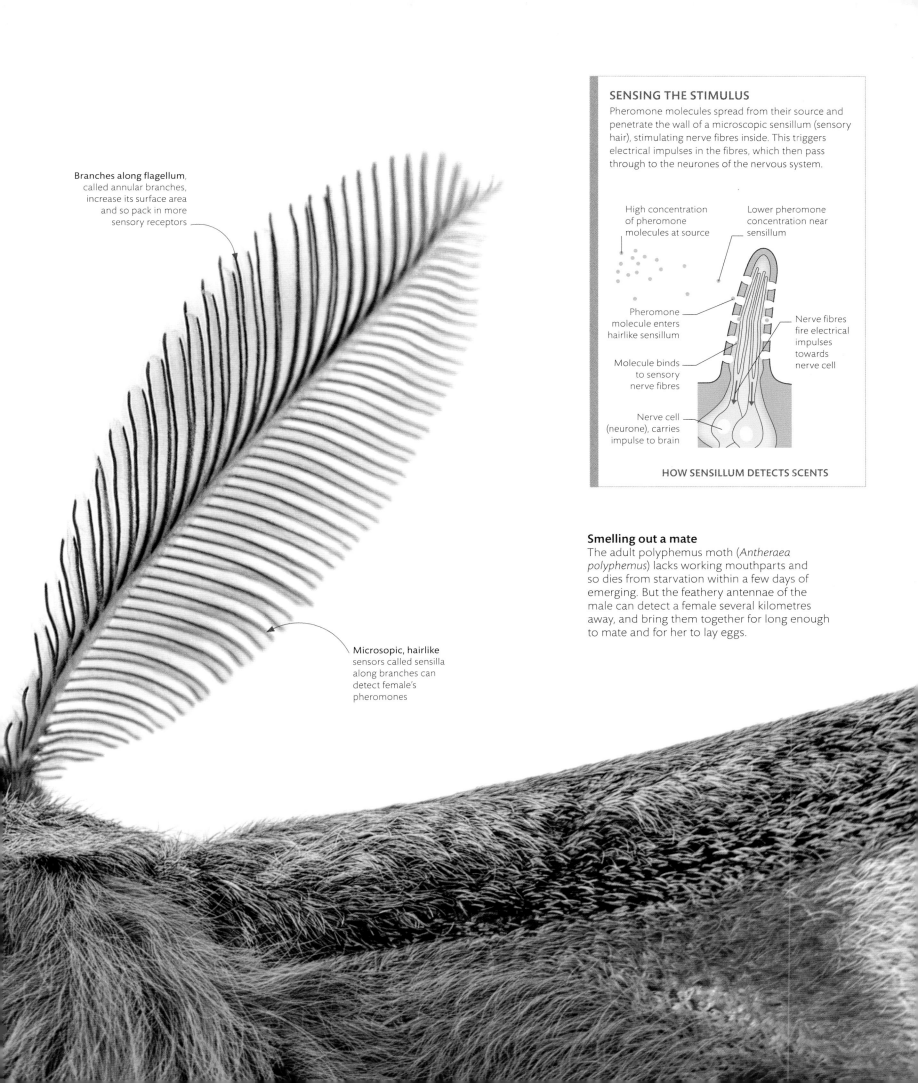

Branches along flagellum, called annular branches, increase its surface area and so pack in more sensory receptors

Microsopic, hairlike sensors called sensilla along branches can detect female's pheromones

SENSING THE STIMULUS
Pheromone molecules spread from their source and penetrate the wall of a microscopic sensillum (sensory hair), stimulating nerve fibres inside. This triggers electrical impulses in the fibres, which then pass through to the neurones of the nervous system.

High concentration of pheromone molecules at source

Lower pheromone concentration near sensillum

Pheromone molecule enters hairlike sensillum

Nerve fibres fire electrical impulses towards nerve cell

Molecule binds to sensory nerve fibres

Nerve cell (neurone), carries impulse to brain

HOW SENSILLUM DETECTS SCENTS

Smelling out a mate
The adult polyphemus moth (*Antheraea polyphemus*) lacks working mouthparts and so dies from starvation within a few days of emerging. But the feathery antennae of the male can detect a female several kilometres away, and bring them together for long enough to mate and for her to lay eggs.

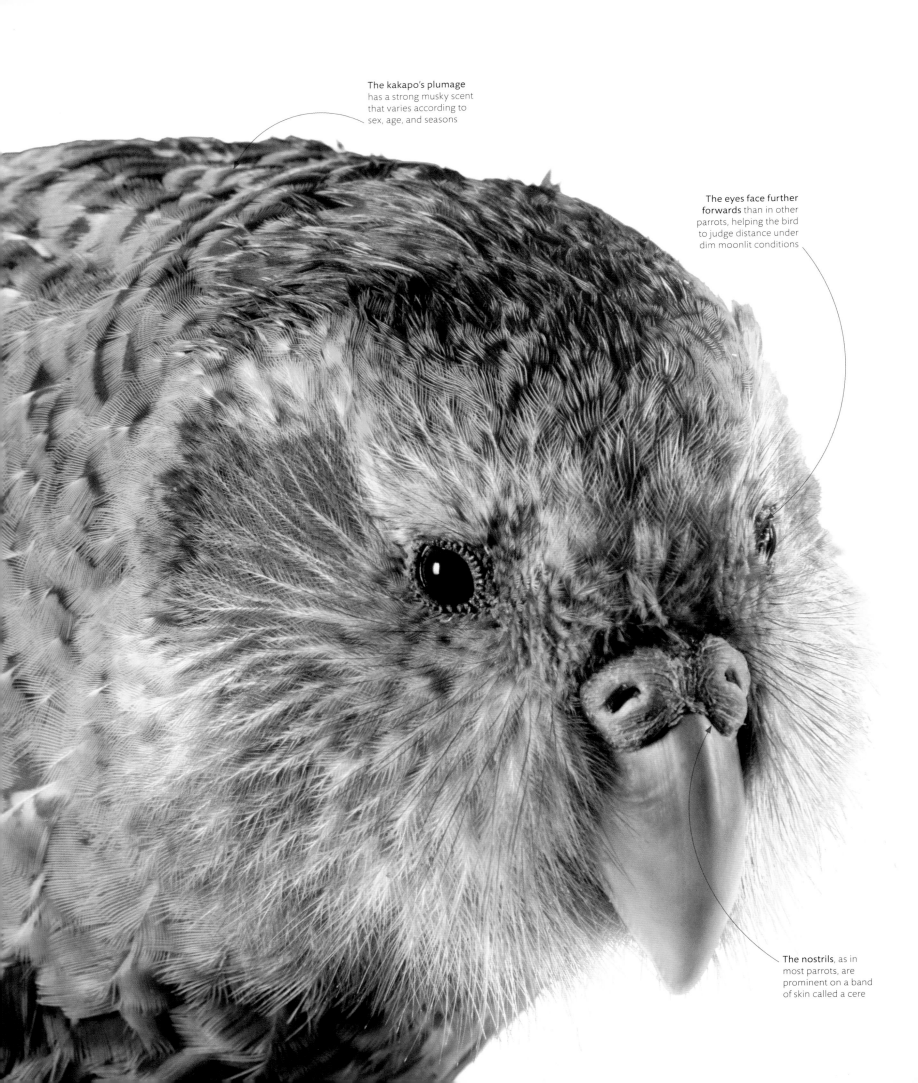

The kakapo's plumage has a strong musky scent that varies according to sex, age, and seasons

The eyes face further forwards than in other parrots, helping the bird to judge distance under dim moonlit conditions

The nostrils, as in most parrots, are prominent on a band of skin called a cere

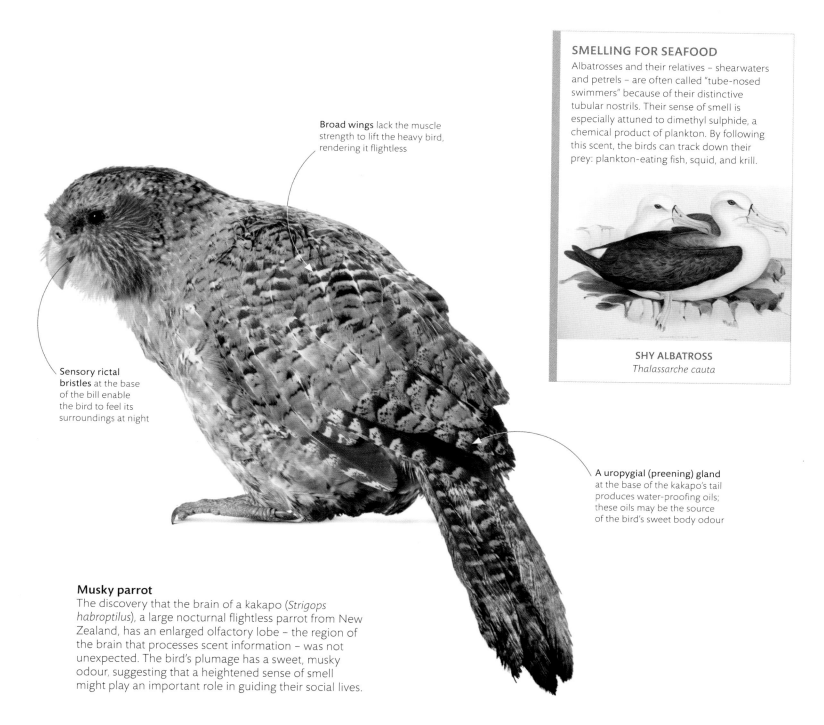

Broad wings lack the muscle strength to lift the heavy bird, rendering it flightless

SMELLING FOR SEAFOOD
Albatrosses and their relatives – shearwaters and petrels – are often called "tube-nosed swimmers" because of their distinctive tubular nostrils. Their sense of smell is especially attuned to dimethyl sulphide, a chemical product of plankton. By following this scent, the birds can track down their prey: plankton-eating fish, squid, and krill.

SHY ALBATROSS
Thalassarche cauta

Sensory rictal bristles at the base of the bill enable the bird to feel its surroundings at night

A uropygial (preening) gland at the base of the kakapo's tail produces water-proofing oils; these oils may be the source of the bird's sweet body odour

Musky parrot
The discovery that the brain of a kakapo (*Strigops habroptilus*), a large nocturnal flightless parrot from New Zealand, has an enlarged olfactory lobe – the region of the brain that processes scent information – was not unexpected. The bird's plumage has a sweet, musky odour, suggesting that a heightened sense of smell might play an important role in guiding their social lives.

smell in birds

Most types of bird seem to use vision or hearing more than their sense of smell, but smell still has a role to play in their lives, and many species rely on it. Albatrosses, for example, fly crosswind over the ocean to pick up the scent of prey in the water below, while turkey vultures can smell carrion on the ground from far away. It is now thought that most birds even produce their own unique odour profiles, helping individuals to recognize one another or locate their nests.

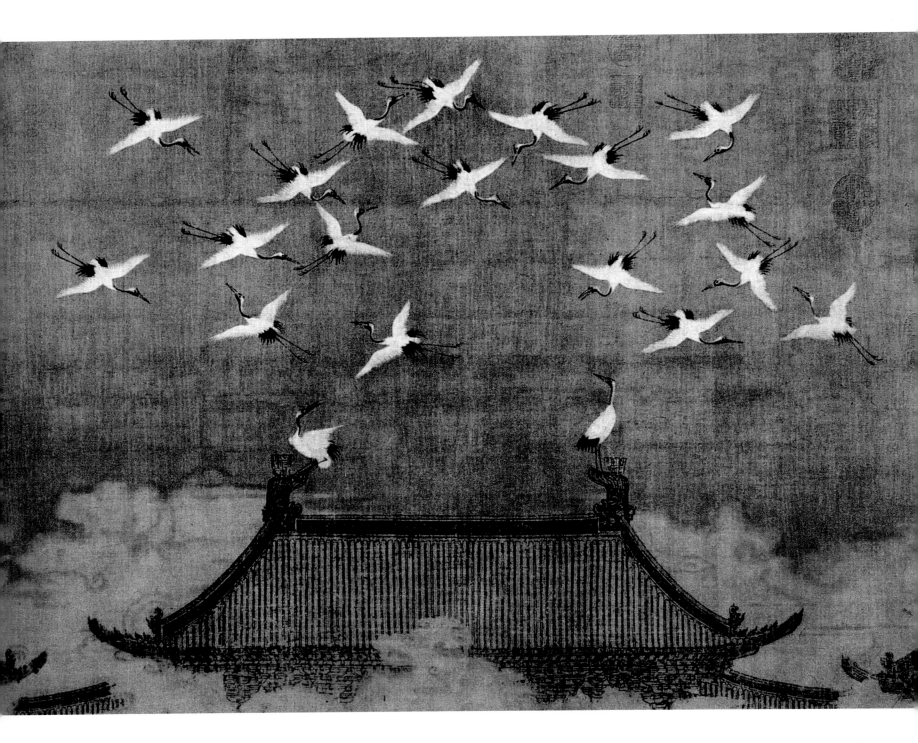

Auspicious Cranes (1112)
A silk handscroll of 20 cranes whirling against
an azure sky commemorates the auspicious
gathering of a cloud of cranes above the roof
of the royal palace. Emperor Huizong (1082–1135),
an artist and poet in his own right, created this
magical piece of work under his own name
Zhao Ji in recognition of the event.

song birds

During a period of extraordinary creativity that became known as the "Chinese Renaissance", artists of the Northern Song Dynasty produced paintings and poetry with a sensitivity and lyricism that would not be equalled elsewhere in the world for the next 400 years. The landscapes, animals, and, in particular, bird paintings produced during the 12th century are regarded by experts as the finest in the history of Chinese art.

At the heart of the enterprise was Zhao Ji, an unprepared heir to the throne who had spent his childhood indulging his passions for the arts with little regard for state affairs. He was crowned Emperor Huizong but finished his disastrous 26-year reign in penury as a common prisoner after the collapse of the Northern Song Dynasty and defeat by Jurchen forces. During his years as ruler, Huizong was a patron who brought the finest artists to his court and who was himself gifted in painting, calligraphy, poetry, music, and architecture.

The intertwined arts of poetry, calligraphy, and painting were used mainly on silk hand scrolls that were designed to be unfurled and read in sections from right to left. Some vertical scrolls were displayed briefly on walls, but most were personal to the viewer rather than public exhibits. The traditional subjects were landscapes and intimate naturalistic animal and bird portraits. Birds as subjects were steeped in the tradition of augury: Mandarin ducks represented happiness and loyalty because they mate for life and were said to pine to death at the loss of a mate; doves denoted love and fidelity; owls were an ill omen; cranes symbolized longevity and wisdom. A flock of 20 cranes descending through sunlit clouds over the royal palace in 1112 was seen as an auspicious sign; Zhao Ji depicted the event with a painting and poem of breathtaking beauty.

The legacy of the Song Dynasty is visible in the work of late 19th- and early 20th-century artists who began to experiment with a synthesis between traditional landscape, flower, fish, and bird painting and Western styles. Brothers Gao Jianfu and Gao Qifeng studied "Nihonga" fusion painting in Japan before establishing, with Chen Shuren, the Lingnan School in the Guangdong region of China. By the 1920s, the Lingnan style had become distinctive. Characterized by blank space and bright colouring, today it remains a popular fusion of old and new.

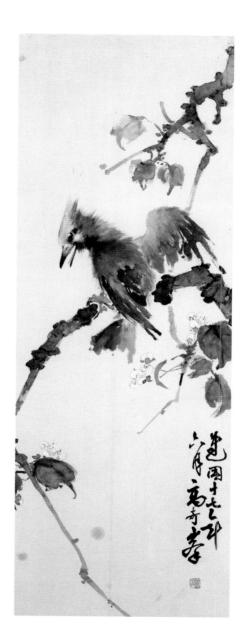

172 · 173 senses

Woodpecker (1927)
Traditional Chinese flower and bird painting is combined with Western styles and Japanese white highlights in Gao Qifeng's woodpecker. The dry brushstrokes on the scroll are typical of the school that he established, which was based on Japanese Nihonga art.

> " Immortal birds, proclaiming good news, suddenly appear with their measured dance. "

ZHAO JI, IN HIS POEM , *AUSPICIOUS CRANES*, 1112

TURBINATES

The rear part of the snout of most mammals is packed with a labyrinth of paper-thin, scroll-like bony walls called turbinates. Turbinates increase the surface area for olfactory epithelium: a lining of sensory cells connected to nerve endings. These detect a huge variety of scent molecules, even at low concentration. Most mammals – except for whales, dolphins, and most primates – have a good sense of smell.

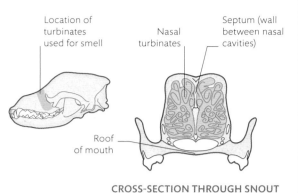

Location of turbinates used for smell

Nasal turbinates

Septum (wall between nasal cavities)

Roof of mouth

CROSS-SECTION THROUGH SNOUT

smell in mammals

The noses of land vertebrates evolved from the simple nasal pits of their fish ancestors into more complex scent-detecting channels. These connect to the back of the mouth, so breathing is possible even with the mouth closed. In mammals, large nasal chambers warm and moisten the incoming air before it reaches the olfactory epithelium (see box). Sensory cells in the lining detect smells from food sources, as well as signalling chemicals, called pheromones, that play an important role in the social lives of mammals.

Rhinarium (a tough nose disc supported by cartilage) is used to push into hard soil and excavate buried food

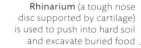

Fleshy tentacles encircle each nostril

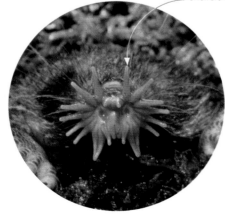

Nostrils can be closed to prevent soil and debris entering the nasal passages

Tip of the snout is moved by small muscles, so the pig can probe for food without moving its head

Smelling in water
The star-nosed mole (*Condylura cristata*) forages for invertebrates in waterlogged ground by blowing out bubbles and breathing them back in to detect the scent of food. It also feels for prey with 25,000 touch receptors on its nasal tentacles.

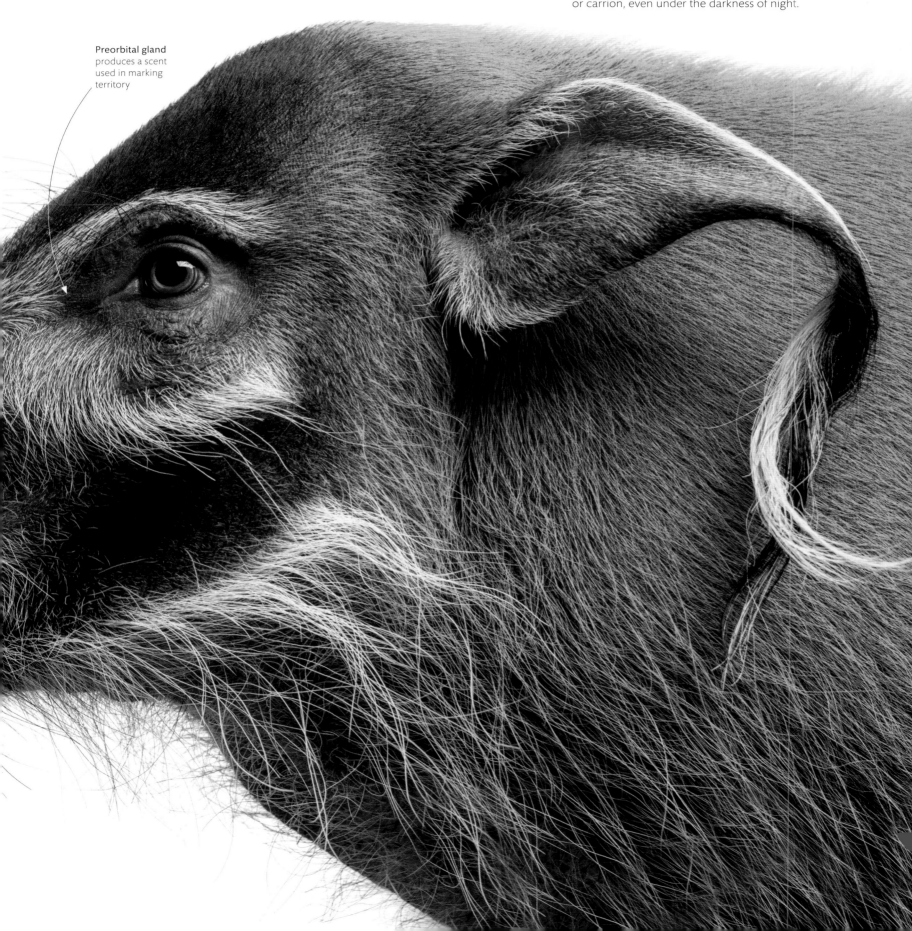

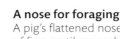

A nose for foraging
A pig's flattened nose, supported by a disc of firm cartilage and tipped with large nostrils, is perfect for rooting for food in the ground. The omnivorous red river hog (*Potamochoerus porcus*) successfully sniffs out roots, bulbs, fruit, or carrion, even under the darkness of night.

Preorbital gland
produces a scent used in marking territory

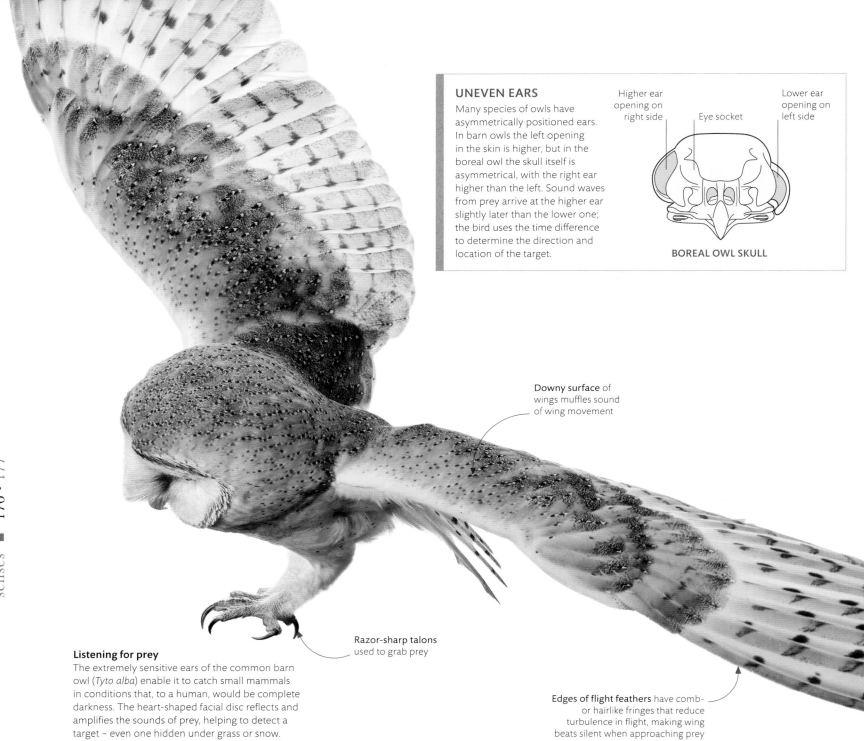

UNEVEN EARS

Many species of owls have asymmetrically positioned ears. In barn owls the left opening in the skin is higher, but in the boreal owl the skull itself is asymmetrical, with the right ear higher than the left. Sound waves from prey arrive at the higher ear slightly later than the lower one; the bird uses the time difference to determine the direction and location of the target.

Higher ear opening on right side

Eye socket

Lower ear opening on left side

BOREAL OWL SKULL

Downy surface of wings muffles sound of wing movement

Razor-sharp talons used to grab prey

Listening for prey
The extremely sensitive ears of the common barn owl (*Tyto alba*) enable it to catch small mammals in conditions that, to a human, would be complete darkness. The heart-shaped facial disc reflects and amplifies the sounds of prey, helping to detect a target – even one hidden under grass or snow.

Edges of flight feathers have comb- or hairlike fringes that reduce turbulence in flight, making wing beats silent when approaching prey

how animals hear

Animals hear by detecting vibrations, or sound waves. Inside their ears, vertebrates have sound-sensitive cells with cilia (microscopic hairs) that deflect when sound waves pass over them, triggering nerve signals that the brain interprets as sound. Land vertebrates, such as birds, have an eardrum (see p.178) that transmits and amplifies sound waves from the air into the inner ear, where the animal discriminates loudness and pitch.

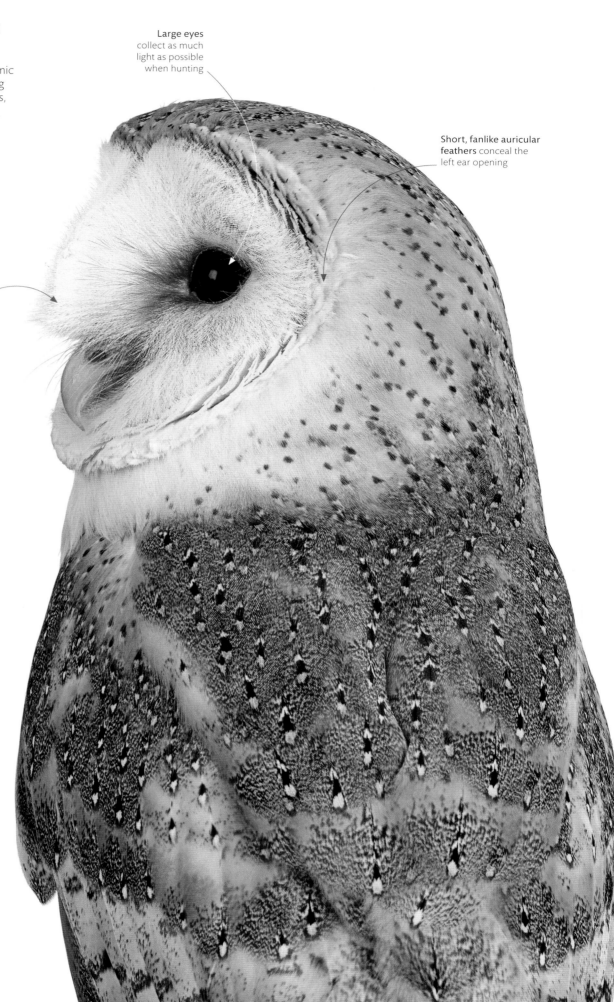

Hidden ears

Instead of the external ear pinnae of mammals (see pp.178–79), the barn owl has a disc of facial feathers that channel sound waves towards its concealed ear openings. The facial disc is more streamlined than pinnae, which is an aerodynamic advantage. Birds have a single bone connecting the eardrum to the inner ear – unlike mammals, which have three bones.

Large eyes collect as much light as possible when hunting

Short, fanlike auricular feathers conceal the left ear opening

Stiff, dense feathers on either side of the face are arranged into a concave pan for intercepting sound waves

Large-eared omnivore
Native to the South American pampas, the maned wolf (*Chrysocyon brachyurus*) uses its large ears to listen for prey among the tall grasses. However, only 20 per cent of its hunting attempts end with a meal, so up to half of its diet is made up of fruit.

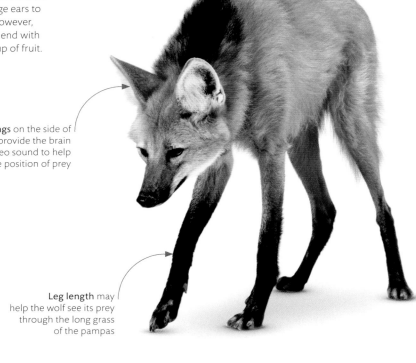

Ear openings on the side of the head provide the brain with stereo sound to help judge the position of prey

Leg length may help the wolf see its prey through the long grass of the pampas

mammal ears

Mammals rely on their hearing more than many other animals, both to listen out for danger – especially at night – and to hunt for prey. Several features have evolved in mammals that improve their sense of hearing. Inside a mammal's head, the ears have evolved three bones, or ossicles, that amplify sound vibrations. On the outside, they have two fleshy pinnae, or natural trumpets, that direct sounds into the ear towards the amplifying system.

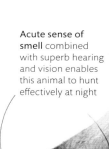

Acute sense of smell combined with superb hearing and vision enables this animal to hunt effectively at night

MAMMALIAN HEARING

The working parts of a mammal's ear lie deep inside its head. Sound waves (see p.180) are channelled into the ear canal, where they vibrate a membranous drum. The vibrations are then transmitted through a chain of tiny bones, or ossicles, towards the inner ear. Here, cells in a fluid-filled coiled tube, the cochlea, detect the vibration and send nerve impulses towards the brain.

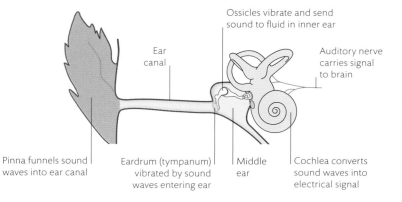

Ossicles vibrate and send sound to fluid in inner ear

Ear canal

Auditory nerve carries signal to brain

Pinna funnels sound waves into ear canal

Eardrum (tympanum) vibrated by sound waves entering ear

Middle ear

Cochlea converts sound waves into electrical signal

MAIN COMPONENTS OF MAMMALIAN EAR

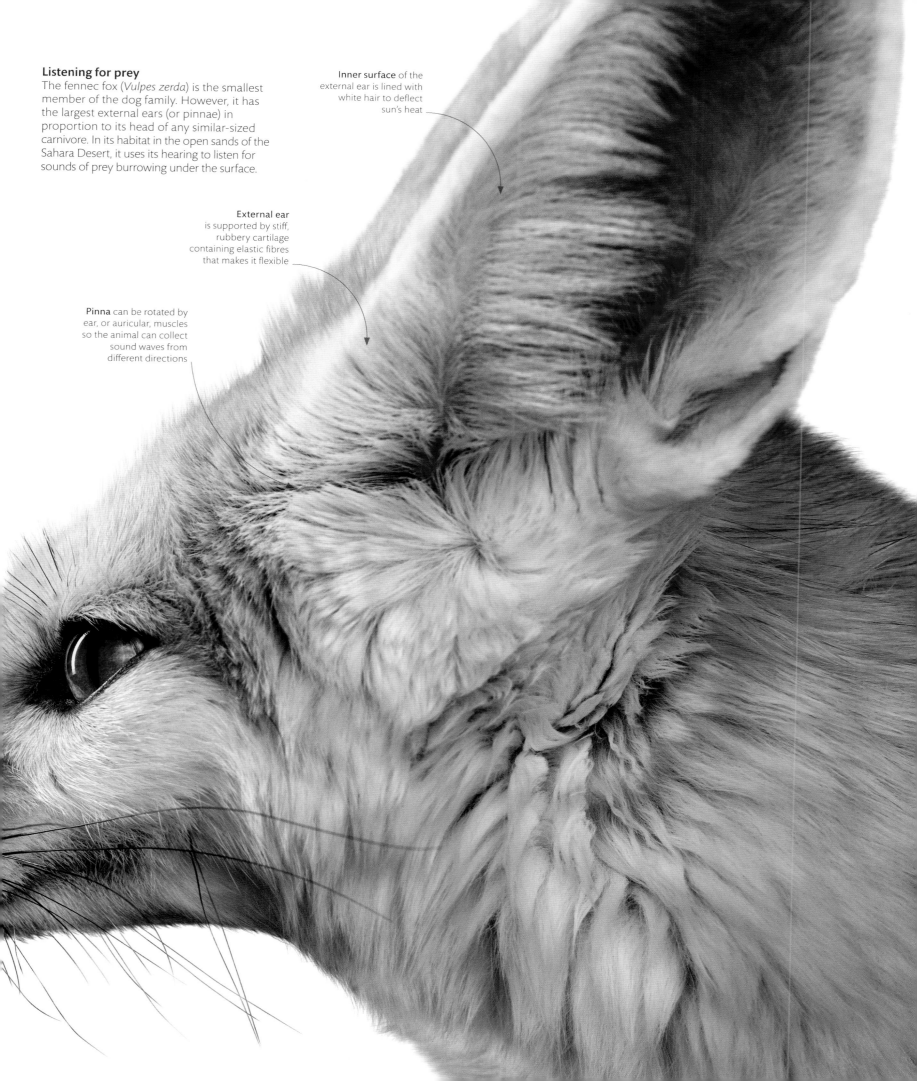

Listening for prey
The fennec fox (*Vulpes zerda*) is the smallest member of the dog family. However, it has the largest external ears (or pinnae) in proportion to its head of any similar-sized carnivore. In its habitat in the open sands of the Sahara Desert, it uses its hearing to listen for sounds of prey burrowing under the surface.

Inner surface of the external ear is lined with white hair to deflect sun's heat

External ear is supported by stiff, rubbery cartilage containing elastic fibres that makes it flexible

Pinna can be rotated by ear, or auricular, muscles so the animal can collect sound waves from different directions

NATURAL SONAR

With each pulse of sound it emits – using the larynx in microbats, but with the tongue in some large fruit bats – the bat disarticulates its inner ear bones to avoid self-deafening, restoring them a fraction of a second later to receive the echo. The pulses have a moth-sized wavelength – any longer, and the sound would fail to reflect strongly. Using high-frequency (short wavelength) sound ensures good reflection from small objects, as well as fine discrimination and resolution.

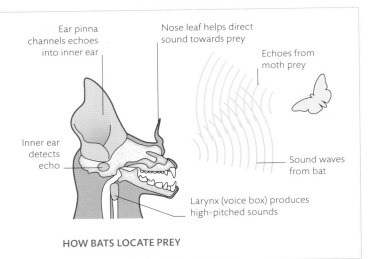

Ear pinna channels echoes into inner ear

Nose leaf helps direct sound towards prey

Echoes from moth prey

Inner ear detects echo

Sound waves from bat

Larynx (voice box) produces high-pitched sounds

HOW BATS LOCATE PREY

Echolocators

Among bats, the microbats possess the most impressive echolocation abilities. A diversity of facial equipment – nose leaves that focus sound pulses and large ears that receive the echoes – has evolved in over 1,000 microbat species. Nose leaves are often highly developed in species that emit calls through the nostrils. However, most microbats actually emit their calls through the mouth, so they lack the elaborate nose leaf or facial ornament.

listening for echoes

When vision is compromized at night or in murky water, some animals rely on a natural form of sonar called echolocation. By listening for the echoes of sounds that the animal emits, it can build up a picture of its surroundings. Bats use echolocation to negotiate obstacles and pinpoint flying insects. The echoes from their high-frequency calls tell the bats the size, shape, position, and distance of objects, and even the texture.

Sound directors, sound detectors
Nose leaves that focus echolocation pulses are found in microbats such as horseshoe bats, false vampire bats, and leaf-nosed bats. Some bats can instantly change the shape of their large ear pinnae to pinpoint the source of echoes from prey.

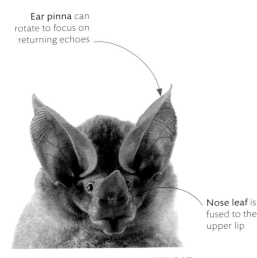

Ear pinna can rotate to focus on returning echoes

Nose leaf is fused to the upper lip

WHITE-THROATED ROUND-EARED BAT
Lophostoma silvicolum

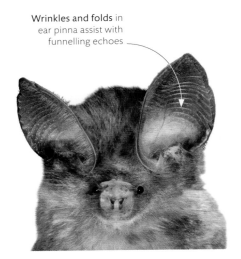

Wrinkles and folds in ear pinna assist with funnelling echoes

POMONA LEAF-NOSED BAT
Hipposideros pomona

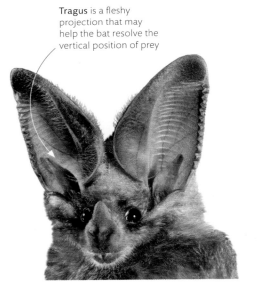

Tragus is a fleshy projection that may help the bat resolve the vertical position of prey

CALIFORNIA LEAF-NOSED BAT
Macrotus californicus

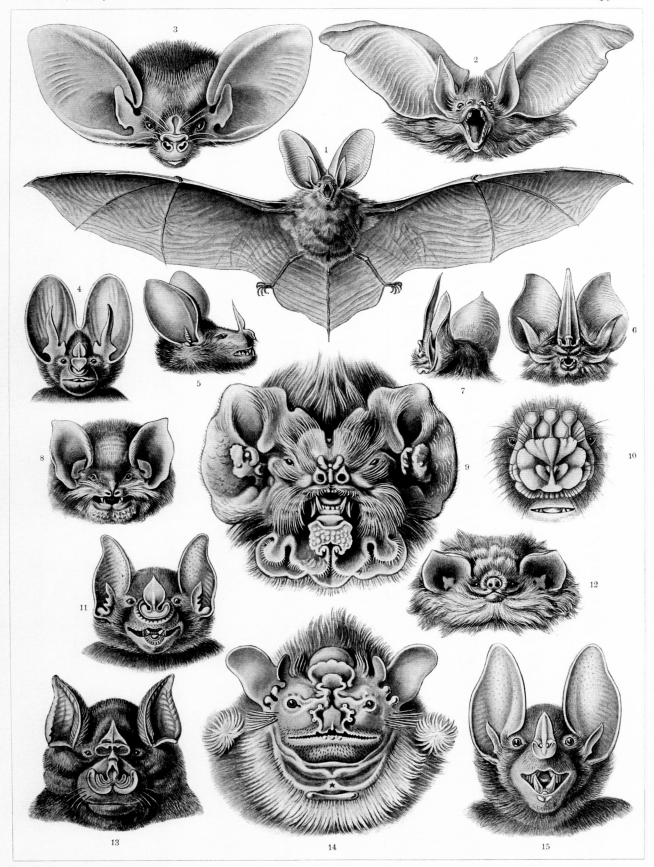

Chiroptera. — Flebertiere.

common dolphin

The intelligence of dolphins is well known and shown by their complex behaviour. As a proportion of body size, their brains are second only to humans. Not only are common dolphins (*Delphinus* sp.) highly skilled at targeting prey, but they even work together to maximize their catch.

Common dolphins swim in warm ocean waters around the world, generally no deeper than about 180 m (600 ft). Like other cetaceans, dolphins are about as well adapted to life in water as it is possible for an air-breathing mammal to be. The entire body is hydrodynamic – it is propelled by a tail fluke (see pp.274–75), steered with front flippers, and its nostrils are fashioned as a blowhole in the top of the head.

A dolphin's brain is a powerful information processor. Sound travels better in water than in air, and dolphins locate prey by echolocation. They beam clicks produced below the blowhole through an oil-filled "melon" in their forehead, and direct the returning echo towards the inner ear through oily channels in the lower jaw. The region of the brain concerned with auditory information is greatly expanded to deal with the data received; only parts of the brain involved in olfaction are reduced compared with other mammals that rely more on a sense of smell.

But a dolphin uses its brain for more than just decoding sensory input. The mammalian neocortex, a folded surface layer of brain tissue – and seat of the highest cognitive skills – is especially well-developed, which explains why they are so good at storing memories, making reasoned decisions, and inventing new behaviours. A pod of common dolphins can communicate so well that they cooperate when hunting fish, herding them into tighter shoals for easier pickings.

Cooperative brains
Long-nosed common dolphins (*Delphinus capensis*) swim in waters over the continental shelves – a pod is seen here forcing sardines into a bait-ball. A second short-nosed species (*D. delphis*) lives further offshore.

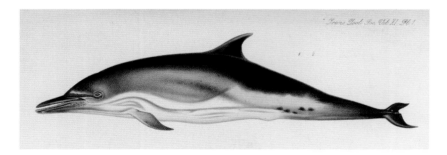

Colourful disguise
A 19th-century illustration depicts the colour patches on common dolphins – among the brightest of all cetaceans – which may serve to break up their outline, helping them avoid detection by bigger predators.

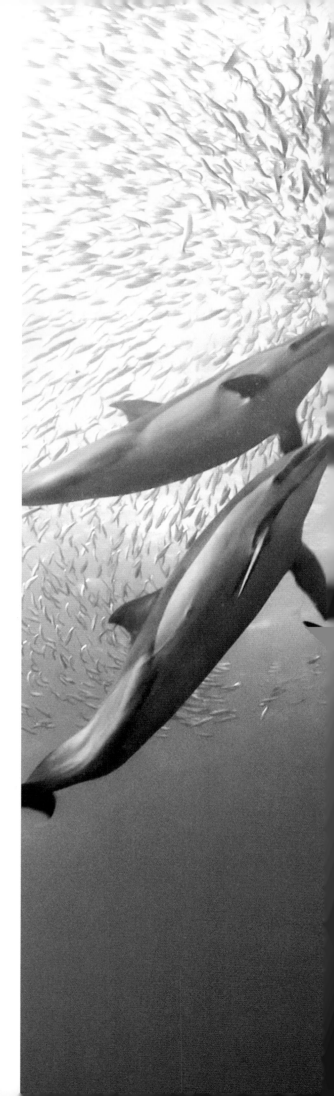

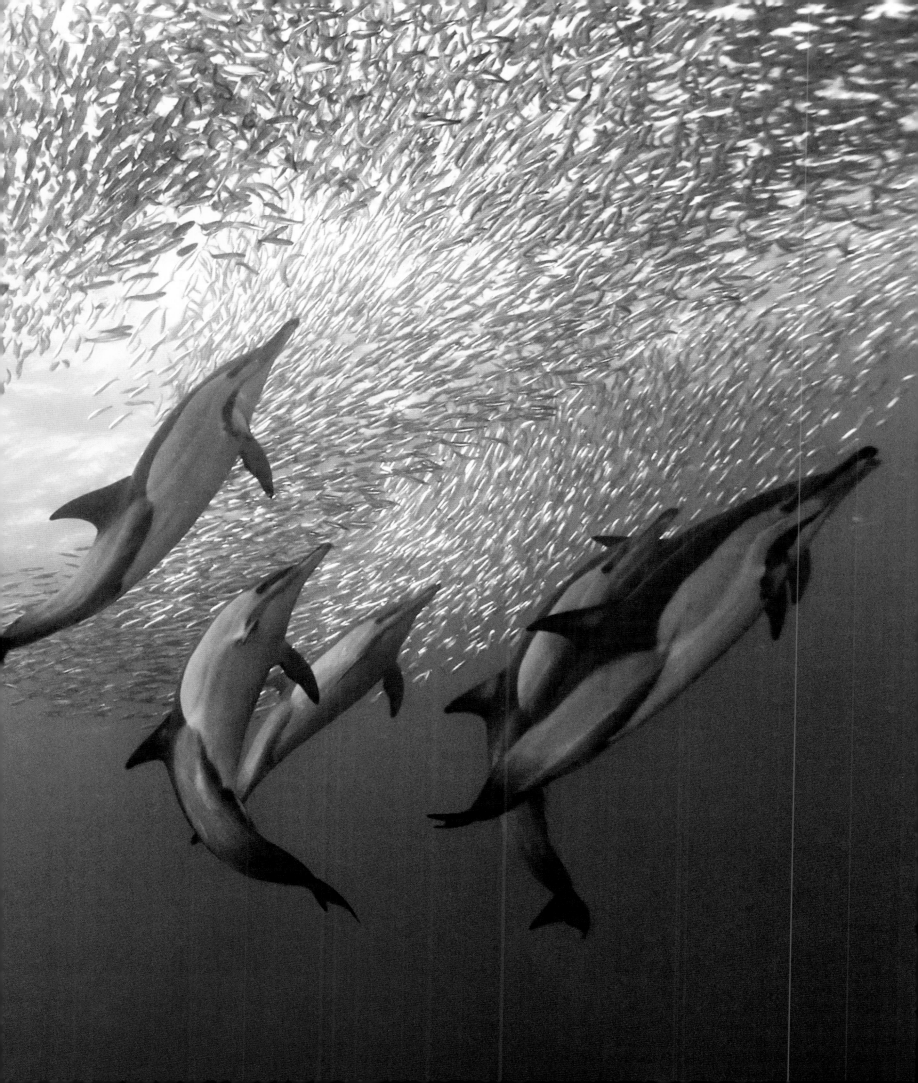

mouths
and jaws

mouth. an opening through which many animals take in food and emit sounds.

jaw. a hinged structure that forms the frame of the mouth in animals that bite and chew, and that is used for manipulating food.

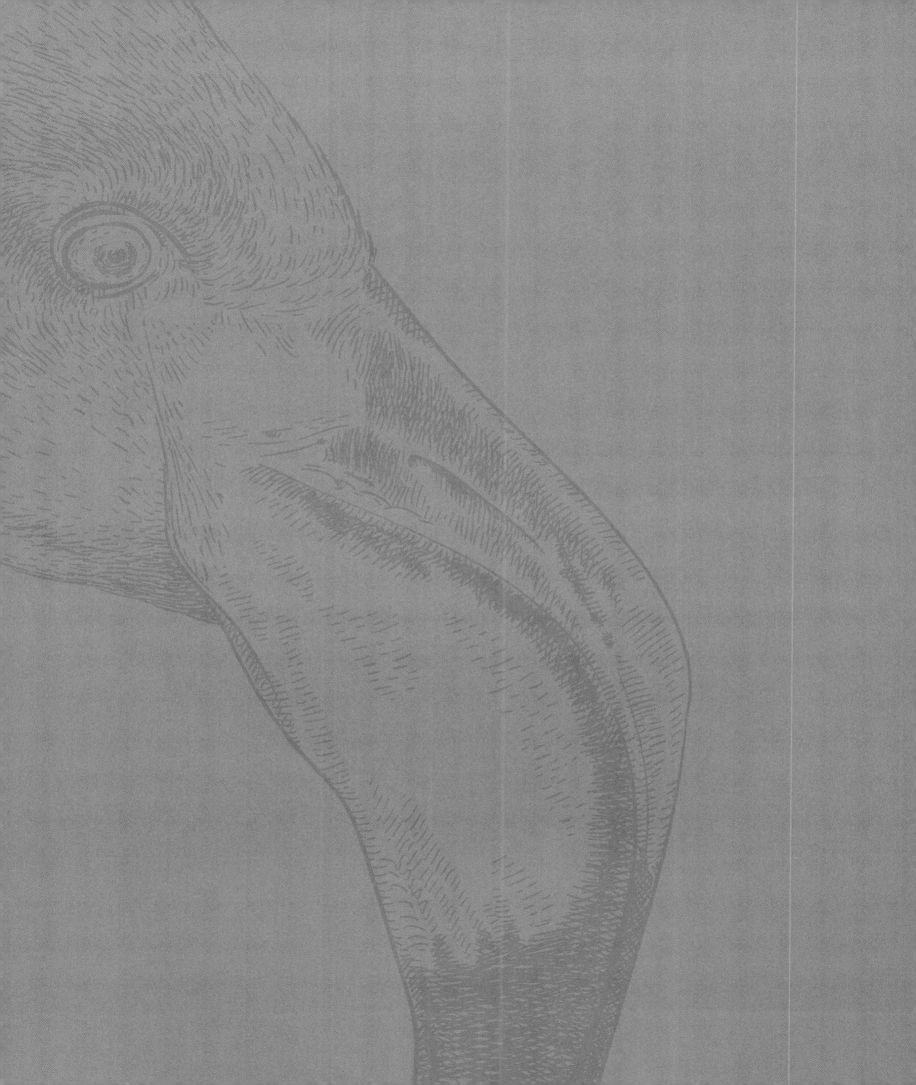

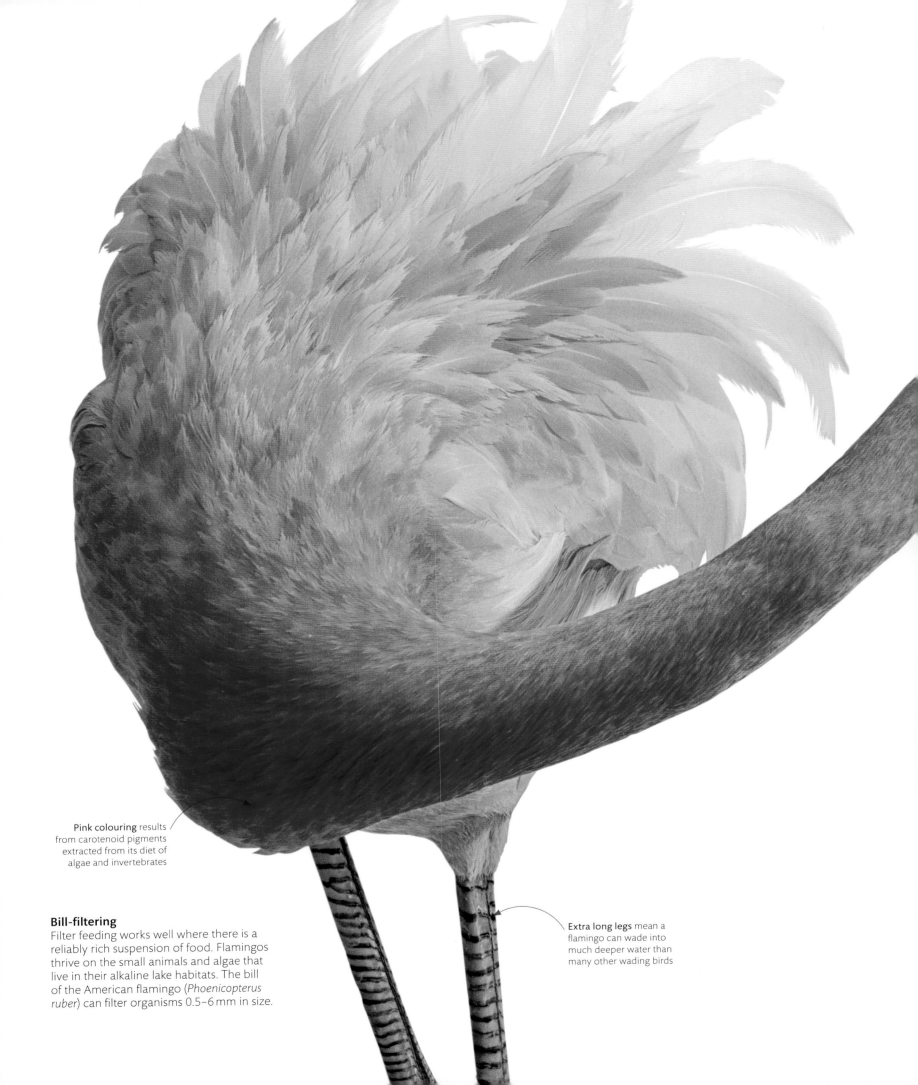

Pink colouring results from carotenoid pigments extracted from its diet of algae and invertebrates

Bill-filtering
Filter feeding works well where there is a reliably rich suspension of food. Flamingos thrive on the small animals and algae that live in their alkaline lake habitats. The bill of the American flamingo (*Phoenicopterus ruber*) can filter organisms 0.5–6 mm in size.

Extra long legs mean a flamingo can wade into much deeper water than many other wading birds

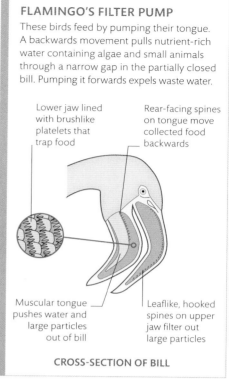

FLAMINGO'S FILTER PUMP
These birds feed by pumping their tongue. A backwards movement pulls nutrient-rich water containing algae and small animals through a narrow gap in the partially closed bill. Pumping it forwards expels waste water.

Lower jaw lined with brushlike platelets that trap food

Rear-facing spines on tongue move collected food backwards

Muscular tongue pushes water and large particles out of bill

Leaflike, hooked spines on upper jaw filter out large particles

CROSS-SECTION OF BILL

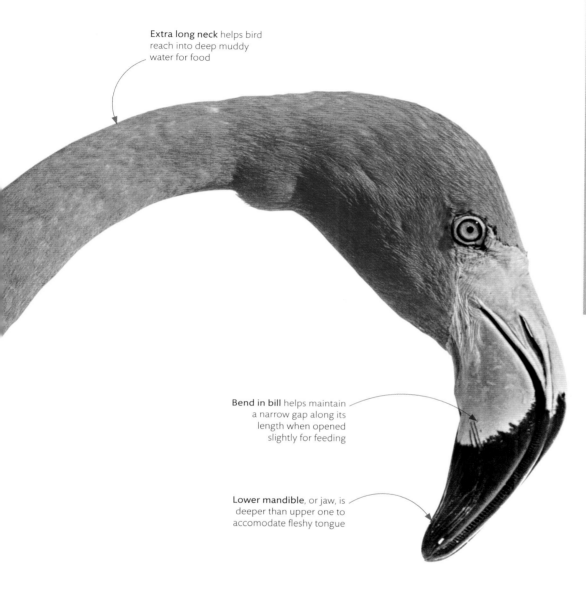

Extra long neck helps bird reach into deep muddy water for food

Bend in bill helps maintain a narrow gap along its length when opened slightly for feeding

Lower mandible, or jaw, is deeper than upper one to accomodate fleshy tongue

Gill rakers collect plankton for swallowing as the fish swims with its mouth wide open

filter feeding

Many animals acquire their nourishment from the tiniest of food particles, often suspended in water, and have evolved efficient methods for collecting them. The smallest invertebrates trap the small particles with mucus, but in bigger, stronger animals, food-rich water is driven through a fine-mesh filter that traps the foodstuff. Known as filter feeding, this is not only exploited by the largest animals of all, the blue whale and whale shark, but also smaller ones like mackerel, as well as water-wading flamingos.

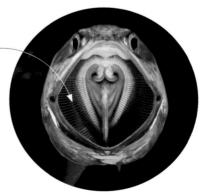

Ocean filter
Many open ocean fish such as this Indian mackerel (*Rastrelliger kanagurta*), have bony extensions in their gill cage, or gill rakers, that filter plankton from the water as the fish swims.

christmas tree worms

Miniature "forests" of these treelike animals not only live on coral reefs, but also help to protect them. Christmas tree worms (*Spirobranchus giganteus*) may look more like plants than animals, but their feathery spirals are in fact modified gill-tentacles through which they feed and breathe.

These remarkable creatures begin life as embryos created by broadcast spawning. Male and female worms simultaneously release sperm and eggs into the water that together produce free-floating larvae. After hours or weeks, a larva settles on a hard coral and metamorphoses into a creature inhabiting a mucus tube – the beginnings of the carbonate tube in which this worm will live for as long as 30 years.

The juvenile worm ingests calcium-rich particles and processes them, via a special gland, into calcium carbonate, which it excretes to form a 20-cm (8-in) tube that extends deep into the coral's hard outer framework. The only visible part of a Christmas tree worm is its prostomium, which takes the form of two spiral "trees" – the rest of the animal remains in its tube. If the worm senses danger, it retracts the trees into the tube, too (see box below).

Each Christmas tree worm spiral consists of 5 to 12 whorls that bristle with tiny tentacles, or radioles, which in turn are covered with filamentous pinnules, which themselves bear microscopic hairlike cilia. The cilia beat, creating a current that pulls in the plankton on which the worm feeds.

At the base of each tree, these animals have two compound eyes, each with up to 1,000 facets. Scientists are not sure what they can see, but experiments show that they retract if predatory fish approach, even when the fish does not cast a shadow.

Gift for the coral reef
The worm feeds by wafting food along a groove running down each tentacle – and eventually to the mouth at the base of the "tree". As it feeds, the currents it creates also bring in nutrients for the coral and help to disperse damaging waste.

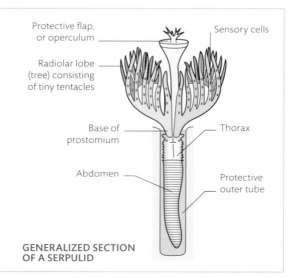

HEADS UP
Life as a tube-building, sedentary filter feeder is possible through some extraordinary modifications to the average annelid head. The highly specialized tentacles around the mouth of serpulid worms form a protective lid, while others have developed into feathery structures that serve as both food-gathering organs and gills for respiration. A Christmas tree worm's head end, or prostomium, stands 1–2 cm (⅓–¾ in) high and can be 3.8 cm (½ in) wide and its 3 cm (1⅕ in) body is safely protected by its tube. If danger is detected the prostomium can retract its inside this tube, too.

Protective flap, or operculum

Sensory cells

Radiolar lobe (tree) consisting of tiny tentacles

Base of prostomium

Thorax

Abdomen

Protective outer tube

GENERALIZED SECTION OF A SERPULID

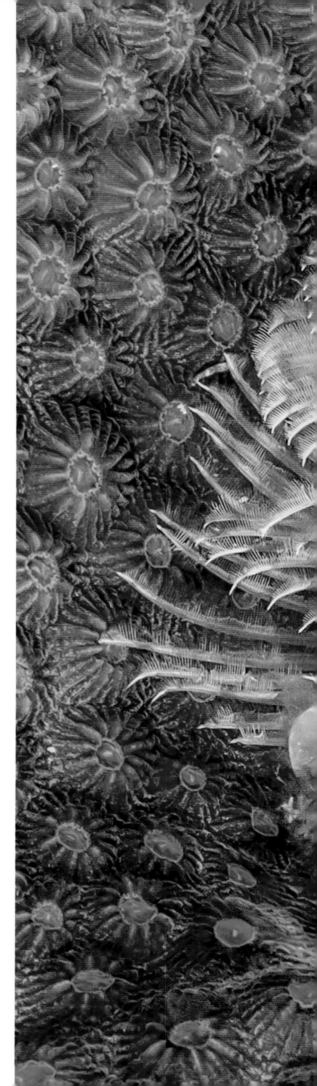

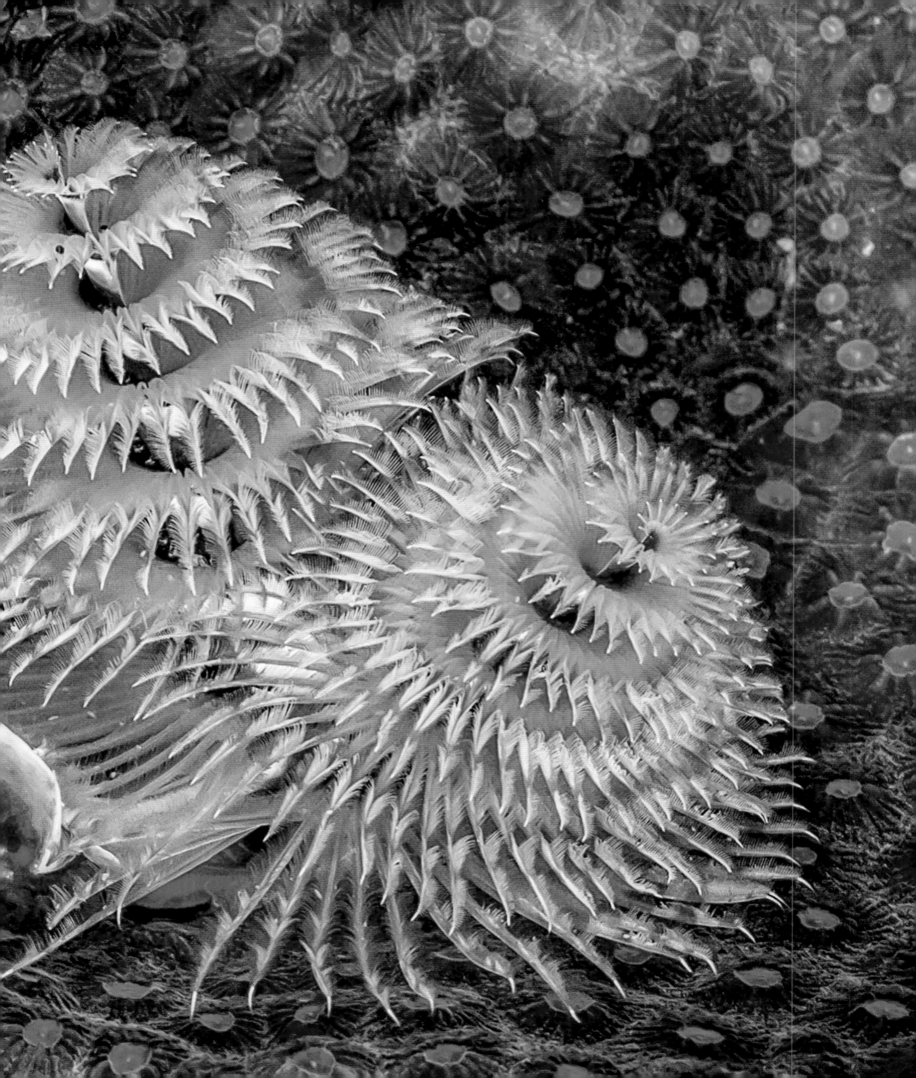

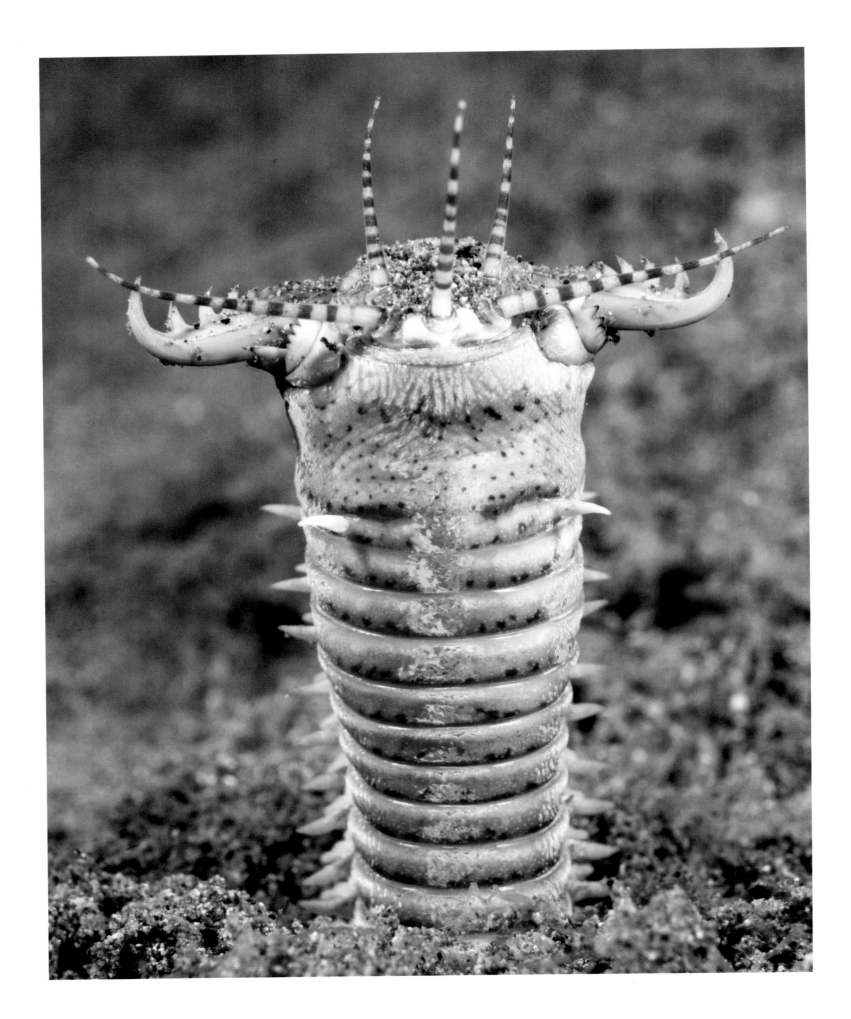

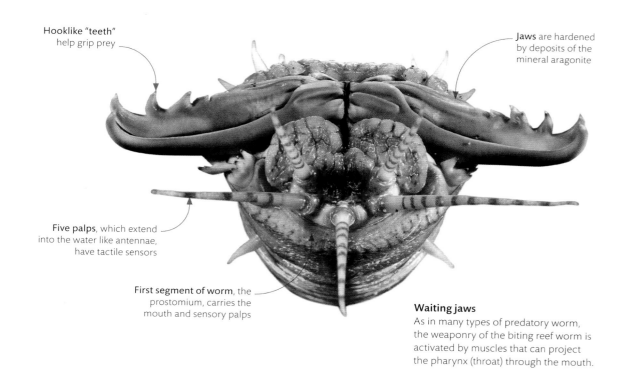

Hooklike "teeth"
help grip prey

Jaws are hardened
by deposits of the
mineral aragonite

Five palps, which extend
into the water like antennae,
have tactile sensors

First segment of worm, the
prostomium, carries the
mouth and sensory palps

Waiting jaws
As in many types of predatory worm,
the weaponry of the biting reef worm is
activated by muscles that can project
the pharynx (throat) through the mouth.

invertebrate jaws

Prior to the evolution of jaws, animals fed on whatever material they could fit
into their mouths and swallow whole, as jellyfish and anemones still do.
Muscle-powered jaws that could slice, grind, or rasp gave early wormlike
animals access to new food sources, enabling animals to break large solid
items of food – such as leaves or flesh – into manageable fragments that could
be swallowed and digested easily. Sharp jaws became equally effective as
defensive weapons or tools for capturing and dispatching prey. Today, jaws
of varying size, form, and complexity are wielded by a wide diversity of
invertebrates, from ants to giant squid and predatory worms.

Hydraulic jaws
Pressure of fluids inside the body pushes the throat of a
biting reef worm (*Eunice aphroditois*) outwards, making
the serrated jaws and sensory palps gape around the
mouth. Any fish that swims close enough stimulates the
palps and releases the pressure, so the jaws snap shut
around the fish's body as the throat retracts back inside.

BLOODSUCKER JAWS

Many jaws work like scissors, pivoting blades
together as they close, but those of a leech are
more like scalpels. After clamping an oral sucker
to the skin of a victim, the leech's three jaws slice
through, leaving a Y-shaped wound. Salivary
glands release their secretions over the sharp
blades. Anticoagulant in the saliva ensures that
the blood keeps flowing freely from the wound,
while powerful muscles in the leech's pharynx
(throat) suck up the liquid meal.

Three toothed
blades, arranged
in a Y shape

LEECH JAWS

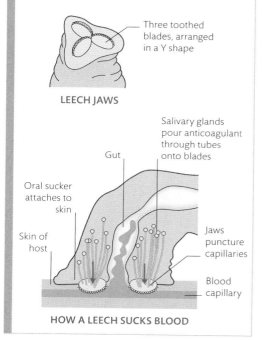

Salivary glands
pour anticoagulant
through tubes
onto blades

Gut

Oral sucker
attaches to
skin

Skin of
host

Jaws
puncture
capillaries

Blood
capillary

HOW A LEECH SUCKS BLOOD

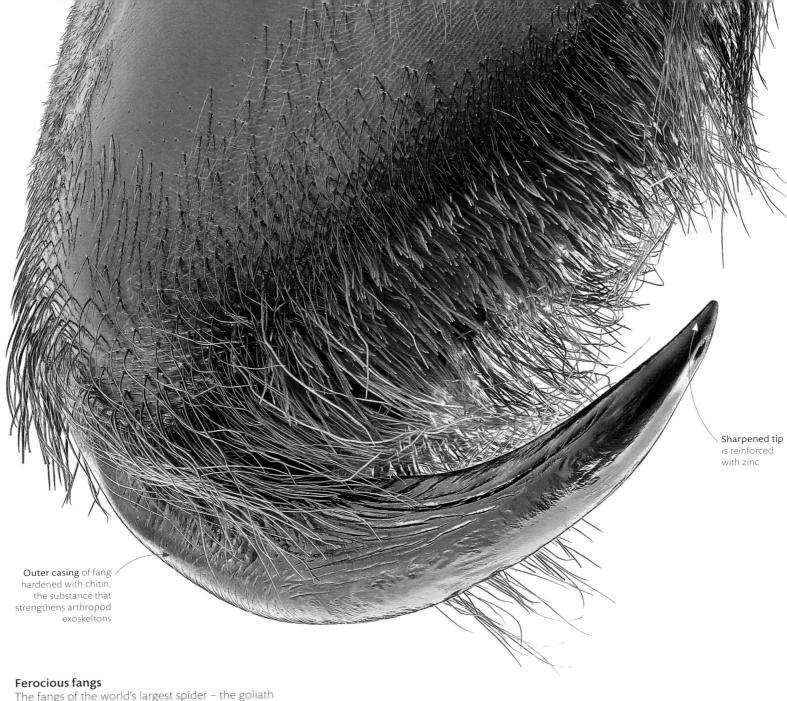

Sharpened tip
is reinforced
with zinc

Outer casing of fang
hardened with chitin,
the substance that
strengthens arthropod
exoskeltons

Ferocious fangs
The fangs of the world's largest spider – the goliath
birdeater (*Theraphosa blondi*) – retain their menacing
gleam even in their discarded moulted skin, as seen
here. These fangs, turned inwards, reveal the tiny
apertures that deliver the venom. Despite its size
this spider is not life-threatening to humans.

injecting venom

Spiders are the dominant invertebrate predators almost everywhere on land. For billions of
insects, and a few lizards or birds, spiders are killers. Except for a few hundred non-venomous
types, most of the 50,000 or so spider species kill with venom that they inject into their victim
with fangs. Spiders cannot ingest solid food, so while the venom paralyzes prey to stop it from
struggling, they spew digestive enzymes into the body to liquidize it. Only a few – black widows,
the Brazilian wolf spider, and the Australian funnel-web – have venom that is dangerous to humans.

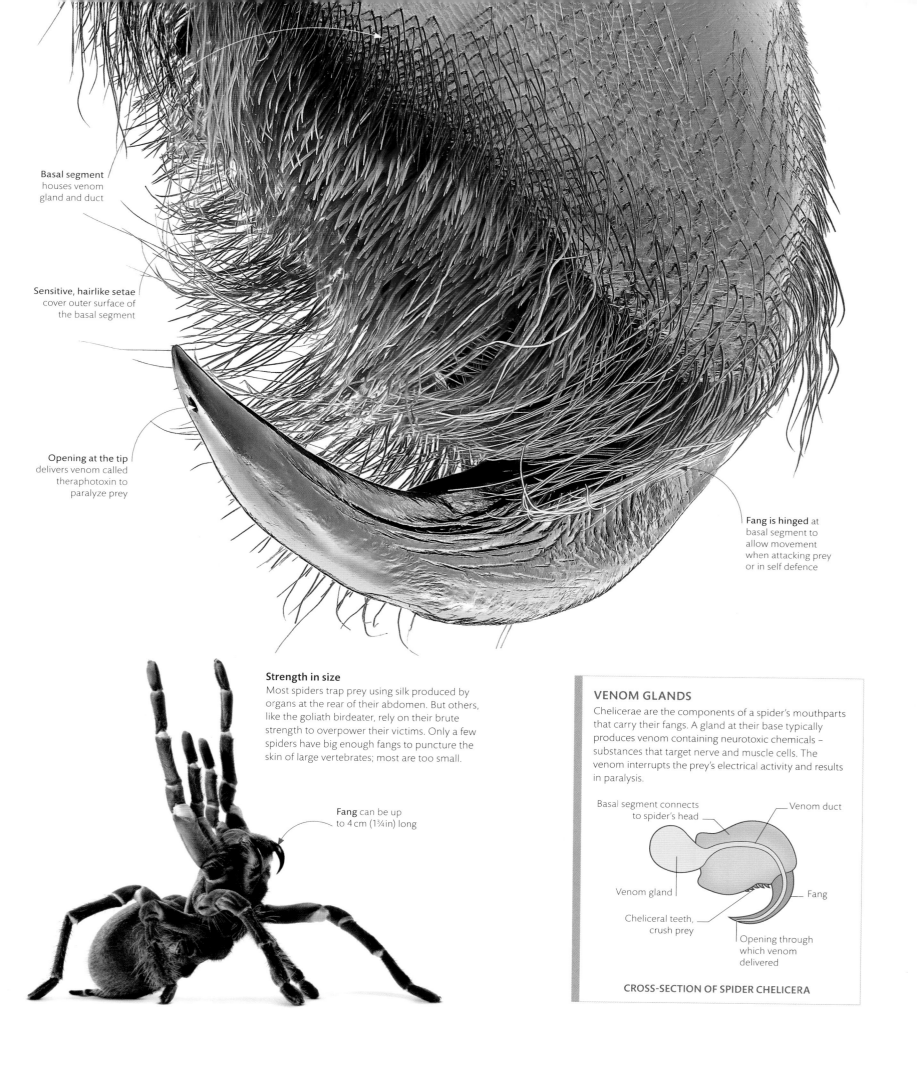

Basal segment
houses venom
gland and duct

Sensitive, hairlike setae
cover outer surface of
the basal segment

Opening at the tip
delivers venom called
theraphotoxin to
paralyze prey

Fang is hinged at
basal segment to
allow movement
when attacking prey
or in self defence

Strength in size

Most spiders trap prey using silk produced by
organs at the rear of their abdomen. But others,
like the goliath birdeater, rely on their brute
strength to overpower their victims. Only a few
spiders have big enough fangs to puncture the
skin of large vertebrates; most are too small.

Fang can be up
to 4 cm (1¾ in) long

VENOM GLANDS

Chelicerae are the components of a spider's mouthparts
that carry their fangs. A gland at their base typically
produces venom containing neurotoxic chemicals –
substances that target nerve and muscle cells. The
venom interrupts the prey's electrical activity and results
in paralysis.

Basal segment connects
to spider's head

Venom duct

Venom gland

Fang

Cheliceral teeth,
crush prey

Opening through
which venom
delivered

CROSS-SECTION OF SPIDER CHELICERA

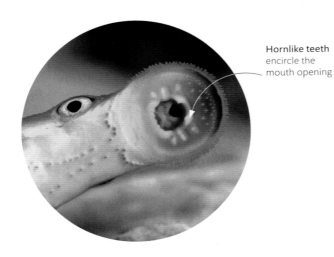

Hornlike teeth encircle the mouth opening

Life without jaws
Lampreys and hagfish are the only living vertebrates that lack biting jaws. The European brook lamprey (*Lampetra planeri*) uses its suckerlike mouth to cling to rocks; only its larvae, which lack the suction disc, feed. Many other lamprey species are parasitic as adults, attaching to a fish by the mouth and feeding on the host's blood and tissues.

FROM GILL ARCHES TO JAWS
Studies of embryological development and fossil evidence suggest that the jaws of vertebrates evolved from the gill arches of fish. Gill arches are skeletal struts between the gill openings in the side of the head. It is likely that the front-most arch became the upper and lower mandibles of biting jaws.

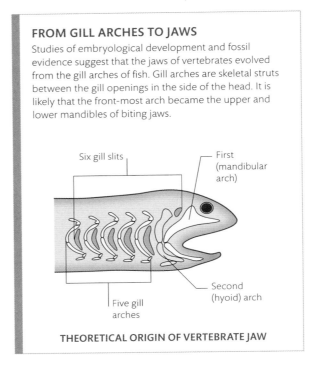

Six gill slits

First (mandibular arch)

Five gill arches

Second (hyoid) arch

THEORETICAL ORIGIN OF VERTEBRATE JAW

vertebrate jaws

The first fish that swam in prehistoric oceans half a billion years ago had jawless mouths that probably rasped food from sea-floor sediment – until the evolution of jaws enabled vertebrates to diversify. Opening and closing jaws can pump oxygen-rich water across the gills, and this may have been their first role in the earliest jawed fish. Since jaws can also bite, their evolution inevitably revolutionized the way that vertebrates could feed. Biting jaws opened the way for herbivorous vertebrates to chew on plants and for carnivores to kill prey.

Jaw bones
As in all mammals, the lower jaw of the pygmy hippopotamus is formed from a single bone called a dentary. The reptilian ancestors of mammals hinged their jaws with different bones, but during evolution, those bones have been repurposed as the tiny middle-ear ossicles that improve mammal hearing.

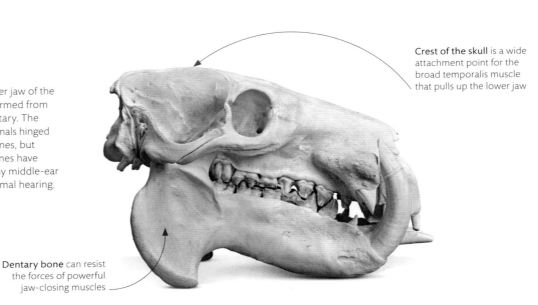

Crest of the skull is a wide attachment point for the broad temporalis muscle that pulls up the lower jaw

Dentary bone can resist the forces of powerful jaw-closing muscles

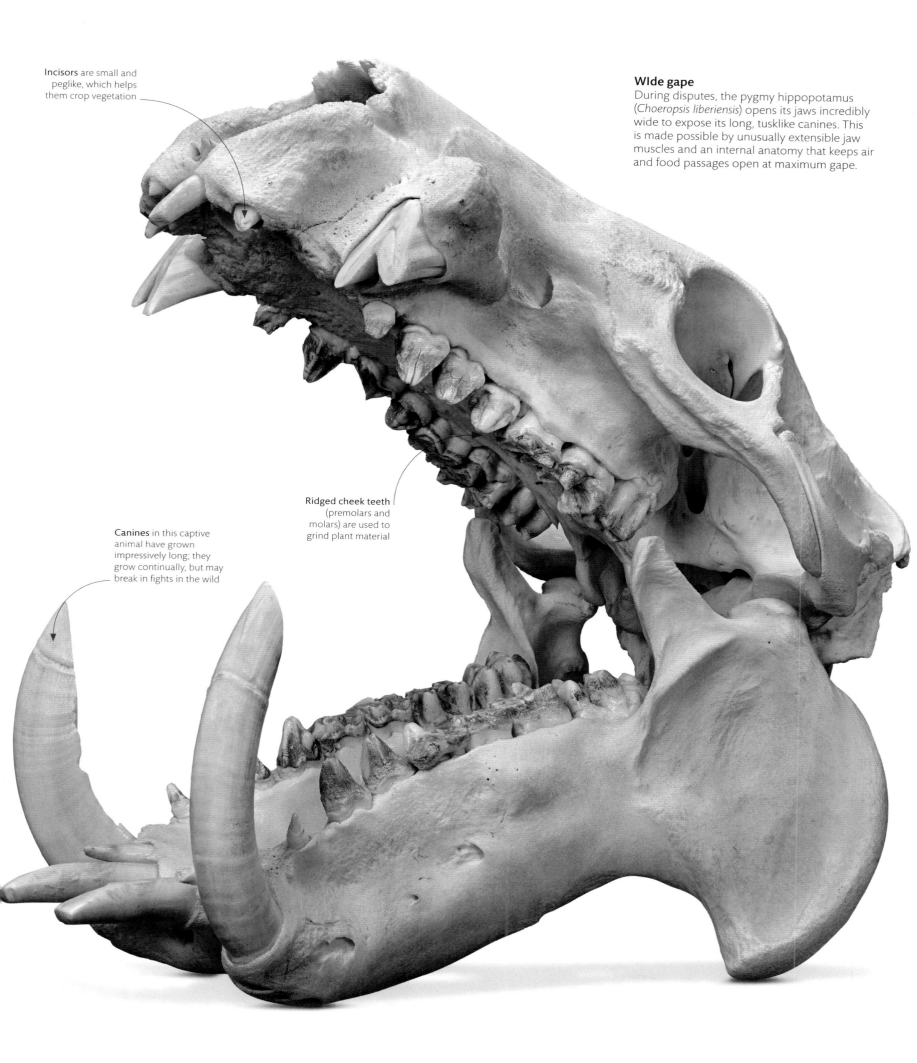

Incisors are small and peglike, which helps them crop vegetation

Wide gape
During disputes, the pygmy hippopotamus (*Choeropsis liberiensis*) opens its jaws incredibly wide to expose its long, tusklike canines. This is made possible by unusually extensible jaw muscles and an internal anatomy that keeps air and food passages open at maximum gape.

Ridged cheek teeth (premolars and molars) are used to grind plant material

Canines in this captive animal have grown impressively long; they grow continually, but may break in fights in the wild

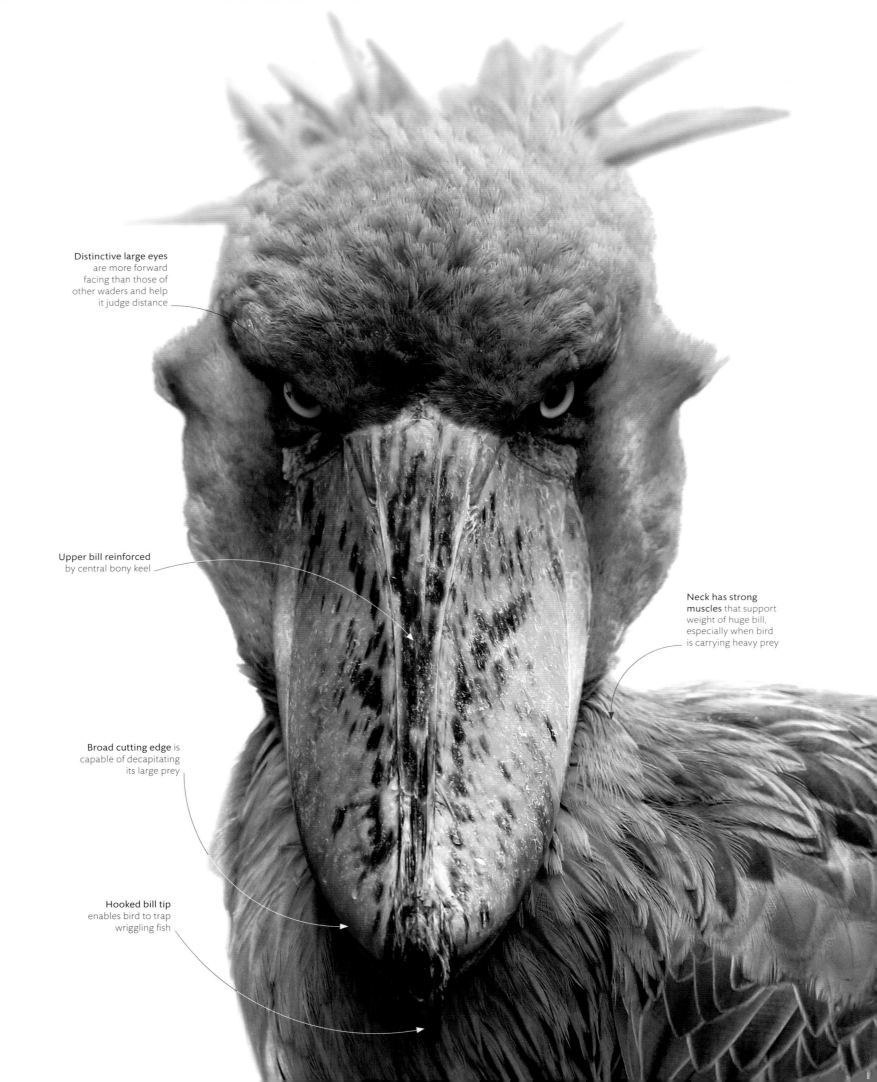

Distinctive large eyes are more forward facing than those of other waders and help it judge distance

Upper bill reinforced by central bony keel

Neck has strong muscles that support weight of huge bill, especially when bird is carrying heavy prey

Broad cutting edge is capable of decapitating its large prey

Hooked bill tip enables bird to trap wriggling fish

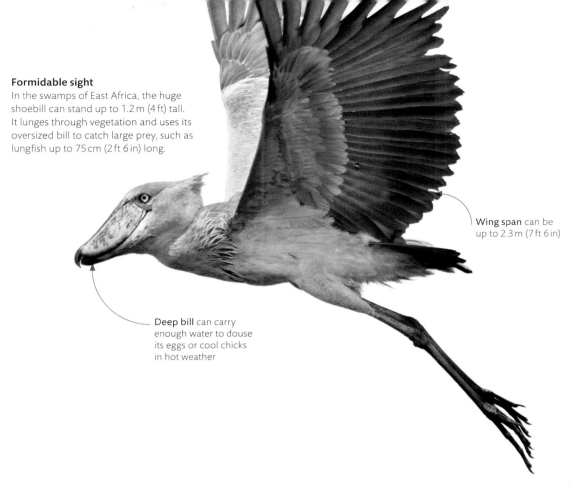

Formidable sight
In the swamps of East Africa, the huge shoebill can stand up to 1.2 m (4 ft) tall. It lunges through vegetation and uses its oversized bill to catch large prey, such as lungfish up to 75 cm (2 ft 6 in) long.

Wing span can be up to 2.3 m (7 ft 6 in)

Deep bill can carry enough water to douse its eggs or cool chicks in hot weather

bird bills

During evolution, birds lost the feature that gave their reptilian ancestors such a formidable bite – teeth. Instead, they gained a versatile horny beak that varies from a cutting-edged weapon to a delicate tool that can crush seeds or probe for nectar (see pp.198–99). Thanks to the flexibility of their jaws, birds can open their mouth to manipulate food not just by dropping the lower mandible, but also by hinging the top mandible upwards.

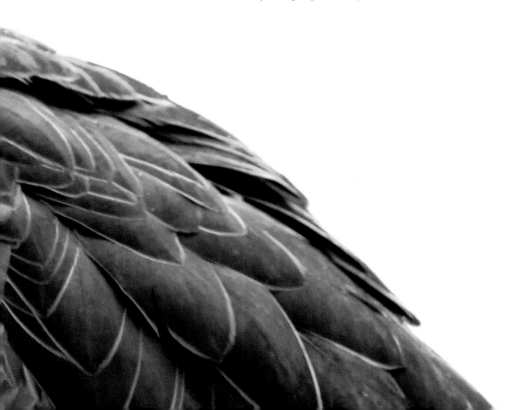

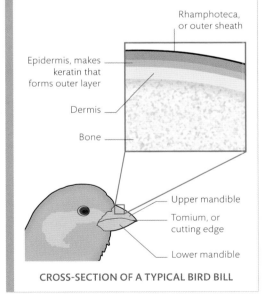

STRUCTURE OF A BIRD'S BILL
The bony core of a bird's bill has a tough, often hard and horny, outer coating, or rhamphotheca, made of the same tissue as claws and nails – keratin. This layer is supplied with blood vessels and nerves, making the bill sensitive to touch.

Rhamphotheca, or outer sheath

Epidermis, makes keratin that forms outer layer

Dermis

Bone

Upper mandible

Tomium, or cutting edge

Lower mandible

CROSS-SECTION OF A TYPICAL BIRD BILL

Fearsome wader
Once called the whale-headed stork, the shoebill (*Balaeniceps rex*) is a wading bird, but it is more closely related to the pelican according to its DNA. Although lacking a pouch, its upper bill is hooked and reinforced with a keel, like that of a pelican.

bird bill shapes

Birds' bills have evolved in response to the food available in their habitats, resulting in a wide variety of beak shapes that reflect what and how they eat. While many birds are omnivores, consuming both live prey and vegetation depending on the season, most specialize in a particular food source, such as seeds, nectar, or invertebrates.

fruit, seed, and nut eating

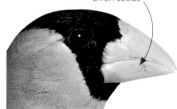

Cone-shape helps bird peck out and crush cedar, pine, and birch seeds

JAPANESE GROSBEAK
Eophona personata

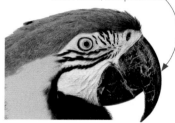

Strong beak can break nutshells, crush seeds, and peel fruit

BLUE-AND-YELLOW MACAW
Ara ararauna

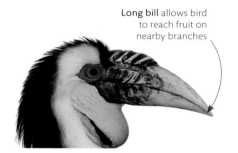

Long bill allows bird to reach fruit on nearby branches

WREATHED HORNBILL
Rhyticeros undulatus

nectar drinking

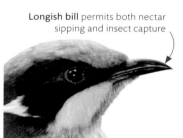

Longish bill permits both nectar sipping and insect capture

YELLOW-TUFTED HONEYEATER
Lichenostomus melanops

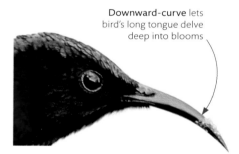

Downward-curve lets bird's long tongue delve deep into blooms

BLACK-THROATED SUNBIRD
Aethopyga saturata

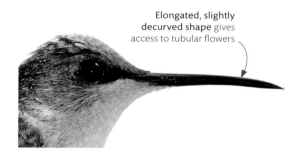

Elongated, slightly decurved shape gives access to tubular flowers

VIOLET-CROWNED WOODNYMPH
Thalurania colombica colombica

mud and soil probing

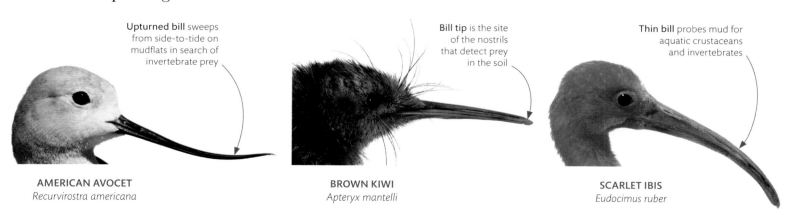

Upturned bill sweeps from side-to-tide on mudflats in search of invertebrate prey

AMERICAN AVOCET
Recurvirostra americana

Bill tip is the site of the nostrils that detect prey in the soil

BROWN KIWI
Apteryx mantelli

Thin bill probes mud for aquatic crustaceans and invertebrates

SCARLET IBIS
Eudocimus ruber

Matching bill to food

The form and size of a bird's bill is a good clue as to what makes up the bulk of its diet. For example, a short, thick, conical bill – such as that of a finch – indicates a seed-eater, while carnivorous birds, such as eagles and other raptors, typically have a curved, razor-sharp beak tip that rips flesh into manageable chunks.

flesh ripping and gobbling

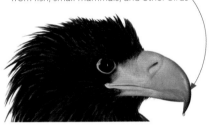

Hooked, deeply curved tip tears flesh from fish, small mammals, and other birds

STELLER'S SEA EAGLE
Haliaeetus pelagicus

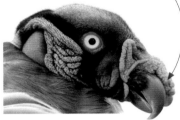

Sharp bill rips skin and hard tissue from carcasses

KING VULTURE
Sarcoramphus papa

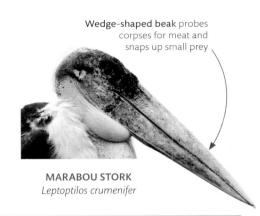

Wedge-shaped beak probes corpses for meat and snaps up small prey

MARABOU STORK
Leptoptilos crumenifer

insect catching

Short beak set on wide mouth that scoops up insects

EUROPEAN NIGHTJAR
Caprimulgus europaeus

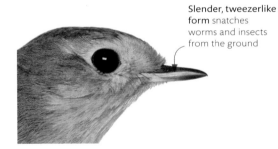

Slender, tweezerlike form snatches worms and insects from the ground

EUROPEAN ROBIN
Erithacus rubecula

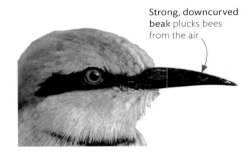

Strong, downcurved beak plucks bees from the air

GREEN BEE-EATER
Merops orientalis

fishing

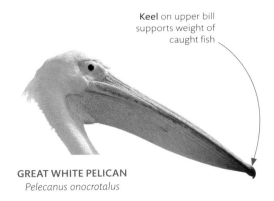

Keel on upper bill supports weight of caught fish

GREAT WHITE PELICAN
Pelecanus onocrotalus

Colourful beak holds several small fish at once

ATLANTIC PUFFIN
Fratercula arctica

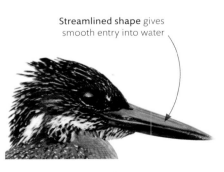

Streamlined shape gives smooth entry into water

GIANT KINGFISHER
Megaceryle maxima

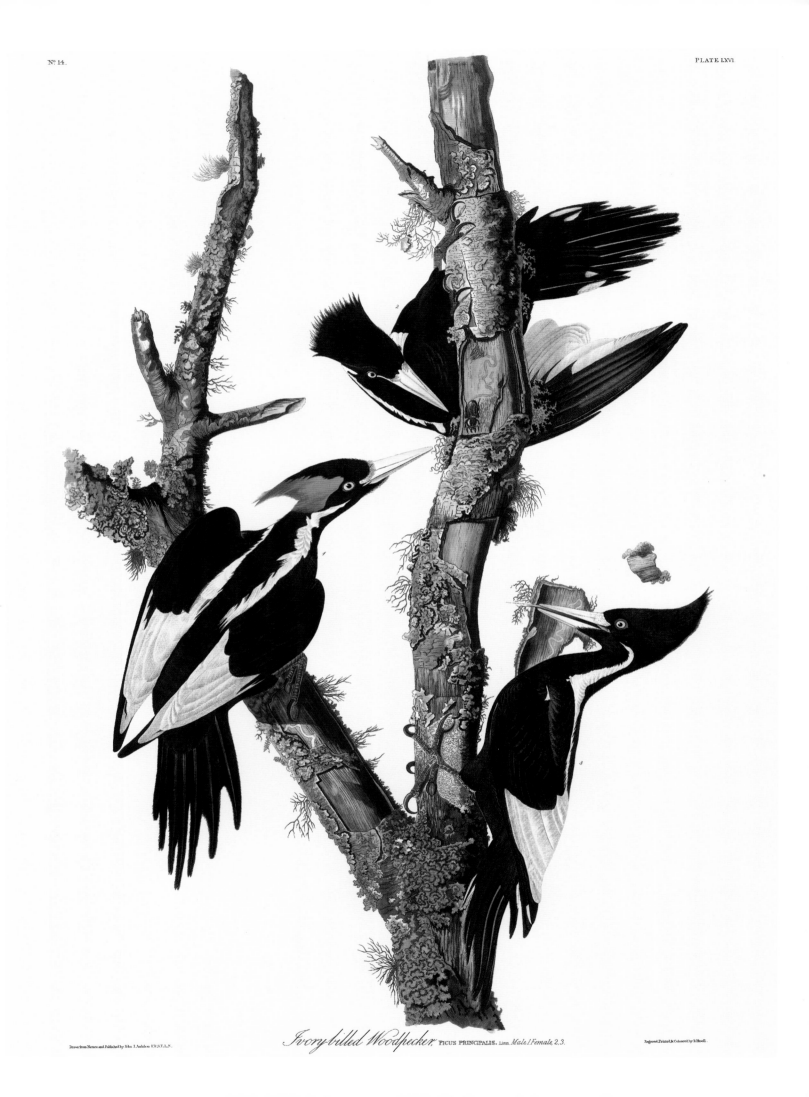

Ivory-billed Woodpecker, PICUS PRINCIPALIS. Linn. *Male 1 Female, 2, 3.*

Drawn from Nature and Published by John J. Audubon F.R.S.F.L.S.

Engraved, Printed & Coloured by R. Havell.

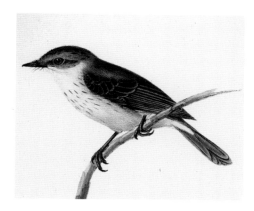

Darwin's flycatcher
John Gould worked extensively with the preserved specimens brought back from Charles Darwin's voyages. His illustration of a yellow and grey female Darwin's flycatcher (*Pyrocephalus nanus*) from the Galápagos Islands is from *The Zoology of the Voyage of HMS Beagle* (1838).

art of the ornithologist

The number, brilliance, and variety of form, song, and flight of the world's birds was one of the most tantalizing challenges to 19th-century naturalists. This was the age of recording, sketching, and classifying birds, and new printing techniques that produced iconic works of ornithological art.

Enthusiasts travelling the globe brought home accounts of their natural history discoveries and also bird specimens, many of which ended up in the hands of taxidermist and ornithologist John Gould (1804–1881). The first curator and preserver at the London Zoological Society, Gould went on to publish some of the world's finest bird volumes. A new method of printing, lithography, which involved drawing images on limestone blocks, enabled the production of vibrant hand-coloured prints. Gould travelled in Europe to catalogue and sketch birds for his major work, *The Birds of Europe* (1832–33) before embarking on a two-year expedition to Tasmania and Australia. Working with his wife, Elizabeth,

a painter, he produced a seven-volume work on Australian birds that included 328 species new to science.

Lifelike as the birds appear in these vibrant bird books, the majority were killed, dissected, and stuffed before they were painted. In the US, naturalist and hunter John James Audubon wire-rigged the fresh corpses of birds against a backdrop of their habitats to produce dazzlingly realistic images. Shunned by US scientists, he sought subscriptions from noble houses and university libraries in Britain to sponsor monumental aquatint plates of every bird in the Americas. The 200 printed copies of *The Birds of America* took nearly 12 years and $115,000 to complete.

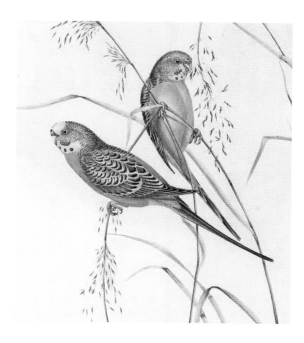

Budgerigars
English ornithologist John Gould produced 600 lithographs in superb detail and colour for his iconic *The Birds of Australia* (1840–48). After he brought the first two budgerigars to England in 1840, they became popular as pets.

Ivory-billed woodpecker
John James Audubon's desire to produce life-sized portraits of every bird in America called for "double elephant paper", 100 cm x 67 cm (39 in x 26 in). The ivory-billed woodpecker, now thought to be extinct, is one of the 435 monumental plates in *The Birds of America* (1827–38).

> " I never for a day gave up listening to the songs of our birds, or watching their peculiar habits, or delineating them in the best way I could. "

JOHN JAMES AUDUBON, *AUDUBON AND HIS JOURNALS*, 1899

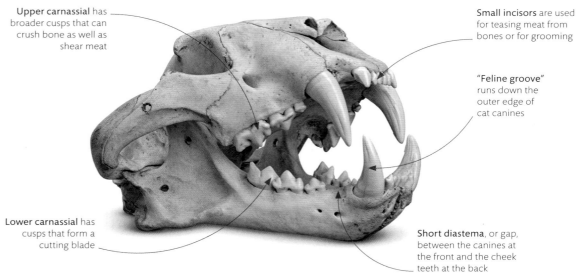

Upper carnassial has broader cusps that can crush bone as well as shear meat

Small incisors are used for teasing meat from bones or for grooming

"Feline groove" runs down the outer edge of cat canines

Lower carnassial has cusps that form a cutting blade

Short diastema, or gap, between the canines at the front and the cheek teeth at the back

Tiger skull
Like those of other carnivorans, the formidably strong jaws of the tiger (*Panthera tigris*) carry deeply-rooted cheek teeth, called carnassials, that slice together like scissors to cut through meat. Long canines that puncture skin are used to grip struggling prey.

carnivore teeth

Any animal that consumes flesh – such as a jellyfish, a tiger beetle, or a crocodile – is described as carnivorous, but some of the most impressive adaptations to a meat-eating lifestyle occur in the mammalian order Carnivora. These so-called carnivorans – including cats, dogs, weasels, and bears – rely on powerful biting jaws equipped with teeth that can stab, kill, and dismember. And like most other mammals, their modified teeth ensure that the mouth can multi-task while processing food.

DIFFERENTIATED TEETH

In most fish, amphibians, and reptiles, the teeth are similar in shape. An alligator's teeth are uniformly conical. This uniformity is known as homodont dentition. But mammals, in contrast, have differentiated, or heterodont, dentition, which enables them to process food in different ways. Typically, there are chisel-shaped incisors at the front that nibble or crop, conical stabbing canines behind the incisors, and ridged cheek teeth (premolars and molars) at the back that crush and grind food.

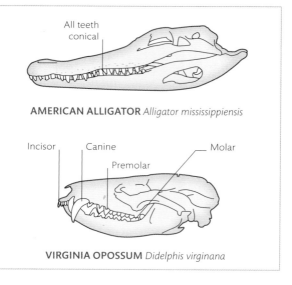

All teeth conical

AMERICAN ALLIGATOR *Alligator mississippiensis*

Incisor — Canine — Molar
Premolar

VIRGINIA OPOSSUM *Didelphis virginiana*

A killer's teeth
Although still fearsome, the canine teeth of a cheetah (*Acinonys jubatus*) are proportionately smaller than those of other cats; smaller canine roots leave more space in the nasal cavity to "pant" through the nostrils while gripping prey after a high-speed chase. The daggerlike canines, used for clamping a victim's throat to suffocate, are unmistakably those of a carnivore.

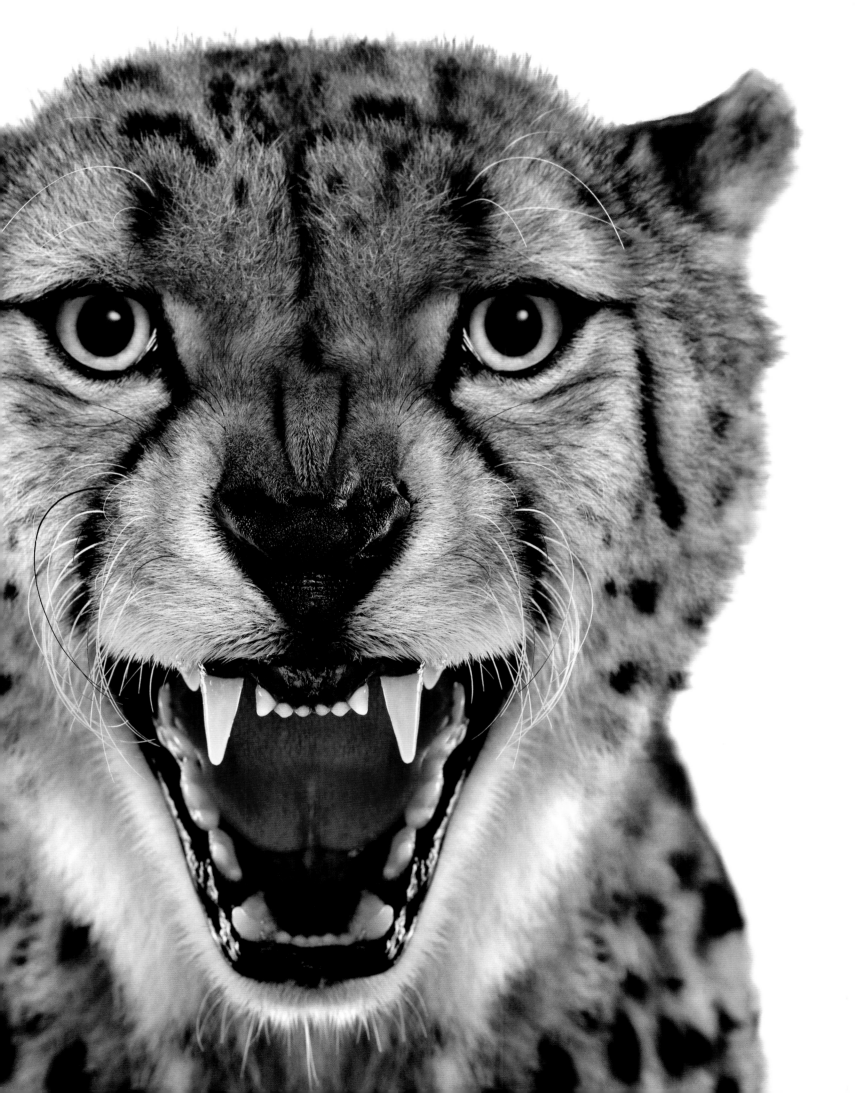

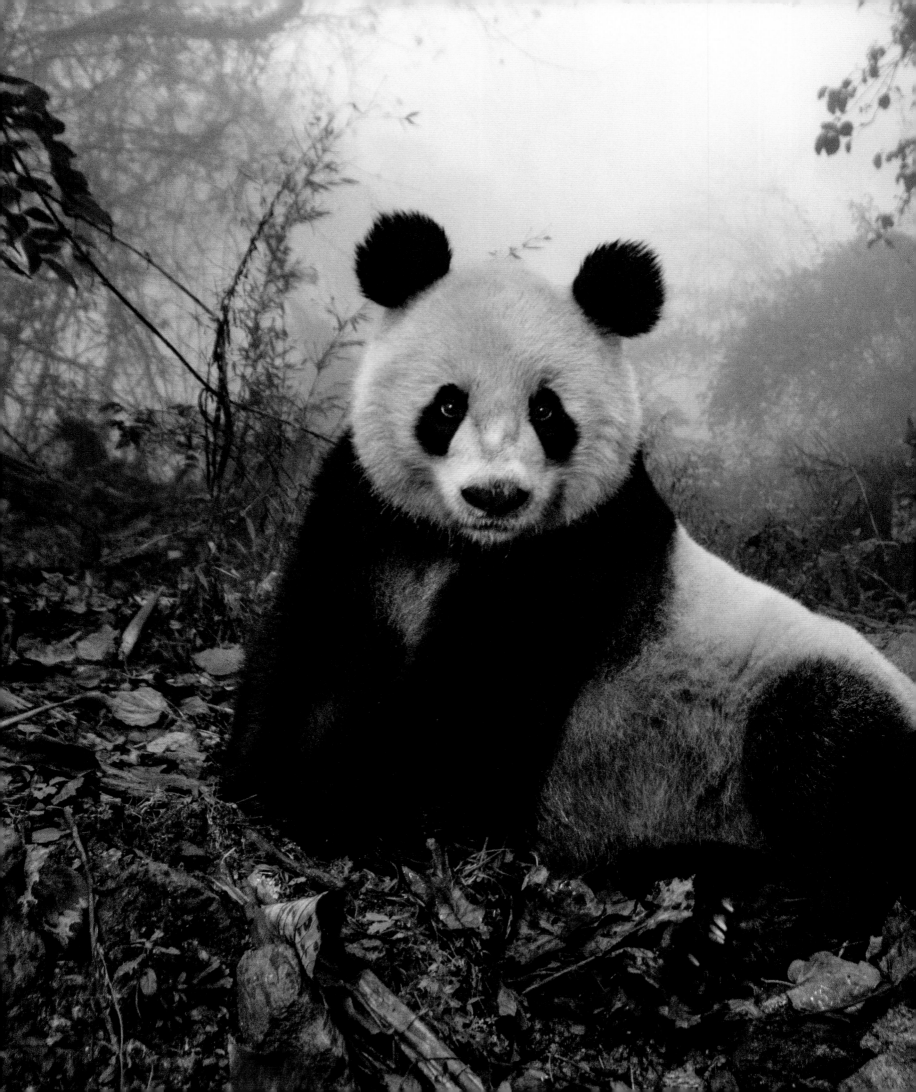

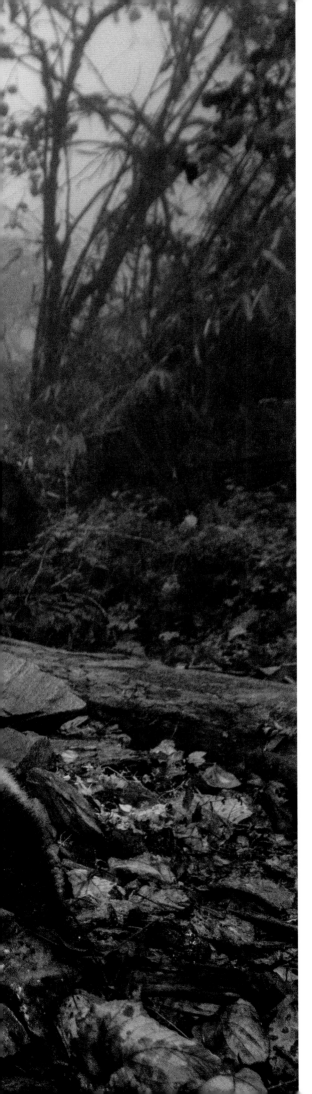

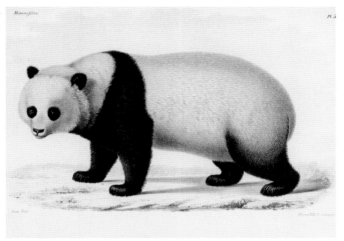

spotlight species

giant panda

With its rounded face, mainly vegetarian diet, and black-and-white coat, the giant panda (*Ailuropoda melanoleuca*) at first seems more like a large raccoon than a bear. While DNA testing places this mammal firmly in the bear family, its anatomy and dietary habits continue to intrigue scientists.

The giant panda is one of the world's most vulnerable mammals and it is estimated that only 1,500–2,000 remain in the wild. Adults can be 1.2–1.8 m (4–6 ft) long and can weigh up to 136 kg (300 lb). Although slow-moving, they can swim and wade through water, and are agile climbers, and use an enlarged wrist-bone, called a pseudo-thumb, to manipulate bamboo. Mature by the age of 5 or 6, the normally solitary males and females spend a few days or weeks together during the mating season, usually between March and May. Cubs are born 3–6 months later, and stay with their mothers for up to two years.

This animal's diet is an enigma. A panda spends up to 16 hours a day eating 9–18 kg (20–40 lb) of food, then rests for hours. Its heritage as well as its canine teeth and short digestive tract suggest that it is a carnivore. But 99 per cent of its food intake is nutritionally poor bamboo, and it has ultra-strong jaw muscles and broad, flat molars that can grind this tough plant material. Most carnivores lack the gut bacteria needed to digest grasses, but giant pandas have enough to break down some of the cellulose they eat; they only receive 17–20 per cent of a meal's energy – just enough to survive. Their diet means they cannot build up the fat reserves that allow a deep sleep (torpor) through winter.

Mountain dweller
Once also found in lowland areas, human encroachment has restricted the animal to higher ground. Giant pandas now inhabit only a few dense mountain forests of central China.

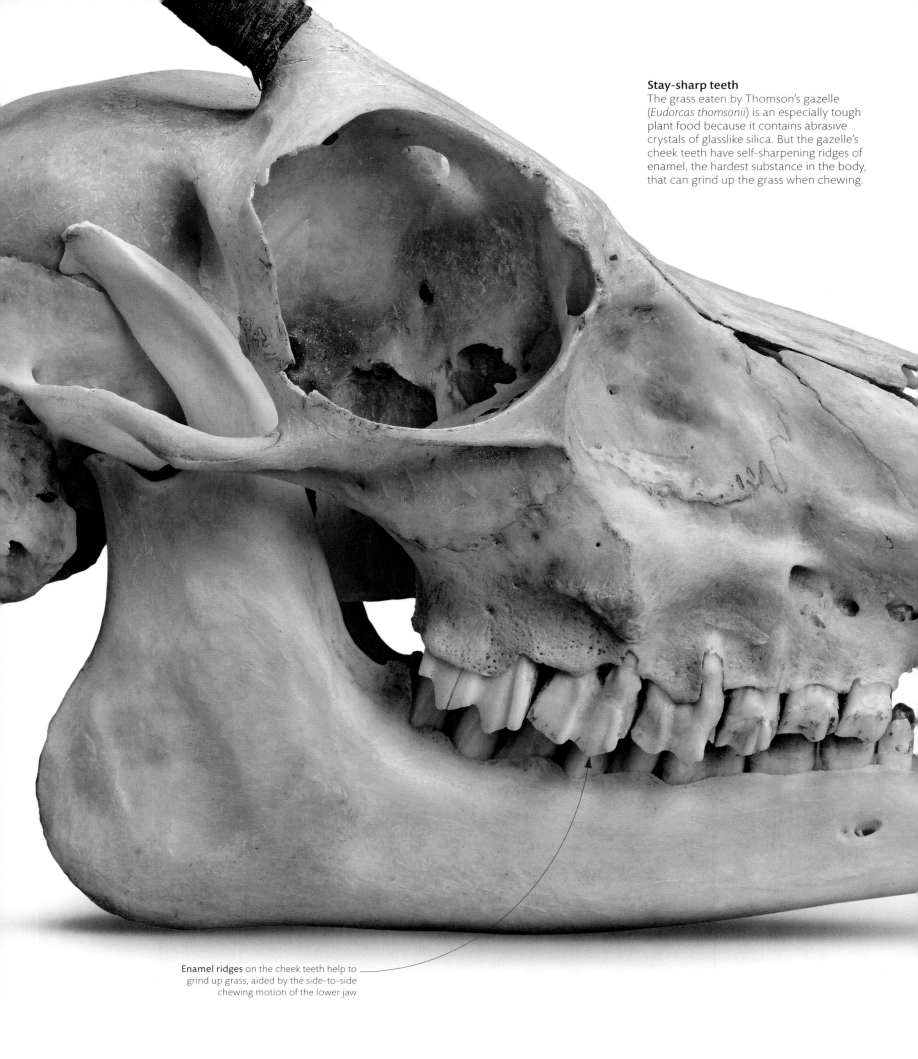

Stay-sharp teeth
The grass eaten by Thomson's gazelle (*Eudorcas thomsonii*) is an especially tough plant food because it contains abrasive crystals of glasslike silica. But the gazelle's cheek teeth have self-sharpening ridges of enamel, the hardest substance in the body, that can grind up the grass when chewing.

Enamel ridges on the cheek teeth help to grind up grass, aided by the side-to-side chewing motion of the lower jaw

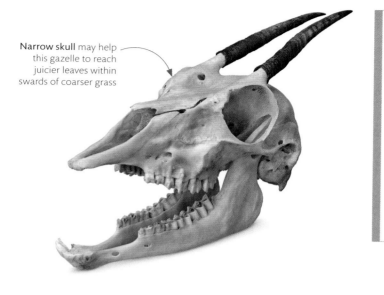

Narrow skull may help this gazelle to reach juicier leaves within swards of coarser grass

USING MICROBES

The microbes inside the gut of herbivores produce the enzyme cellulase, which digests plant fibre into sugar and fatty acids. In many plant-eating mammals, such as horses, rhinoceroses, and rabbits, the microbes inhabit enlarged parts of the intestine, but ruminants (including cows and antelopes) house the microbes in a multi-chambered stomach.

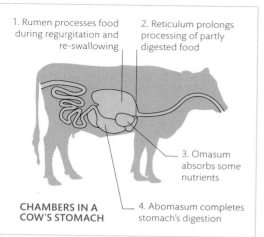

1. Rumen processes food during regurgitation and re-swallowing

2. Reticulum prolongs processing of partly digested food

3. Omasum absorbs some nutrients

4. Abomasum completes stomach's digestion

CHAMBERS IN A COW'S STOMACH

A skull for grazing

Like other grazers that eat abrasive grass, Thomson's gazelle has a long row of grinding cheek teeth. Browsers – herbivores that feed on the soft leaves of low shrubs – often have a shorter row of cheek teeth.

eating plants

An animal that relies on plants for its nutrition faces a problem: plant material is packed with fibrous cellulose – the constituent of plant cell walls – that is difficult to digest. Herbivores need the means not only to cut and chew tough leaves, but also to extract nutrients from the resulting pulp. As well as having a specialized dentition, herbivores harbour cultures of living microbes in their digestive system. The microbes provide the enzymes necessary to digest plant fibre into sugars that can be assimilated by the body.

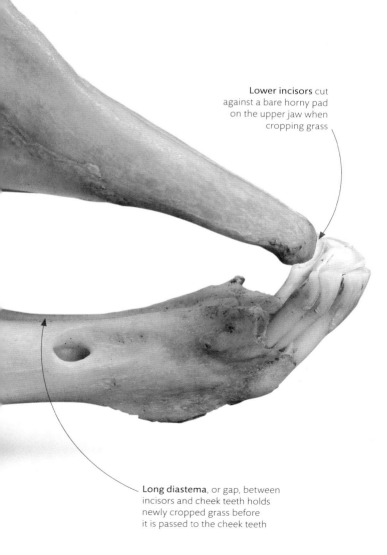

Lower incisors cut against a bare horny pad on the upper jaw when cropping grass

Long diastema, or gap, between incisors and cheek teeth holds newly cropped grass before it is passed to the cheek teeth

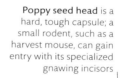

Poppy seed head is a hard, tough capsule; a small rodent, such as a harvest mouse, can gain entry with its specialized gnawing incisors

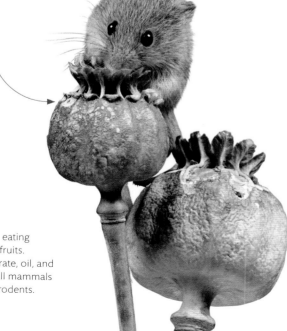

Eating seeds

Some herbivores concentrate on eating more easily digestible seeds and fruits. Many seeds are rich in carbohydrate, oil, and protein – ideal for sustaining small mammals with a high metabolism, such as rodents.

flexible faces

Eating is the primary purpose of the mouth and jaws of any animal. But in mammals – especially higher primates – the extraordinary flexibility of the face, which is controlled by hundreds of tiny muscles, can be also commandeered to communicate signals that are important in organizing social groups. For animals that depend heavily on vision and visual displays, the face has become a way of advertising mood and intent.

EXPRESSIVE FACES

Chimpanzees live in complex social groups, and use facial expressions to advertise their feelings and emotions. Expressions like a play-face indicating happiness, a tooth-bearing grin showing fear, or a pout meaning they are craving reassurance, effect responses in other members of the group. They might even trigger feelings of empathy that help reinforce social bonding.

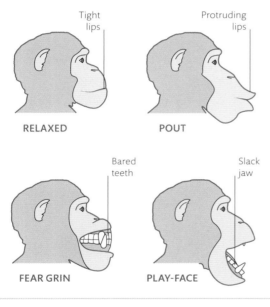

Tight lips

RELAXED

Protruding lips

POUT

Bared teeth

FEAR GRIN

Slack jaw

PLAY-FACE

Look at me

The Bornean orangutan (*Pongo pygmaeus*) collects fruit from trees and uses its flexible lips to help it separate the flesh from skin and seeds. Although orangutans are generally more solitary than African apes, they are social animals and share the capacity for facial expression.

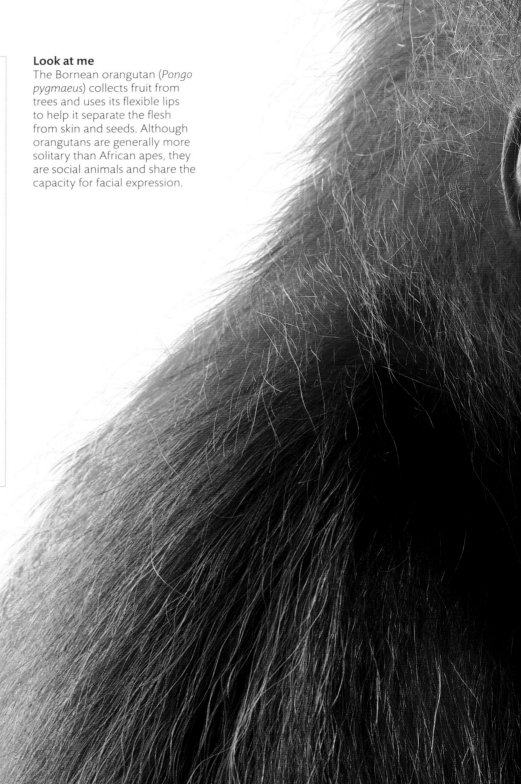

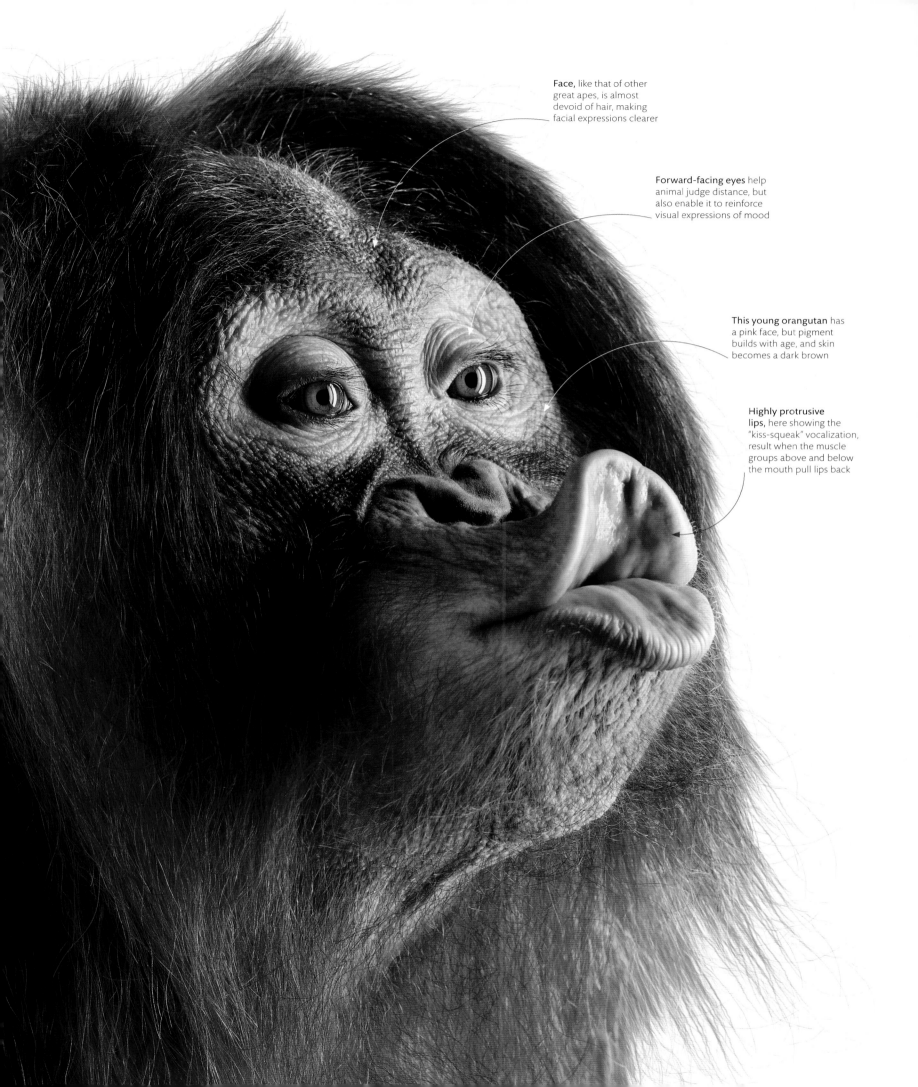

Face, like that of other great apes, is almost devoid of hair, making facial expressions clearer

Forward-facing eyes help animal judge distance, but also enable it to reinforce visual expressions of mood

This young orangutan has a pink face, but pigment builds with age, and skin becomes a dark brown

Highly protrusive lips, here showing the "kiss-squeak" vocalization, result when the muscle groups above and below the mouth pull lips back

legs, arms, tentacles, and tails

leg. a weight-bearing limb used for
locomotion and support.

arm. a vertebrate forelimb typically used
for grasping, or the appendage of an octopus.

tentacle. a flexible appendage used for
locomotion, grasping, feeling, or feeding.

tail. an elongated, flexible appendage
at the hindmost part of an animal.

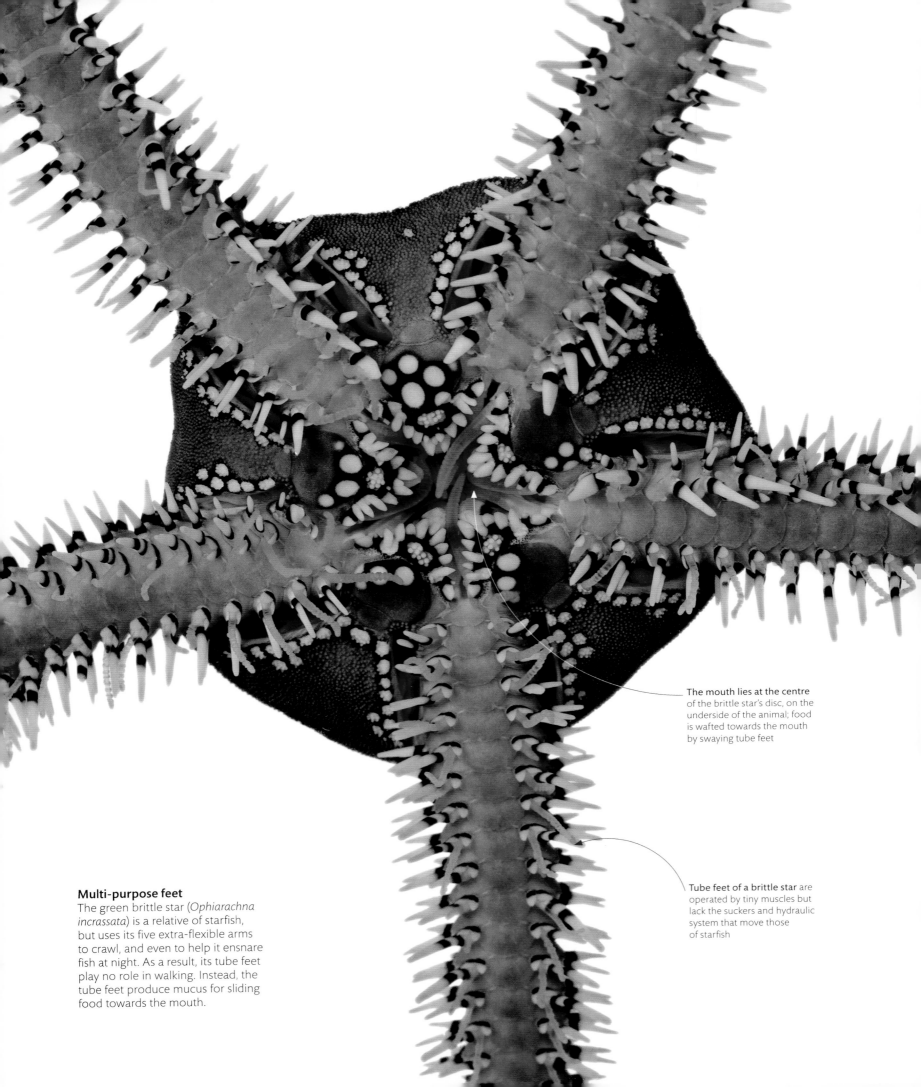

Multi-purpose feet
The green brittle star (*Ophiarachna incrassata*) is a relative of starfish, but uses its five extra-flexible arms to crawl, and even to help it ensnare fish at night. As a result, its tube feet play no role in walking. Instead, the tube feet produce mucus for sliding food towards the mouth.

The mouth lies at the centre of the brittle star's disc, on the underside of the animal; food is wafted towards the mouth by swaying tube feet

Tube feet of a brittle star are operated by tiny muscles but lack the suckers and hydraulic system that move those of starfish

tube feet

Starfish and brittle stars have a similar kind of radial symmetry to anemones, but while anemones are fixed to the sea bed and rely on food drifting by, the other two animals can move to hunt or graze. Through the coordinated efforts of hundreds of fleshy projections – called tube feet – on their underside, starfish seem to glide along the ocean floor. Meanwhile, brittle stars can wriggle and even grip potential prey with their long flexible arms, leaving their tube feet for probing and sensing.

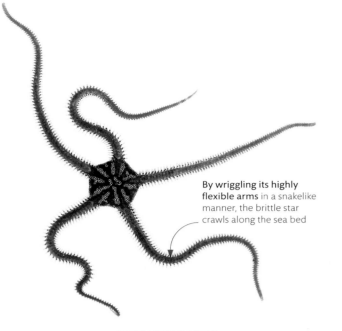

By wriggling its highly **flexible arms** in a snakelike manner, the brittle star crawls along the sea bed

GREEN BRITTLE STAR
Ophiarachna incrassata

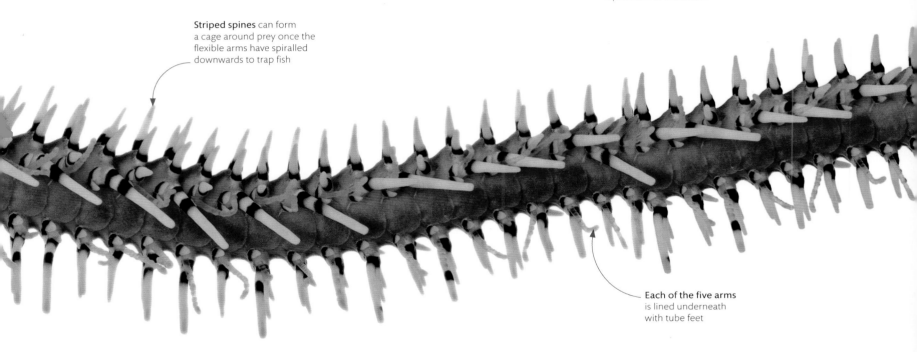

Striped spines can form a cage around prey once the flexible arms have spiralled downwards to trap fish

Each of the five arms is lined underneath with tube feet

STARFISH HYDRAULICS

In starfish, a sac of sea water – called an ampulla – which is located above each tube foot, is filled by a system of water channels unique to these kinds of animals. When muscles in the ampulla contract, they squeeze the sea water down into the tube foot. The tube foot extends and sticks to the sea bed by its sucker, before muscles in the foot contract to pull the animal forwards.

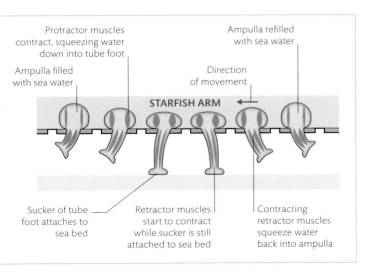

Protractor muscles contract, squeezing water down into tube foot

Ampulla refilled with sea water

Ampulla filled with sea water

Direction of movement

STARFISH ARM

Sucker of tube foot attaches to sea bed

Retractor muscles start to contract while sucker is still attached to sea bed

Contracting retractor muscles squeeze water back into ampulla

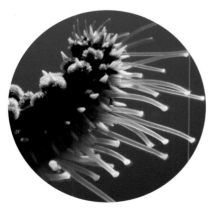

Suckered feet
Starfish tube feet are tipped with suckers that help them walk. These suckers can be strong enough to pull open mussel shells.

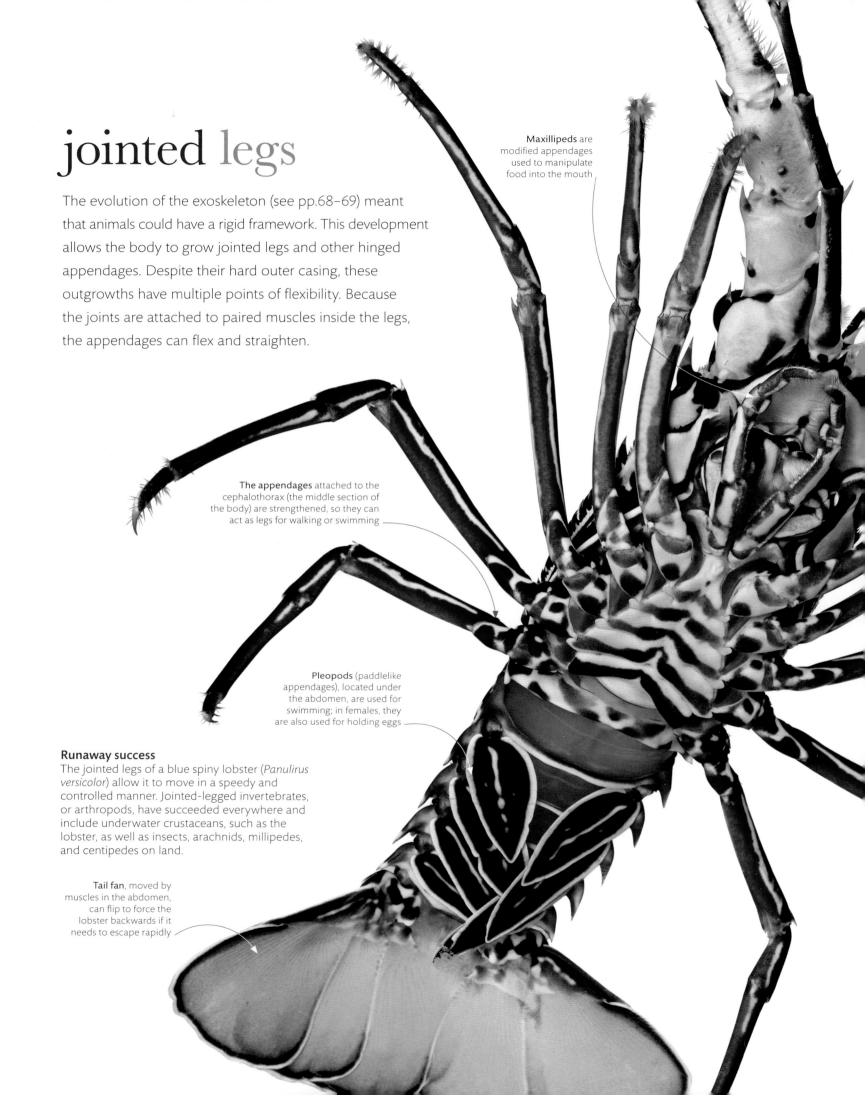

jointed legs

The evolution of the exoskeleton (see pp.68–69) meant that animals could have a rigid framework. This development allows the body to grow jointed legs and other hinged appendages. Despite their hard outer casing, these outgrowths have multiple points of flexibility. Because the joints are attached to paired muscles inside the legs, the appendages can flex and straighten.

Maxillipeds are modified appendages used to manipulate food into the mouth

The appendages attached to the cephalothorax (the middle section of the body) are strengthened, so they can act as legs for walking or swimming

Pleopods (paddlelike appendages), located under the abdomen, are used for swimming; in females, they are also used for holding eggs

Runaway success
The jointed legs of a blue spiny lobster (*Panulirus versicolor*) allow it to move in a speedy and controlled manner. Jointed-legged invertebrates, or arthropods, have succeeded everywhere and include underwater crustaceans, such as the lobster, as well as insects, arachnids, millipedes, and centipedes on land.

Tail fan, moved by muscles in the abdomen, can flip to force the lobster backwards if it needs to escape rapidly

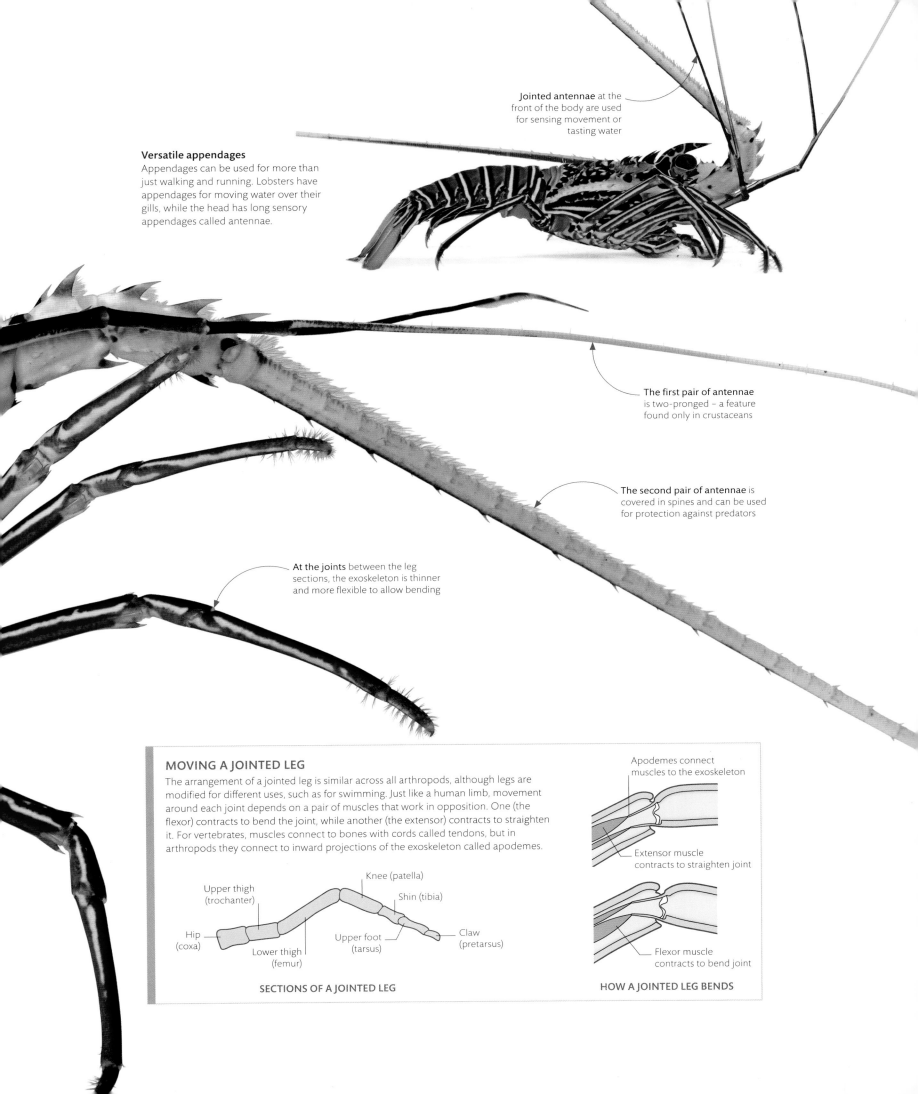

Versatile appendages
Appendages can be used for more than just walking and running. Lobsters have appendages for moving water over their gills, while the head has long sensory appendages called antennae.

Jointed antennae at the front of the body are used for sensing movement or tasting water

The first pair of antennae is two-pronged – a feature found only in crustaceans

The second pair of antennae is covered in spines and can be used for protection against predators

At the joints between the leg sections, the exoskeleton is thinner and more flexible to allow bending

MOVING A JOINTED LEG

The arrangement of a jointed leg is similar across all arthropods, although legs are modified for different uses, such as for swimming. Just like a human limb, movement around each joint depends on a pair of muscles that work in opposition. One (the flexor) contracts to bend the joint, while another (the extensor) contracts to straighten it. For vertebrates, muscles connect to bones with cords called tendons, but in arthropods they connect to inward projections of the exoskeleton called apodemes.

Upper thigh (trochanter)

Knee (patella)

Shin (tibia)

Hip (coxa)

Lower thigh (femur)

Upper foot (tarsus)

Claw (pretarsus)

SECTIONS OF A JOINTED LEG

Apodemes connect muscles to the exoskeleton

Extensor muscle contracts to straighten joint

Flexor muscle contracts to bend joint

HOW A JOINTED LEG BENDS

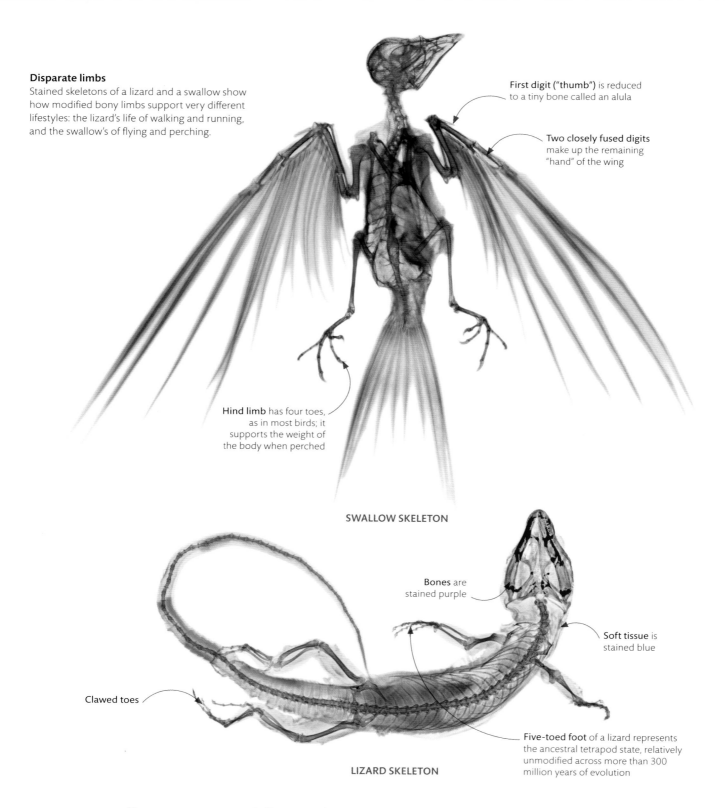

Disparate limbs
Stained skeletons of a lizard and a swallow show how modified bony limbs support very different lifestyles: the lizard's life of walking and running, and the swallow's of flying and perching.

First digit ("thumb") is reduced to a tiny bone called an alula

Two closely fused digits make up the remaining "hand" of the wing

Hind limb has four toes, as in most birds; it supports the weight of the body when perched

SWALLOW SKELETON

Bones are stained purple

Soft tissue is stained blue

Clawed toes

Five-toed foot of a lizard represents the ancestral tetrapod state, relatively unmodified across more than 300 million years of evolution

LIZARD SKELETON

vertebrate limbs

Four-limbed vertebrates, or tetrapods, evolved from fish with fleshy fins that became modified for walking on land. The tetrapods inherited bony limbs consisting of a single upper limb bone hinged to two parallel lower limb bones, in turn articulated with a foot carrying no more than five digits. From this common "pentadactyl" (five-toed) arrangement, tetrapods evolved wings and flippers; some, like snakes, lost their limbs altogether.

Aerial frog
This X-ray image of Wallace's flying frog (*Rhacophorus nigropalmatus*) reveals the five-toed feet and four-toed hands typical of most living limbed amphibians. Elongated digits on the frog's hands and feet support wide webbings that are used to parachute down from high tree branches and even glide short distances.

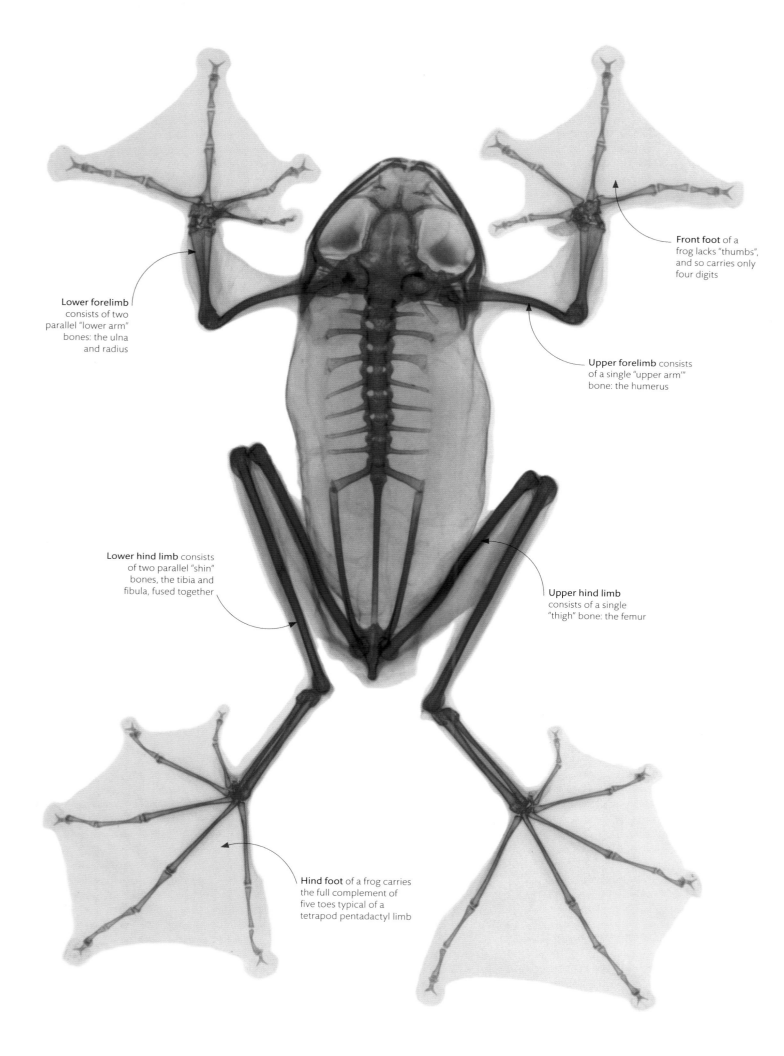

Front foot of a frog lacks "thumbs", and so carries only four digits

Upper forelimb consists of a single "upper arm" bone: the humerus

Lower forelimb consists of two parallel "lower arm" bones: the ulna and radius

Lower hind limb consists of two parallel "shin" bones, the tibia and fibula, fused together

Upper hind limb consists of a single "thigh" bone: the femur

Hind foot of a frog carries the full complement of five toes typical of a tetrapod pentadactyl limb

legs, arms, tentacles, and tails

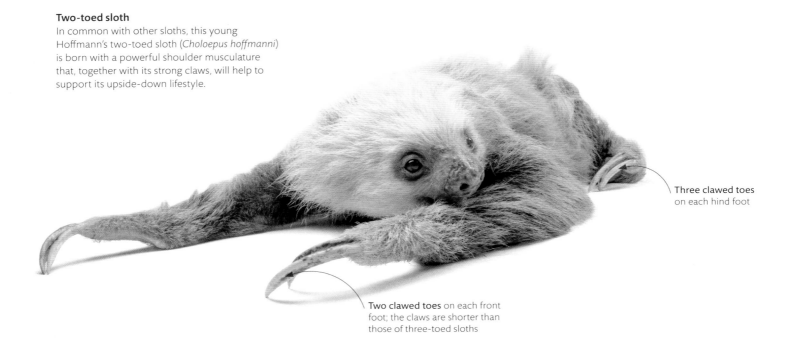

Two-toed sloth
In common with other sloths, this young Hoffmann's two-toed sloth (*Choloepus hoffmanni*) is born with a powerful shoulder musculature that, together with its strong claws, will help to support its upside-down lifestyle.

Three clawed toes on each hind foot

Two clawed toes on each front foot; the claws are shorter than those of three-toed sloths

vertebrate claws

The curved claws at the end of land vertebrates' digits are used for a variety of tasks, from grooming and improving traction, to use as weapons. They occur in all birds, most reptiles and mammals, and even some amphibians. Claws are composed of keratin – the same hard protein found in horns – produced by special cells at their base. They grow continuously, nourished by a central blood vessel; only wear prevents them from growing too long.

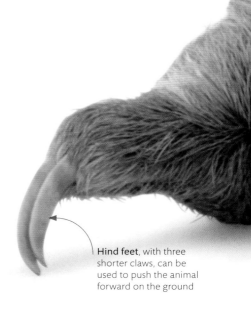

Hind feet, with three shorter claws, can be used to push the animal forward on the ground

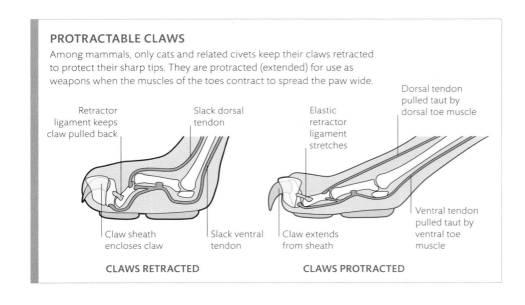

PROTRACTABLE CLAWS
Among mammals, only cats and related civets keep their claws retracted to protect their sharp tips. They are protracted (extended) for use as weapons when the muscles of the toes contract to spread the paw wide.

Retractor ligament keeps claw pulled back

Slack dorsal tendon

Elastic retractor ligament stretches

Dorsal tendon pulled taut by dorsal toe muscle

Claw sheath encloses claw

Slack ventral tendon

Claw extends from sheath

Ventral tendon pulled taut by ventral toe muscle

CLAWS RETRACTED

CLAWS PROTRACTED

Claws as hooks
Like the other sloths of Central and South America, a brown-throated three-toed sloth (*Bradypus variegatus*) cannot grasp branches to climb because its digits are partially fused. Instead, it uses its enlarged claws as hooks to haul itself up and hang from trees. On the ground, where the claws are a hindrance, the sloth is forced to crawl on its forearms.

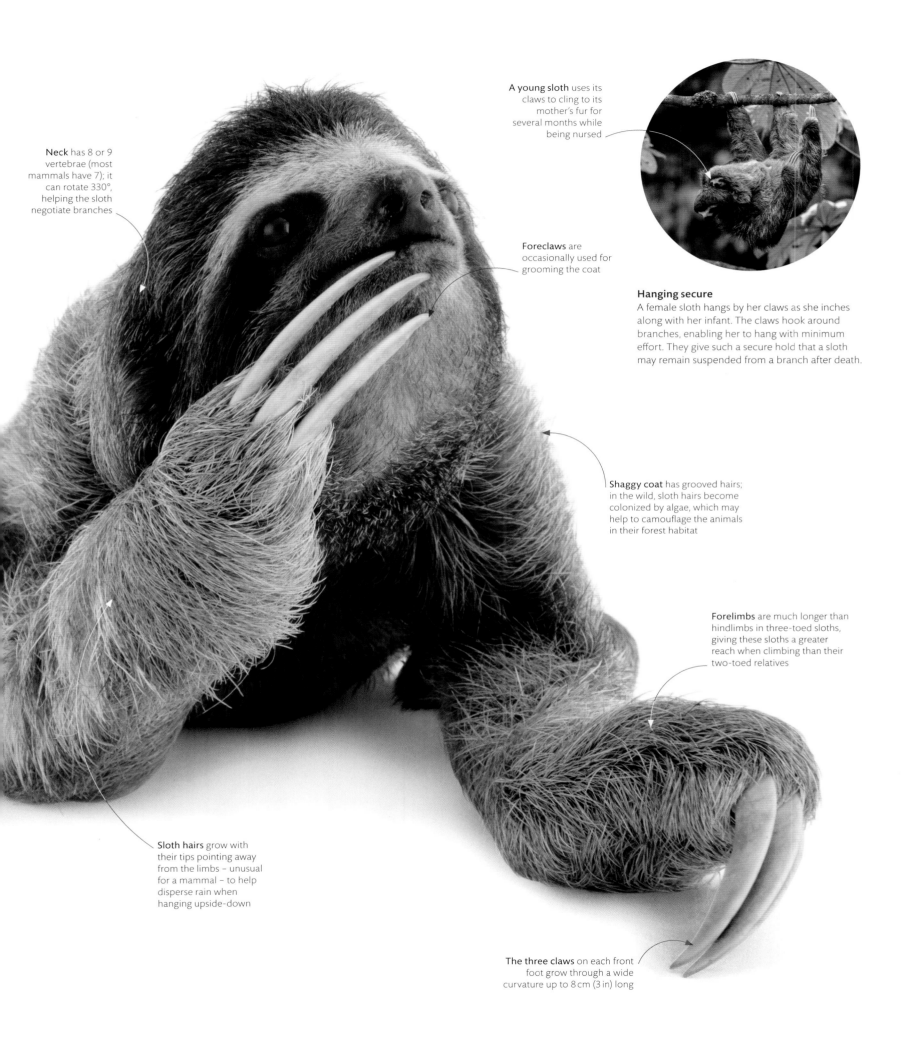

Neck has 8 or 9 vertebrae (most mammals have 7); it can rotate 330°, helping the sloth negotiate branches

A young sloth uses its claws to cling to its mother's fur for several months while being nursed

Foreclaws are occasionally used for grooming the coat

Hanging secure
A female sloth hangs by her claws as she inches along with her infant. The claws hook around branches, enabling her to hang with minimum effort. They give such a secure hold that a sloth may remain suspended from a branch after death.

Shaggy coat has grooved hairs; in the wild, sloth hairs become colonized by algae, which may help to camouflage the animals in their forest habitat

Forelimbs are much longer than hindlimbs in three-toed sloths, giving these sloths a greater reach when climbing than their two-toed relatives

Sloth hairs grow with their tips pointing away from the limbs – unusual for a mammal – to help disperse rain when hanging upside-down

The three claws on each front foot grow through a wide curvature up to 8 cm (3 in) long

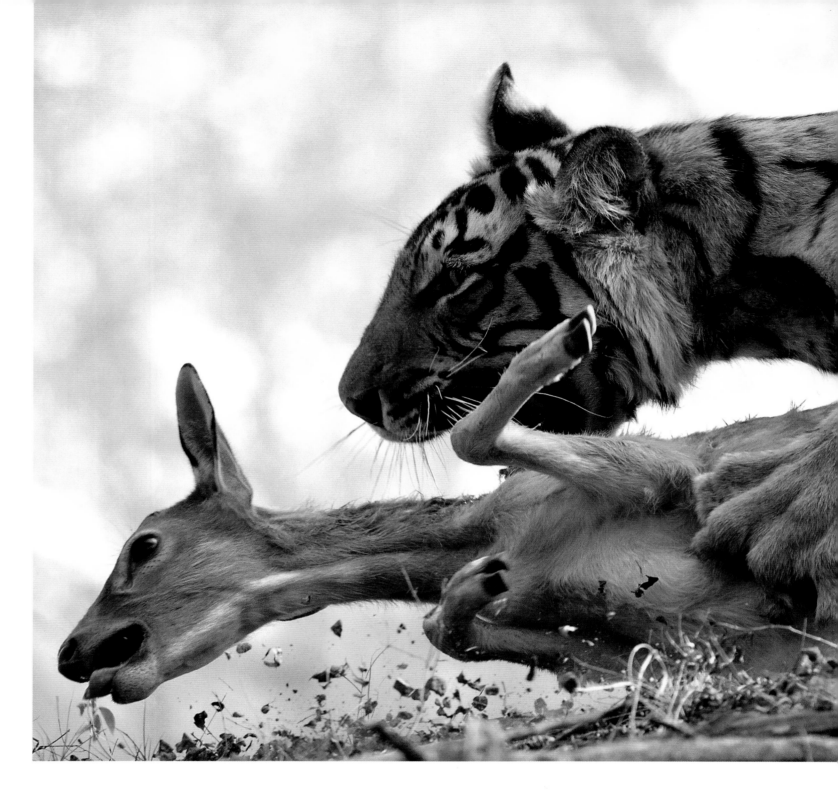

spotlight species

tiger

The tiger (*Panthera tigris*) is the largest of the cat family and a consummate predator, with almost every aspect of its anatomy geared towards the capture, killing, and devouring of prey. Solitary, mostly nocturnal hunters, tigers rely on finely tuned senses, speed, and raw power to ambush or stalk animals ranging in size from pigs and small deer to Asian water buffalo.

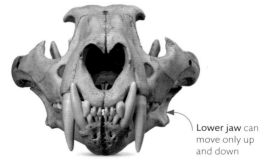

Lower jaw can move only up and down

Mighty bite
The tiger skull is short, broad, and robust and the lower jaw sacrifices chewing ability for pure bite power.

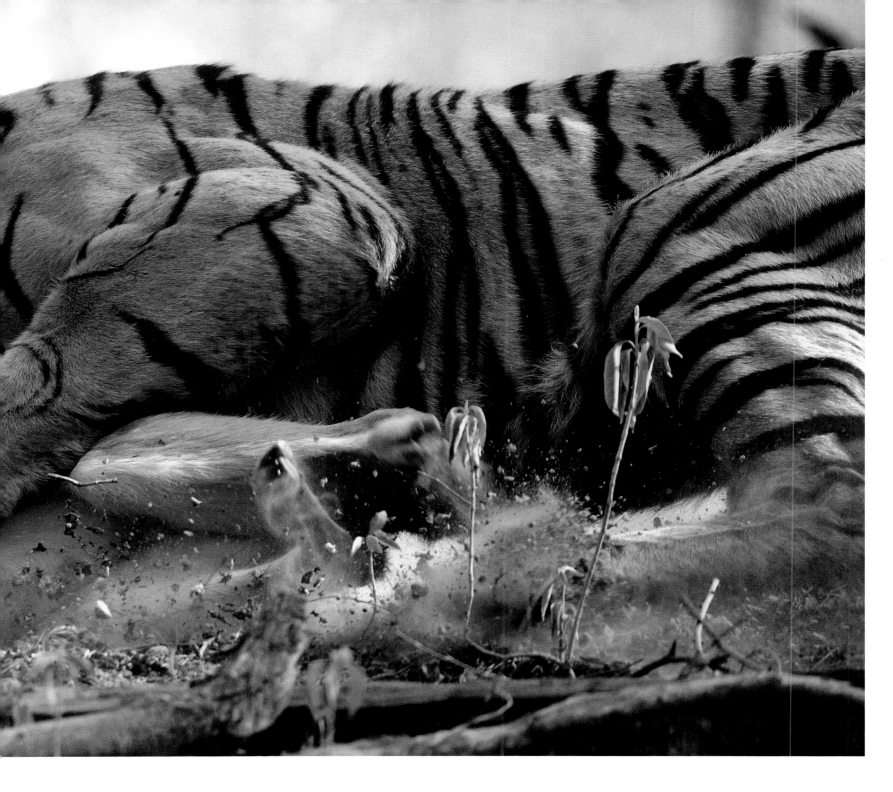

Not all tigers are giants – they vary considerably in size. Siberian tigers are up to five times heavier than those found in the tropical forests of Indonesia, where large body size would encumber movement and increase the chances of overheating. But all exhibit immense athleticism, thanks to a flexible spine and long, muscular limbs. A large adult can leap 10 m (32 ft) from a standing start, and its large, forward-facing eyes aid stereoscopic vision enabling it to judge distance accurately. Their forelimbs and shoulders are immense and the claws are fully retractable, so they do not become blunt through abrasion. Claws are deployed in climbing, for marking territory with scratch marks, and in fighting, as well as to snag, grip, and bring down their prey. The jaws and teeth come into their own once the prey is downed – clamping vicelike on to the windpipe of large prey, or dispatching smaller animals with a bone-splitting bite to the neck.

For such large animals, tigers can be remarkably inconspicuous. The boldly patterned coat provides startlingly effective camouflage in strong or dappled light, in forests, on grassland, or even rocky ground. In rare variants, the orange may be replaced by white, but the stripes remain, providing the same disruptive camouflage.

The huge distribution and the variety of habitats in which tigers once thrived hint at the adaptability of a species that belies its shrunken range and endangered status. This results from exclusively manmade activites, and three of the nine described subspecies are extinct.

Raw power
Tigers typically attack from behind and, using bodyweight, can overwhelm prey in seconds. A deer this size can be eaten in one meal, while larger prey may be cached and consumed over several days.

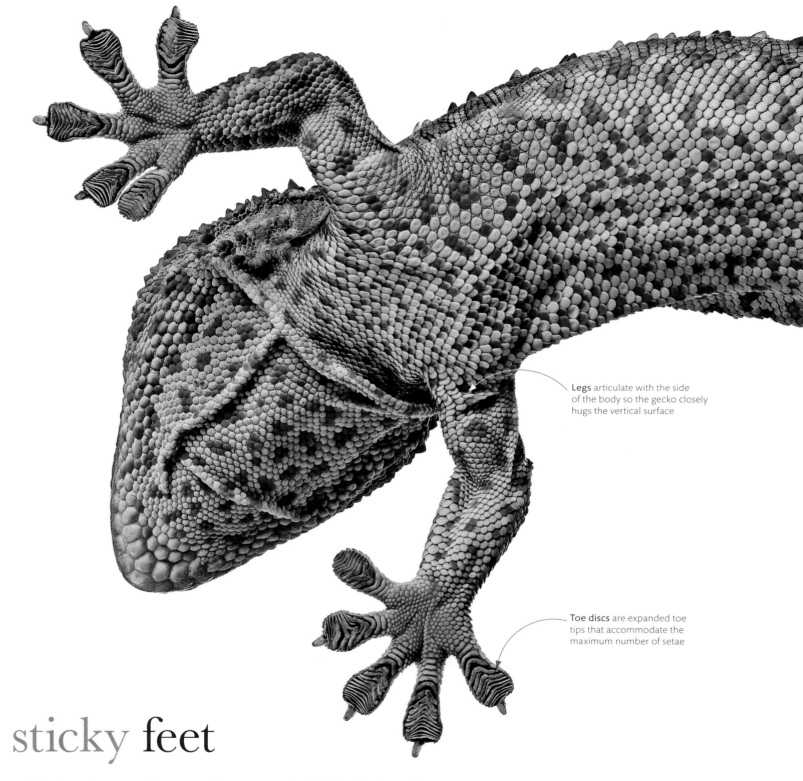

Legs articulate with the side of the body so the gecko closely hugs the vertical surface

Toe discs are expanded toe tips that accommodate the maximum number of setae

sticky feet

While clawed toes and grasping feet are invaluable for climbing, for small animals, the tiny forces of attraction at play between atoms and molecules add up to give sufficient grip on even the smoothest of surfaces. The toe pads of geckos are covered in millions of microscopic hairs called setae. They all stick to a surface, and can collectively contribute enough force to hold the weight of a 300 g (11 oz) lizard, even upside down.

Wall climber
Many species of geckos use their climbing ability to grip smooth leaves or rockfaces, but for some, the walls and ceilings of buildings serve just as well for hunting insects and other invertebrates. The tokay gecko (*Gecko gecko*) originally lived in rainforests, but has adapted to life alongside humans and is seen inside homes in the tropics.

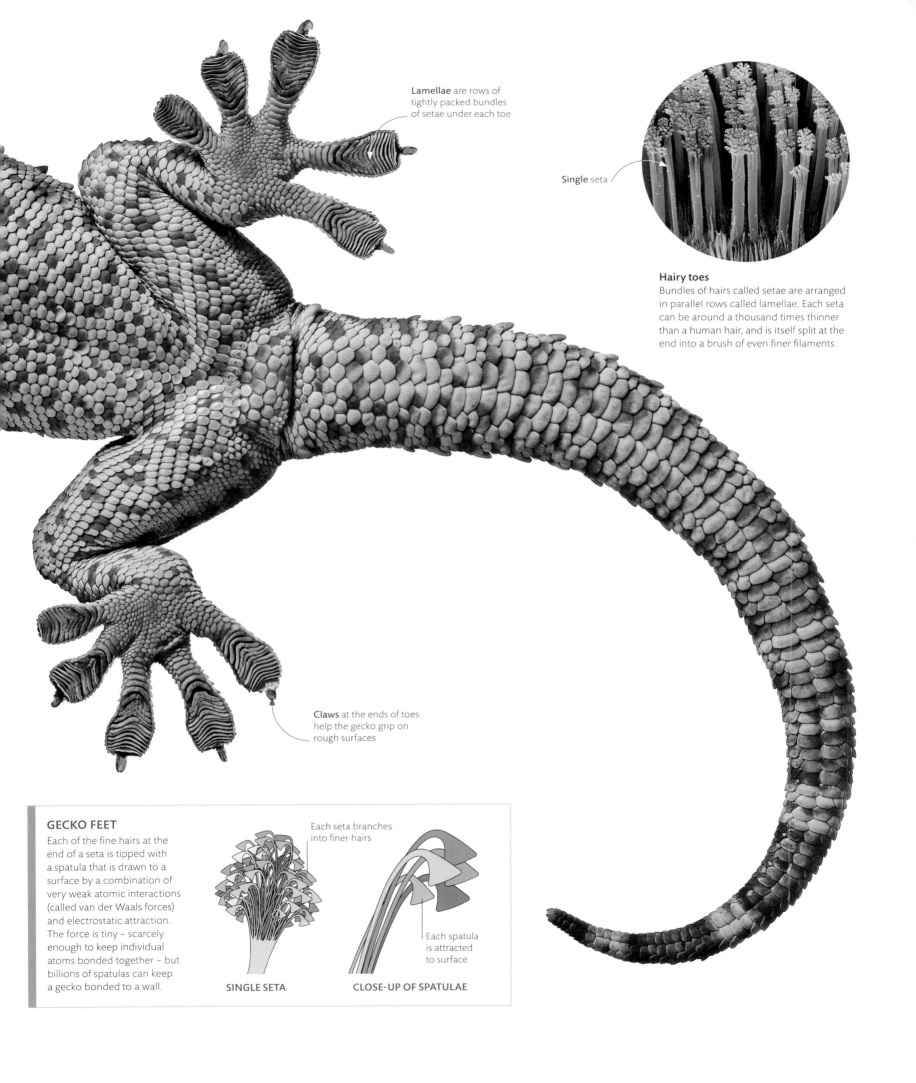

Lamellae are rows of tightly packed bundles of setae under each toe

Single seta

Hairy toes
Bundles of hairs called setae are arranged in parallel rows called lamellae. Each seta can be around a thousand times thinner than a human hair, and is itself split at the end into a brush of even finer filaments.

Claws at the ends of toes help the gecko grip on rough surfaces

GECKO FEET

Each of the fine hairs at the end of a seta is tipped with a spatula that is drawn to a surface by a combination of very weak atomic interactions (called van der Waals forces) and electrostatic attraction. The force is tiny – scarcely enough to keep individual atoms bonded together – but billions of spatulas can keep a gecko bonded to a wall.

Each seta branches into finer hairs

Each spatula is attracted to surface

SINGLE SETA

CLOSE-UP OF SPATULAE

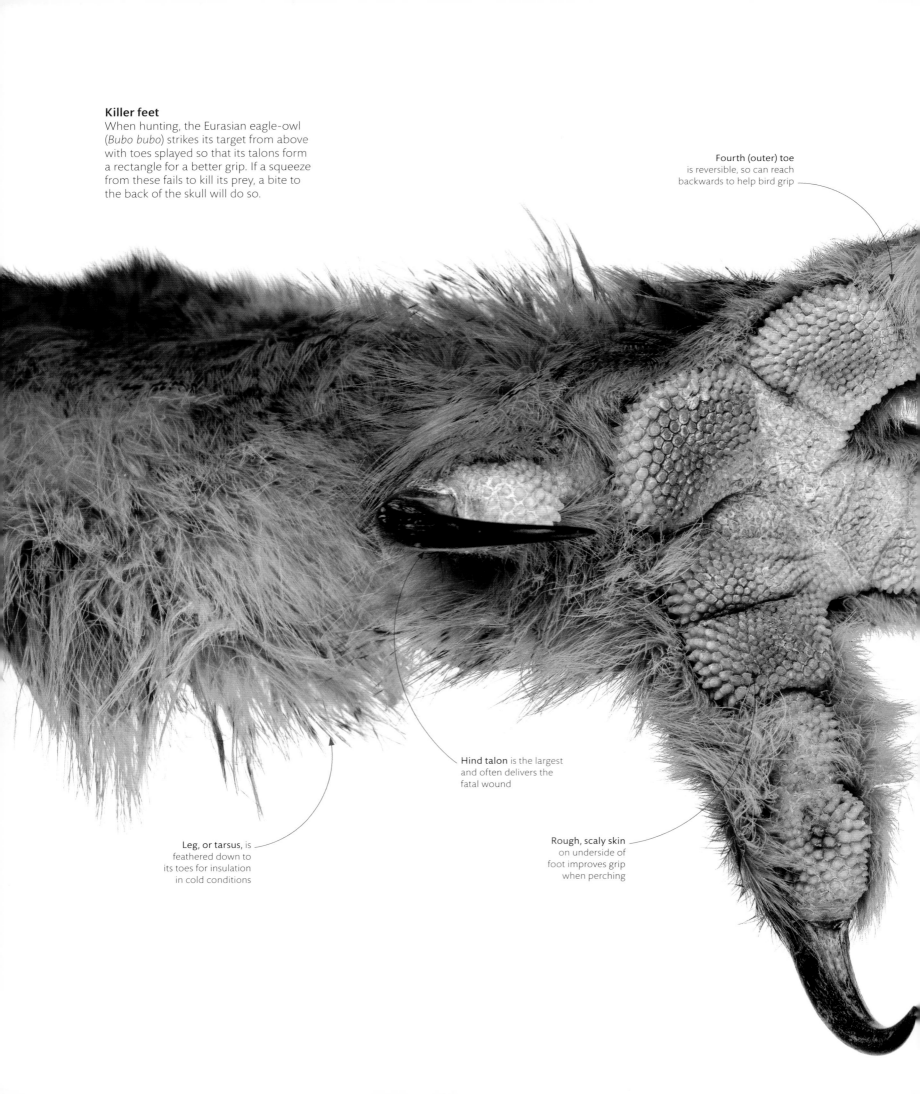

Killer feet
When hunting, the Eurasian eagle-owl (*Bubo bubo*) strikes its target from above with toes splayed so that its talons form a rectangle for a better grip. If a squeeze from these fails to kill its prey, a bite to the back of the skull will do so.

Fourth (outer) toe is reversible, so can reach backwards to help bird grip

Hind talon is the largest and often delivers the fatal wound

Leg, or tarsus, is feathered down to its toes for insulation in cold conditions

Rough, scaly skin on underside of foot improves grip when perching

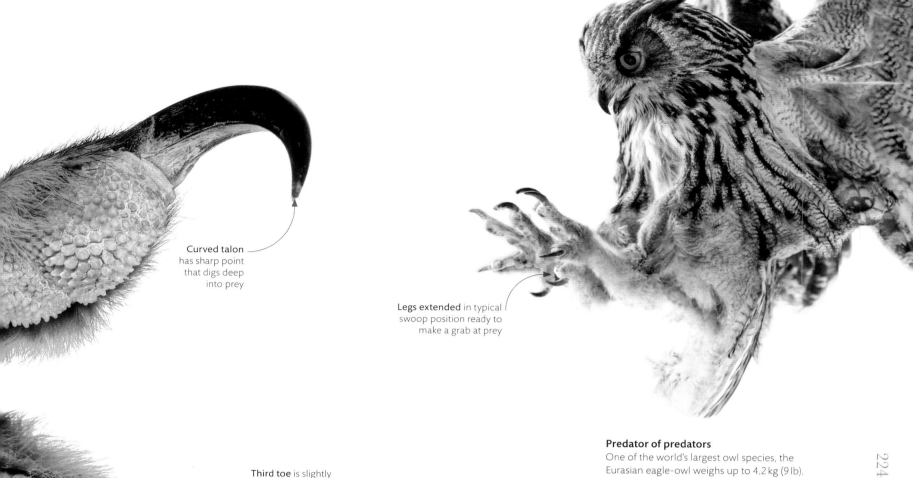

Curved talon has sharp point that digs deep into prey

Legs extended in typical swoop position ready to make a grab at prey

Third toe is slightly longer than the rest

Predator of predators
One of the world's largest owl species, the Eurasian eagle-owl weighs up to 4.2 kg (9 lb). It can kill prey up to the size of young deer and is even known to tackle other predators such foxes or buzzards.

FISHING FEET
Like owls, fishing ospreys lift their heavy prey with feet that have two toes pointing forwards and two back. In addition, they are equipped with long, curved talons and spikey skin on the underside of their toes to ensure they can grip their slippery catch securely as they lift it from the water.

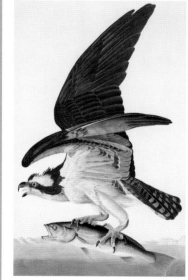

OSPREY *Pandion Haliaetus*

raptor feet

Predatory birds, or raptors, are armed with formidable weaponry – a bill and talons. Both function like daggers to puncture flesh, and are brandished with impressive muscle power, but it is the clawed feet that usually deliver the killing blow. Raptors squeeze prey in a vicelike grip and the talons puncture vital organs, to ensure the body is lifeless when the bird is ready to feed.

Hanging lock

Bats such as this Indian fruit bat (*Pteropus giganteus*) also rely on specially roughened tendons that can lock into position when their claws grip a perch, letting them hang with little muscular effort. Unlike birds, bats do not have opposable digits on their feet.

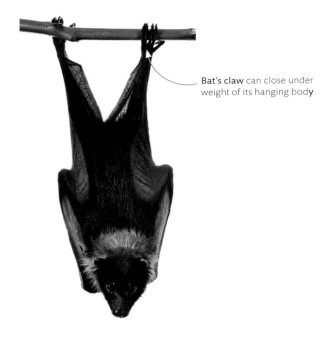

Bat's claw can close under weight of its hanging body

climbing and perching

When four-legged, backboned animals evolved flight, their forelimbs were commandeered as wings, leaving the hindlimbs to support their bodies when they landed. Today, bats still have claws on their wings, which can help with gripping, but birds rely entirely on their two legs to both run and climb. But both of these expert aeronauts have feet that can support their entire weight while they hold tightly onto perches without becoming fatigued – even when they are sleeping. As a result, both bats and birds are perfectly at home far from the ground.

LOCKED TENDONS

When a bird perches, its thigh muscles contract to bend the legs, pulling on the flexor tendons that extend to the tips of the toes. The increase in tension in the tendons causes the bird's toes to grasp the perch. The tendons run through sheaths and both sheaths and tendons have corresponding corrugated surfaces that lock together under the weight of the bird.

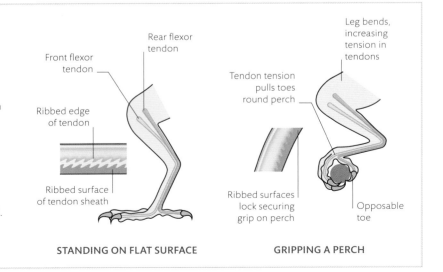

Rear flexor tendon

Front flexor tendon

Ribbed edge of tendon

Ribbed surface of tendon sheath

Leg bends, increasing tension in tendons

Tendon tension pulls toes round perch

Ribbed surfaces lock securing grip on perch

Opposable toe

STANDING ON FLAT SURFACE

GRIPPING A PERCH

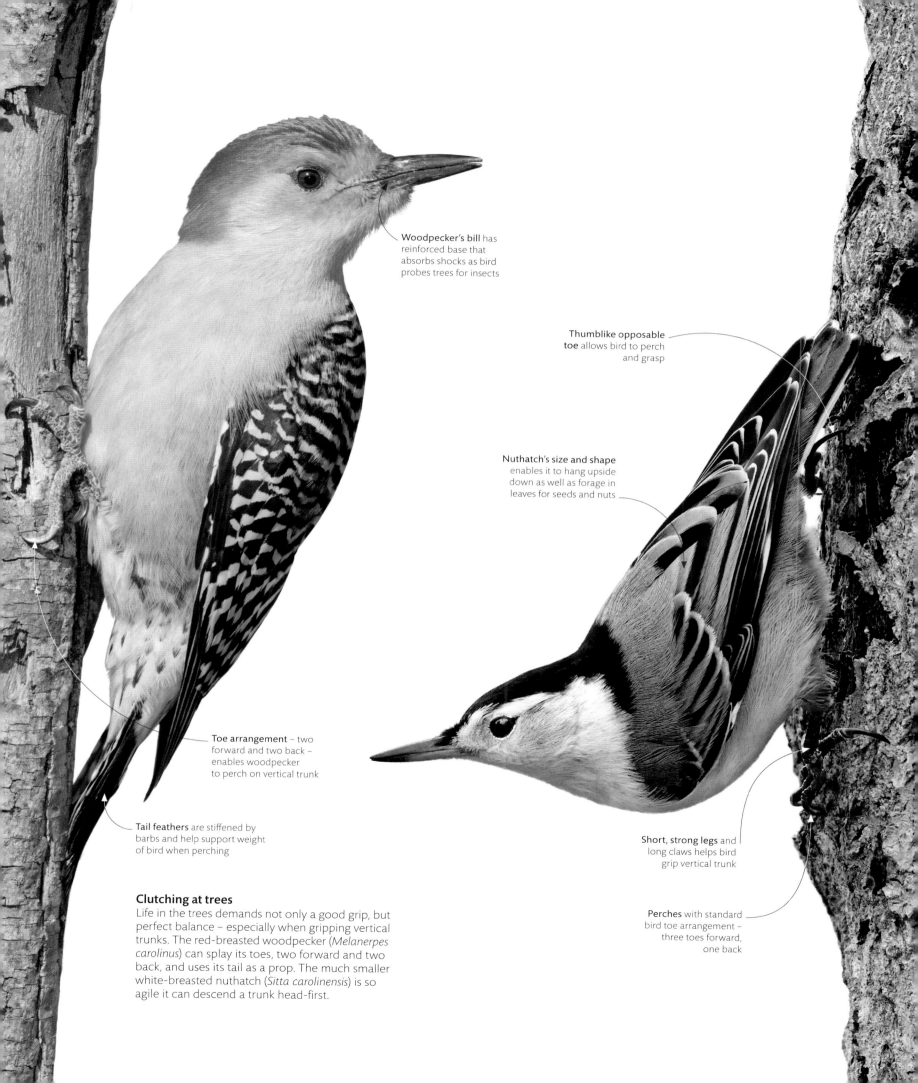

Woodpecker's bill has reinforced base that absorbs shocks as bird probes trees for insects

Thumblike opposable toe allows bird to perch and grasp

Nuthatch's size and shape enables it to hang upside down as well as forage in leaves for seeds and nuts

Toe arrangement – two forward and two back – enables woodpecker to perch on vertical trunk

Tail feathers are stiffened by barbs and help support weight of bird when perching

Short, strong legs and long claws helps bird grip vertical trunk

Perches with standard bird toe arrangement – three toes forward, one back

Clutching at trees

Life in the trees demands not only a good grip, but perfect balance – especially when gripping vertical trunks. The red-breasted woodpecker (*Melanerpes carolinus*) can splay its toes, two forward and two back, and uses its tail as a prop. The much smaller white-breasted nuthatch (*Sitta carolinensis*) is so agile it can descend a trunk head-first.

Even-toed hoofed mammals

Chevrotains are among the smallest hoofed mammals, some little bigger than a rabbit. Like other even-toed hoofed mammals, each limb is supported by the third and fourth hoofed toes.

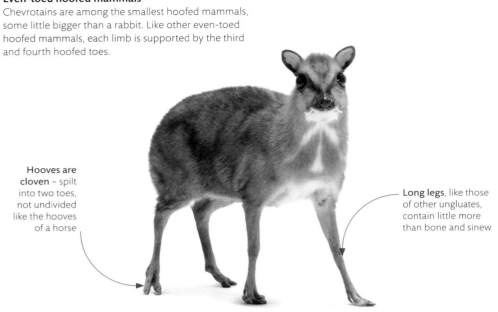

Hooves are cloven – spilt into two toes, not undivided like the hooves of a horse

Long legs, like those of other ungluates, contain little more than bone and sinew

mammal hooves

Many mammals that can run at speed have highly modified feet: they walk on the tips of toes that end in flat-bottomed hooves instead of the curved claws of their ancestors. The number of digits is also reduced, with weight concentrated across a pair of toes in deer, antelope, and other cloven-hoofed mammals, or an odd number of toes in horses, rhinoceroses, and tapirs. Hoof-tipped feet and long, lightweight, slender limbs – with stocky muscles high up near the body's centre of mass – help the legs swing through a bigger stride, maximizing speed when evading predators.

FOOT POSTURES

Plantigrade mammals, such as humans and bears, support their body weight on the flats of their feet; in faster runners, the foot bones are raised, increasing leg length. With digitigrade mammals, such as dogs, the end bones of the digits are flat on the ground, but in unguligrade (hoofed) mammals, the entire digits are raised to the toe-tip position, which maximizes stride length.

KEY

- Femur (thigh)
- Tibula-fibula (lower leg)
- Tarsus (upper foot)
- Metatarsus (lower foot)
- Phalanges (toes)

PLANTIGRADE (BEAR)

DIGITIGRADE (DOG)

UNGULIGRADE (HORSE)

legs, arms, tentacles, and tails

228 • 229

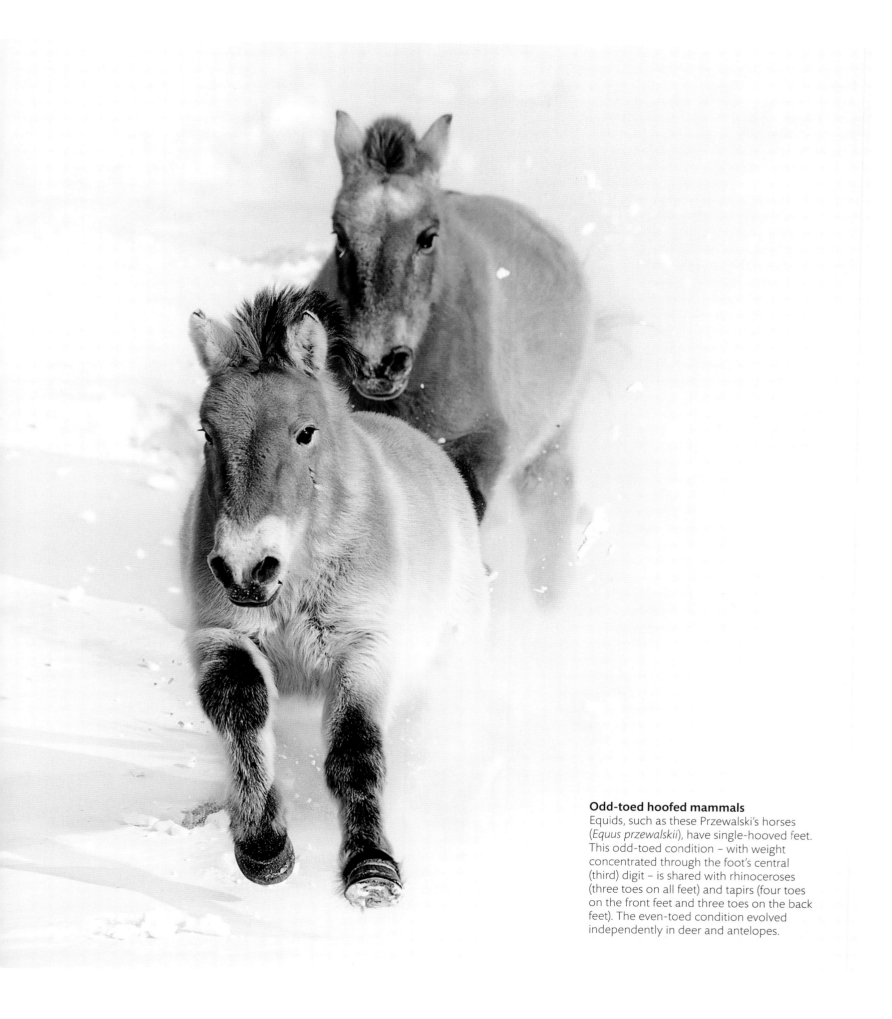

Odd-toed hoofed mammals
Equids, such as these Przewalski's horses (*Equus przewalskii*), have single-hooved feet. This odd-toed condition – with weight concentrated through the foot's central (third) digit – is shared with rhinoceroses (three toes on all feet) and tapirs (four toes on the front feet and three toes on the back feet). The even-toed condition evolved independently in deer and antelopes.

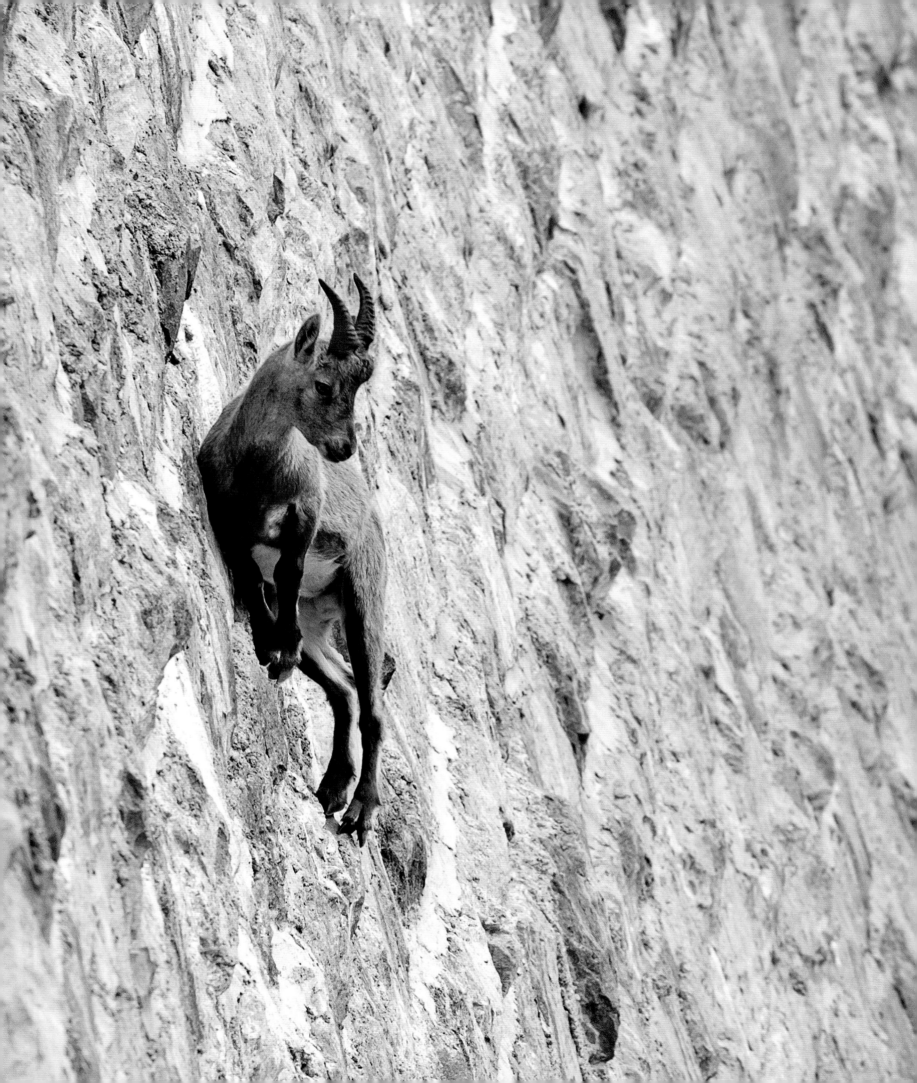

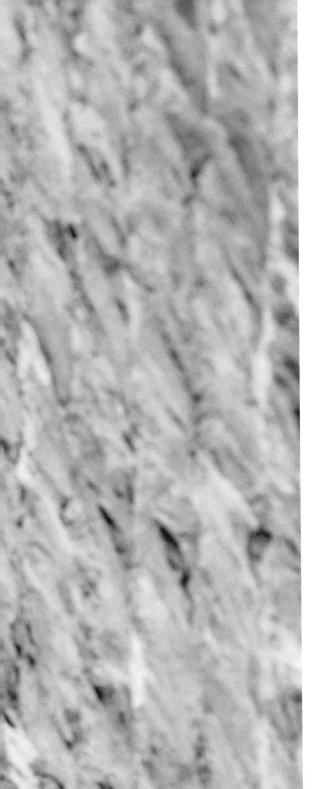

alpine ibex

It takes nerves of steel to balance on a precipitous rock face, but for goats, sheep, and ibex it is a way of life. It is clearly worth the risk as it ensures that they are well out of reach of predators – a considerable benefit for animals that are otherwise exposed in open habitat.

About 11 million years ago – somewhere in Asia – a tribe of sure-footed, hoofed animals evolved into specialized rock-climbers. They developed short, sturdy shin bones and today species of takins, chamois, goats, sheep, and ibex account for more than a third of all horned cloven-hoofed mammals in the mountains across Eurasia and parts of North America.

The Alpine ibex of central Europe became better rock-climbers than most and live at altitudes of up to 3,200 m (10,500 ft), far above the tree-line. They are also seen perching on the walls of dams with gradients of 60 degrees. These

animals have padded hooves for traction (see below) and develop calluses in the backs of their knees to protect them among sharp rocks. Their habitat is free from large predators and in winter the steep rocky slopes and the sides of ravines offer some respite from deep snow. Ewes give birth up there, and their kids quickly learn to keep their footing on the slopes; the mothers wait until spring to lead their young to meadows, where the grazing is better. It takes lighter bodies and shorter legs to negotiate the steep rock faces so, in their 11th year, males move to flatter ground, and never return to the slopes.

Mountain life is difficult and many ibex die in avalanches. They are also at risk of an infectious eye disease that causes blindness, which accounts for nearly 1 in 3 falling to their deaths. But populations are buoyant and when food is abundant, up to 95 per cent of young ibex kids survive into adulthood.

Nimble youth

A young male Alpine ibex is climbing the slope of a manmade dam in northern Italy, attracted to the salt that exudes from the stonework. The ibex follows a zigzag route as it climbs the wall and takes a more linear one on the way down.

HOOVES THAT GRIP

Being cloven-hoofed is key to the Alpine ibex being able to climb a steep rock face. Its toes can splay apart to grip the ground. Beneath each toe, a rubbery sole acts like a suction pad to provide traction between the ibex's feet and the ground that is greater than that between rubber and concrete.

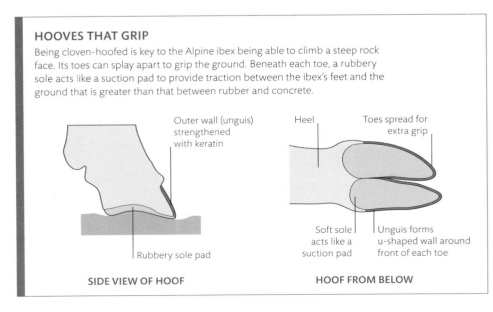

Outer wall (unguis) strengthened with keratin

Rubbery sole pad

SIDE VIEW OF HOOF

Heel

Toes spread for extra grip

Soft sole acts like a suction pad

Unguis forms u-shaped wall around front of each toe

HOOF FROM BELOW

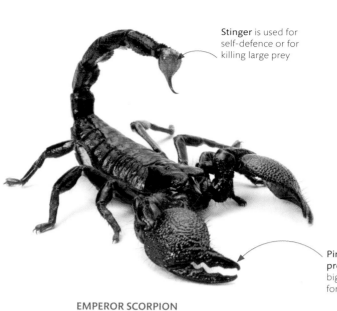

Stinger is used for self-defence or for killing large prey

EMPEROR SCORPION
Pandinus imperator

Pincers crush small prey, or immobilize bigger prey ready for stinging

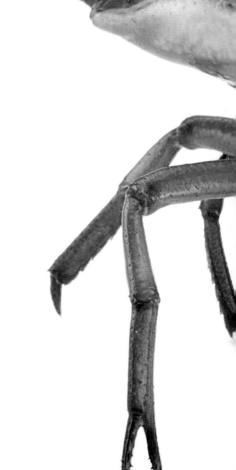

The **dactylus** is the moveable finger of the pincer

The immoveable part of a pincer, the **propodus**, consists of a wide palm packed with muscles

arthropod pincers

Some arthropod mouthparts function like pincers to bite or grasp, but other forms of pincers develop at the ends of the limbs of animals such as crabs, crayfish, and scorpions. These pincers, called chelae, rely on the same kinds of muscles that move walking limbs, except that pincer muscles clamp together clawlike fingers. The smallest pincers deftly manipulate food, while the biggest can be wielded as fearsome defensive weapons.

PINCER MOVEMENT

Like many moving animal parts, pincers contain an antagonistic pair of muscle groups: one to flex the moveable finger closed, another to open it. The muscles pull on extensions of the exoskeleton called apodemes. Large flexor muscles give pincers considerable force: the giant pincer of the tree-climbing robber crab is strong enough to break into a coconut for food.

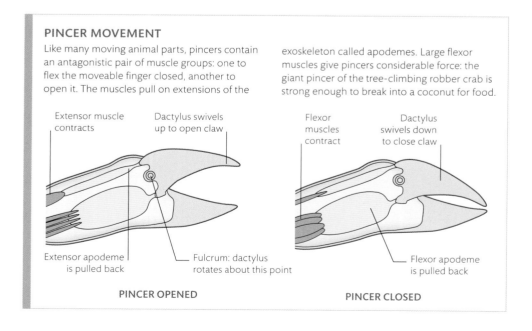

Extensor muscle contracts

Dactylus swivels up to open claw

Extensor apodeme is pulled back

Fulcrum: dactylus rotates about this point

PINCER OPENED

Flexor muscles contract

Dactylus swivels down to close claw

Flexor apodeme is pulled back

PINCER CLOSED

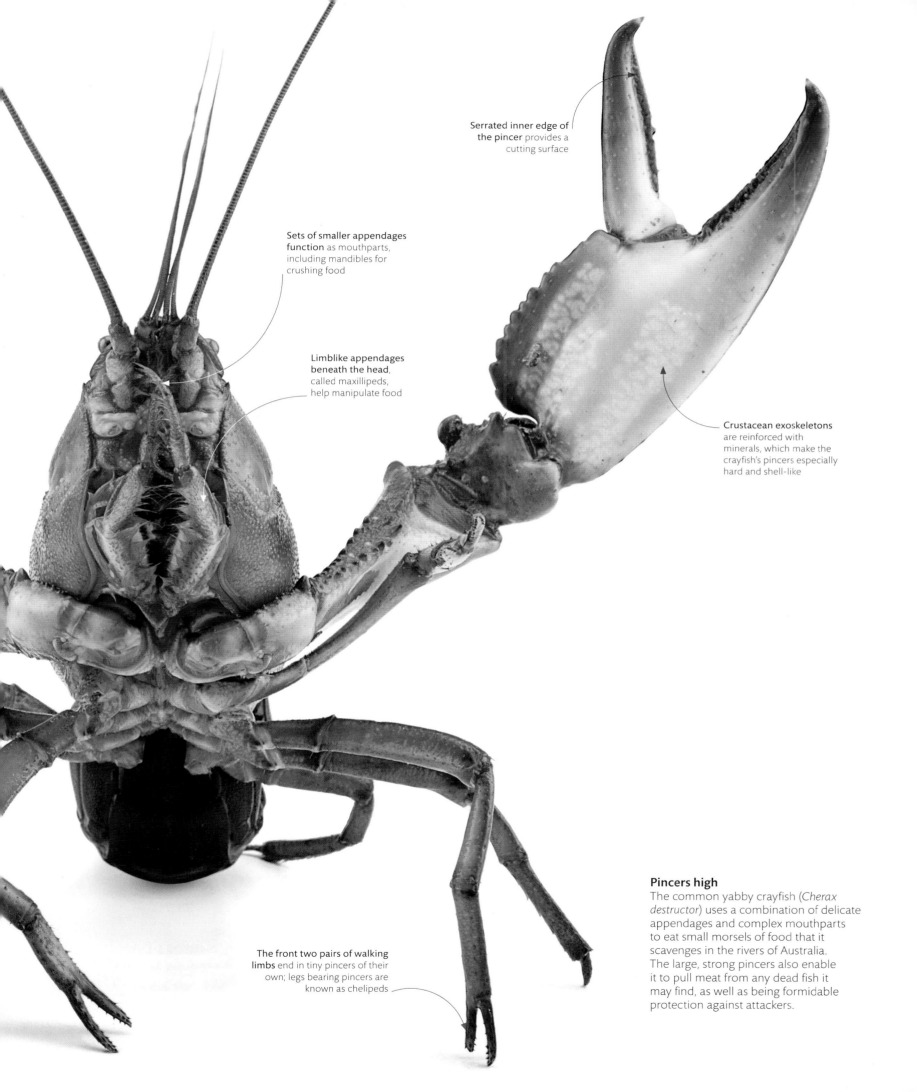

Serrated inner edge of the pincer provides a cutting surface

Sets of smaller appendages function as mouthparts, including mandibles for crushing food

Limblike appendages beneath the head, called maxillipeds, help manipulate food

Crustacean exoskeletons are reinforced with minerals, which make the crayfish's pincers especially hard and shell-like

The front two pairs of walking limbs end in tiny pincers of their own; legs bearing pincers are known as chelipeds

Pincers high
The common yabby crayfish (*Cherax destructor*) uses a combination of delicate appendages and complex mouthparts to eat small morsels of food that it scavenges in the rivers of Australia. The large, strong pincers also enable it to pull meat from any dead fish it may find, as well as being formidable protection against attackers.

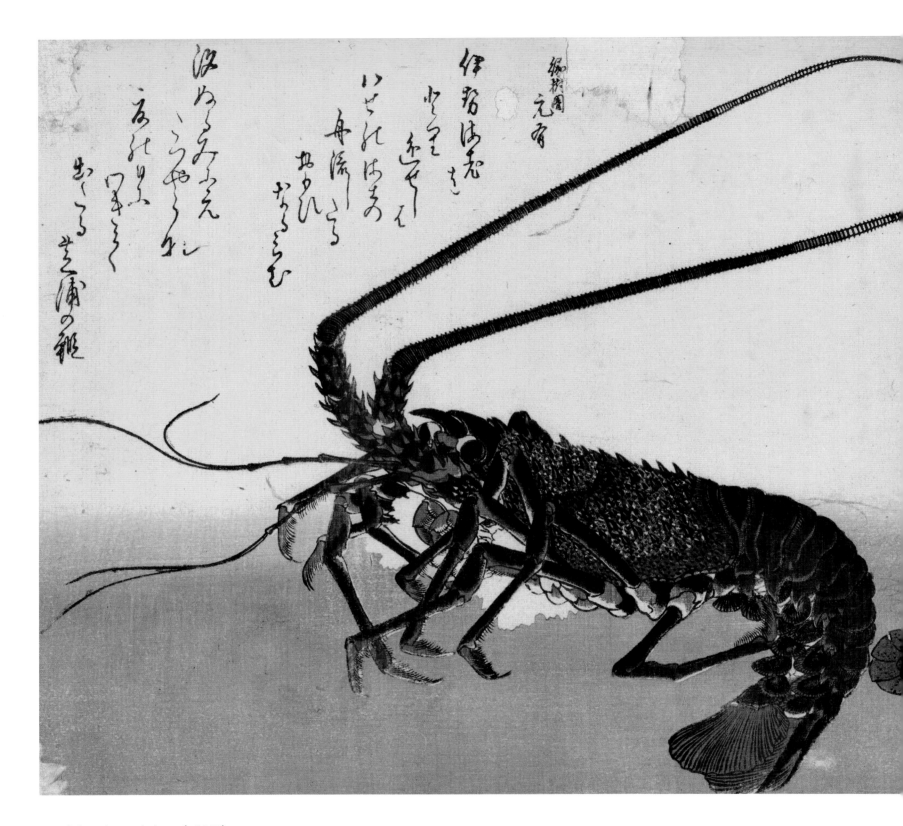

Crayfish and two shrimps (c.1840)
During his country-wide travels in his forties, Utagawa Hiroshige (Ando) was entranced by the splendour of the countryside, producing woodcuts of landscapes and nature portraits that inspired Western artists. The clattering legs and wayward feelers of this crayfish come alive on the page, printed at a time when the technique, colour, and texture of Japanese woodcuts were at a peak.

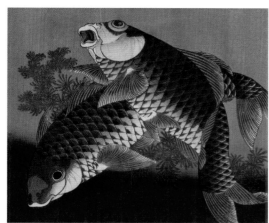

animals in art

artists of the floating world

Woodcut prints of the floating world, called *ukiyo-e*, were not, as one might suppose, nature studies of sea and river life. The "floating world" in Japan referred to the theatres and brothels that were the haunts of wealthy traders in the late 17th century, when prints of city life, erotica, folk tales, and landscapes became immensely popular. Towards the end of the 19th century, domestic tourism opened the country up to its people, and leading artists responded with sublime pictures of scenery and subtle portraits of bird, fish, and other animal life.

At their peak of production, prints were produced in their thousands at fixed prices that made them affordable to all. The publisher had overall control of the editions, working with a designer or artist, a woodblock cutter, and a printer. A copy of the artist's picture was laid face down on a block of fine-grained cherry wood, then the cutter carved through the paper to transfer the design onto this key block. From the mid-18th century, monochrome and simple two-colour prints were replaced with glorious full-colour "brocade" prints that used separate blocks for each colour. Copies were printed by hand, which caused little wear to the blocks and allowed thousands of prints to be produced.

Since ancient times, the bedrock of the Shinto religion endowed every mountain, stream, and tree with spirits known as Kami. In folk tradition, animals were symbolic: for example, rabbits bestowed prosperity, while cranes symbolized a long life. When 19th-century artist Utagawa Hiroshige (Ando) abandoned the *ukiyo-e* subjects of courtesans, kabuki actors, and city life in favour of idealized views of the natural world, it chimed with a national imperative to return to traditional values. Hiroshige's serene landscapes and subtle renderings of birds, animals, and fish also found a wide audience in the West, inspiring Impressionist artists such as Vincent van Gogh, Claude Monet, Paul Cezanne, and James Whistler.

The 70-year-long oeuvre of contemporary artist and printmaker Katsushika Hokusai was similarly celebrated in Japan and in the West. Looking back on his vast production of brush paintings, sea, and landscapes, views of Mount Fuji, and flower, bird, and animal studies he wrote: "At 73, I began to grasp the structures of birds and beasts, insects and fish… If I go on trying, I will surely understand them still better by the time I am 86, so that by 90 I will have penetrated to their essential nature."

> ❝ (Hiroshige) had the knack of vividly bringing home to the mind of an onlooker the beauties of Nature. ❞

SOTARO NAKAI, IN *THE COLOUR-PRINTS OF HIROSHIGE*, EDWARD F. STRANGE, 1925

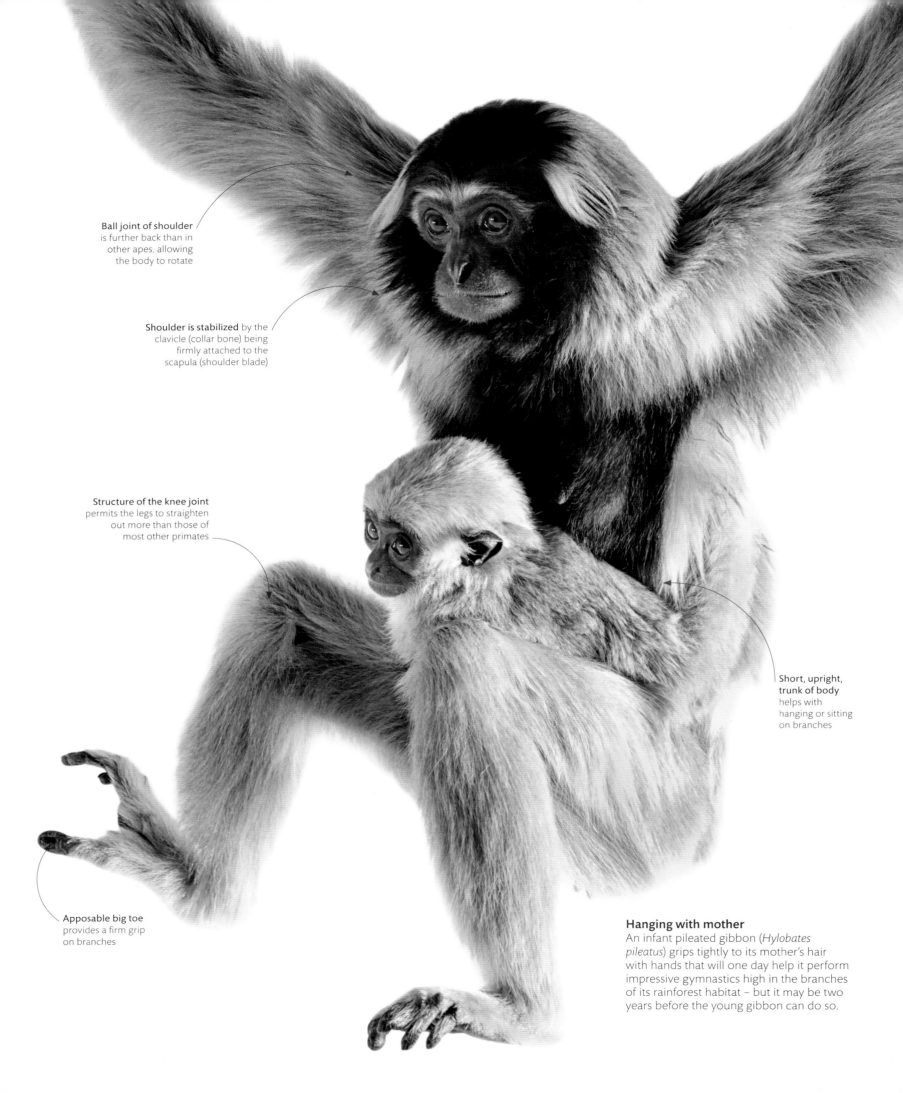

Ball joint of shoulder is further back than in other apes, allowing the body to rotate

Shoulder is stabilized by the clavicle (collar bone) being firmly attached to the scapula (shoulder blade)

Structure of the knee joint permits the legs to straighten out more than those of most other primates

Short, upright, trunk of body helps with hanging or sitting on branches

Apposable big toe provides a firm grip on branches

Hanging with mother

An infant pileated gibbon (*Hylobates pileatus*) grips tightly to its mother's hair with hands that will one day help it perform impressive gymnastics high in the branches of its rainforest habitat – but it may be two years before the young gibbon can do so.

Walking upright
Adaptations to an arboreal lifestlye, such as modified knees, also give gibbons a more rapid stride than other non-human primates when walking on the ground.

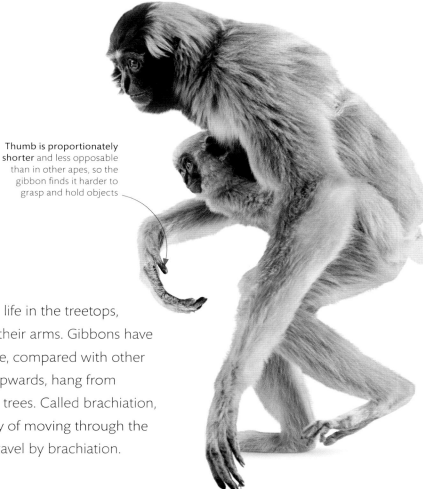

Exceptionally long arms increase the gibbon's reach, maximizing the speed at which it can swing through the trees

Thumb is proportionately shorter and less opposable than in other apes, so the gibbon finds it harder to grasp and hold objects

swinging
through trees

As climbing primates, such as gibbons, adapted to life in the treetops, most of their strong muscular control shifted into their arms. Gibbons have exceptionally long arms, in proportion to body size, compared with other primates. They use their arms to pull themselves upwards, hang from branches, and swing hand-over-hand through the trees. Called brachiation, this arm-swinging propulsion is a fast, efficient way of moving through the canopy. Gibbons and spider monkeys habitually travel by brachiation.

Very long fingers form a reliable hook when hanging from branches

BRACHIATION

The gibbon's flexible wrist joints permit the rotation necessary for arm-swinging. At more leisurely speeds, a gibbon always has at least one hand on a branch, but by swinging completely free it can increase the distance between handholds to move faster through the trees.

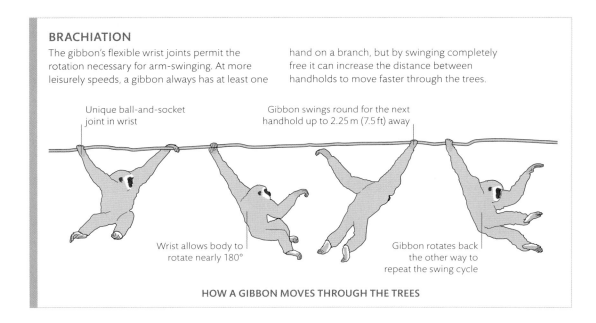

Unique ball-and-socket joint in wrist

Gibbon swings round for the next handhold up to 2.25 m (7.5 ft) away

Wrist allows body to rotate nearly 180°

Gibbon rotates back the other way to repeat the swing cycle

HOW A GIBBON MOVES THROUGH THE TREES

MANUAL DEXTERITY

An opposable thumb – where the thumb is opposite the fingers – helps primates grip. Apes use a power grip for climbing. They can manipulate objects with a precision grip, but their short thumbs have to clamp against the side of their index finger. Only humans – with longer thumbs – can comfortably pinch fingertip and thumb tip together.

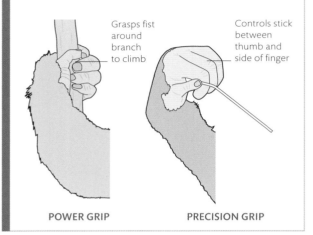

Grasps fist around branch to climb

Controls stick between thumb and side of finger

POWER GRIP

PRECISION GRIP

Versatile tool

The hand of a great ape – such as this bonobo (*Pan paniscus*) – combines strength with extraordinary sensitivity. Like related chimpanzees and humans, most bonobos are right-handed, but they can also use both hands simultaneously.

Long, narrow fingers can move independently of each other

Finger tip pads have more sensory receptor cells than any other part of the body

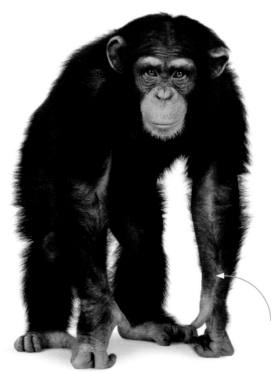

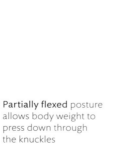

Partially flexed posture allows body weight to press down through the knuckles

"Knuckle" walkers

Grasping hands with sensitive fingertips are useful in trees – but less so when walking on all fours. This common chimpanzee (*Pan troglodytes*), like all African apes, spends much of its time on the ground, and can support its weight on its knuckles. This not only protects the sensitive skin of the palms, but means they can carry objects while moving from place to place.

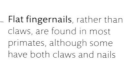

Flat fingernails, rather than claws, are found in most primates, although some have both claws and nails

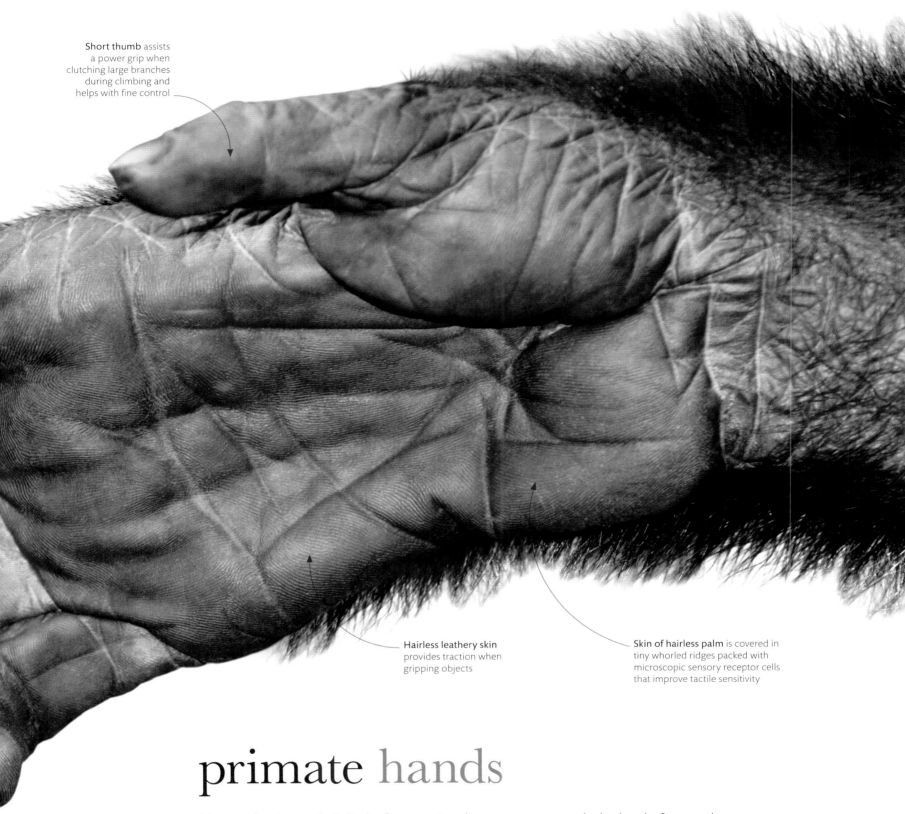

Short thumb assists a power grip when clutching large branches during climbing and helps with fine control

Hairless leathery skin provides traction when gripping objects

Skin of hairless palm is covered in tiny whorled ridges packed with microscopic sensory receptor cells that improve tactile sensitivity

primate hands

Many animals use their limbs for grasping, but none can match the level of manual dexterity found in higher primates. The hands of apes and monkeys have evolved maximum mobility of the digits – and their fingertips are equipped with highly sensitive pads capable of distinguishing countless pinpoints of contact. This, combined with their superior brain power, makes primates expert in manipulating objects around them.

orangutan

The sole surviving members of the *Pongo* genus, orangutans dwell in Asia's dwindling rainforests, and evolutionary adaptations have given them a body ideally suited to climbing through the canopy. These highly intelligent creatures use tools, make rain shelters, and self-medicate with herbs.

Asia's great ape is the largest and heaviest arboreal (or tree-dwelling) mammal on Earth, and one of the most critically endangered. The word orangutan means "person of the forest," and there are three distinct species: Bornean (*Pongo pygmaeus*), Sumatran (*P. abelii*), and the rarer Tapanuli (*P. tapanuliensis*), found exclusively in northern Sumatra's Batang Toru forest. All three are similar-looking, with shaggy hair that varies from orange to reddish depending on the species, relatively short but large torsos, and very long powerful arms.

Adult males weigh more than 90 kg (200 lb) and females range from 30–50 kg (66–110 lb). Their bodies are adapted to be able to navigate their way across highly flexible branches of the forest's trees. Unlike lighter primates, they rely on a combination of movements and physical skills such as walking, climbing, and swinging to cross wider gaps. Orangutans spend most of their lives feeding, foraging, resting, and moving through the rainforest canopy – especially the Sumatran species, which rarely come to ground level because of predators such as tigers. They live on a diet of fruit, supplemented by leaves, non-leafy vegetation, and insects, with the occasional egg or small mammal. Mature animals are mainly solitary, but females, who breed once every 4 or 5 years, can be seen with dependent offspring – usually a single youngster – which can remain with their mother for up to 11 years.

Seeking new heights
With grasping hands and feet, and arm muscles seven times more powerful than a human's, orangutans are master climbers. This Bornean *P. pygmaeus* is scaling a 30-m (98-ft) strangler-fig vine in search of food.

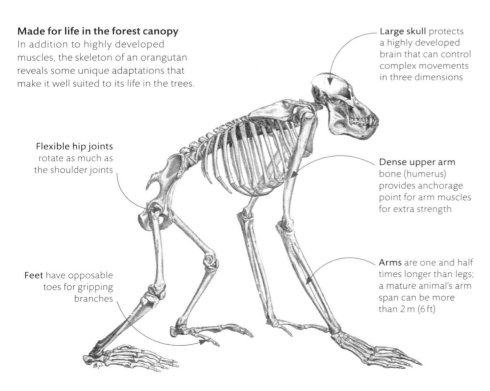

Made for life in the forest canopy
In addition to highly developed muscles, the skeleton of an orangutan reveals some unique adaptations that make it well suited to its life in the trees.

Large skull protects a highly developed brain that can control complex movements in three dimensions

Flexible hip joints rotate as much as the shoulder joints

Dense upper arm bone (humerus) provides anchorage point for arm muscles for extra strength

Feet have opposable toes for gripping branches

Arms are one and half times longer than legs; a mature animal's arm span can be more than 2 m (6 ft)

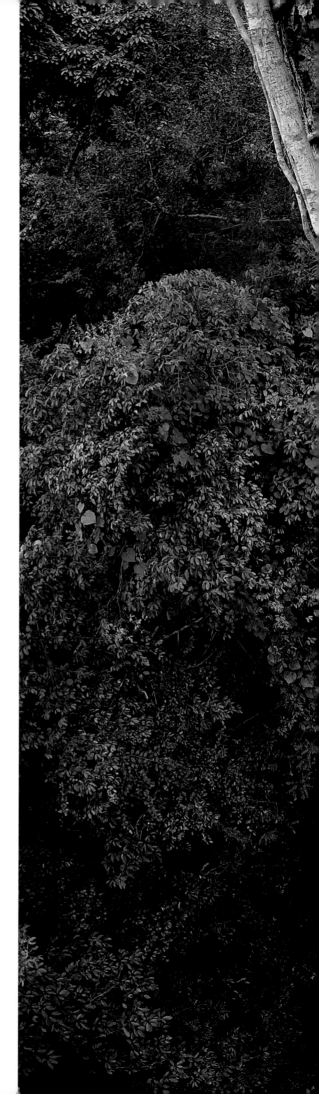

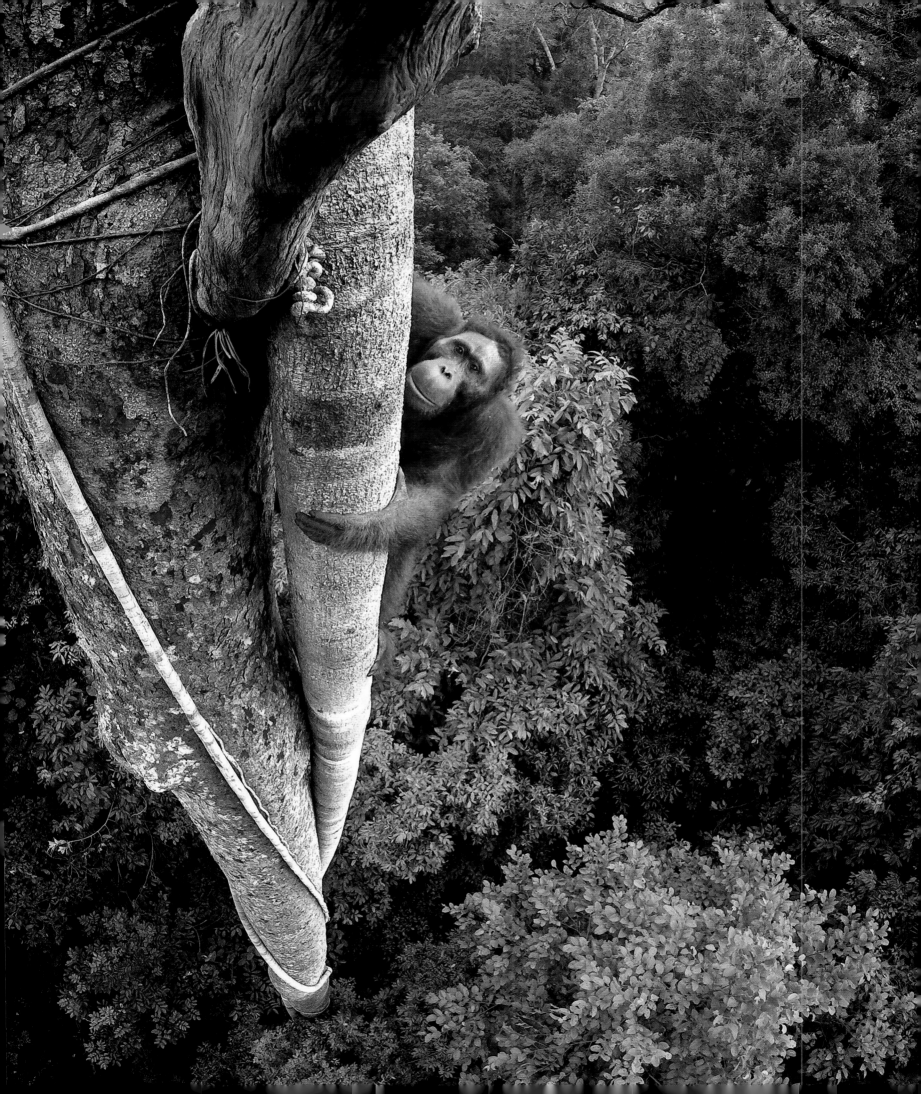

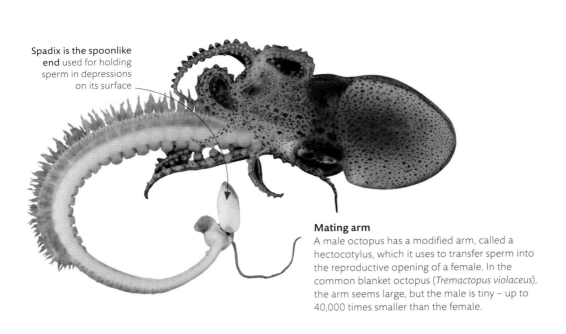

Spadix is the spoonlike end used for holding sperm in depressions on its surface

Mating arm
A male octopus has a modified arm, called a hectocotylus, which it uses to transfer sperm into the reproductive opening of a female. In the common blanket octopus (*Tremactopus violaceus*), the arm seems large, but the male is tiny – up to 40,000 times smaller than the female.

octopus arms

Some molluscs – including octopuses and squids – have abandoned the slow crawling life of their relatives, the slugs and snails, to become agile hunters. They have a large head with big eyes, and around their mouth is a circle of muscular arms. Octopuses have eight suckered arms joined by a web of skin, making a highly effective mechanism for grabbing prey.

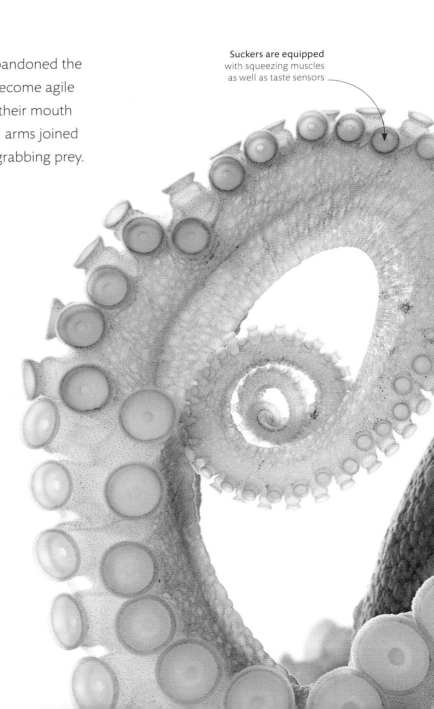

Suckers are equipped with squeezing muscles as well as taste sensors

MUSCLE-PACKED ARM

Composed almost entirely of muscle, an octopus arm works as a muscular hydrostat: the muscles squeeze against each other, rather than against a skeleton. Longitudinal, horizontal, and vertical muscles move the arm in different directions, as in other fleshy hydrostats, such as tongues or elephant trunks.

Epidermis
Dermis
Horizontal muscle
Suction cup muscle
Vein
Vertical muscle
Longitudinal muscle
Artery
Nerve cord
Sucker cup

CROSS-SECTION OF OCTOPUS ARM AND SUCKER

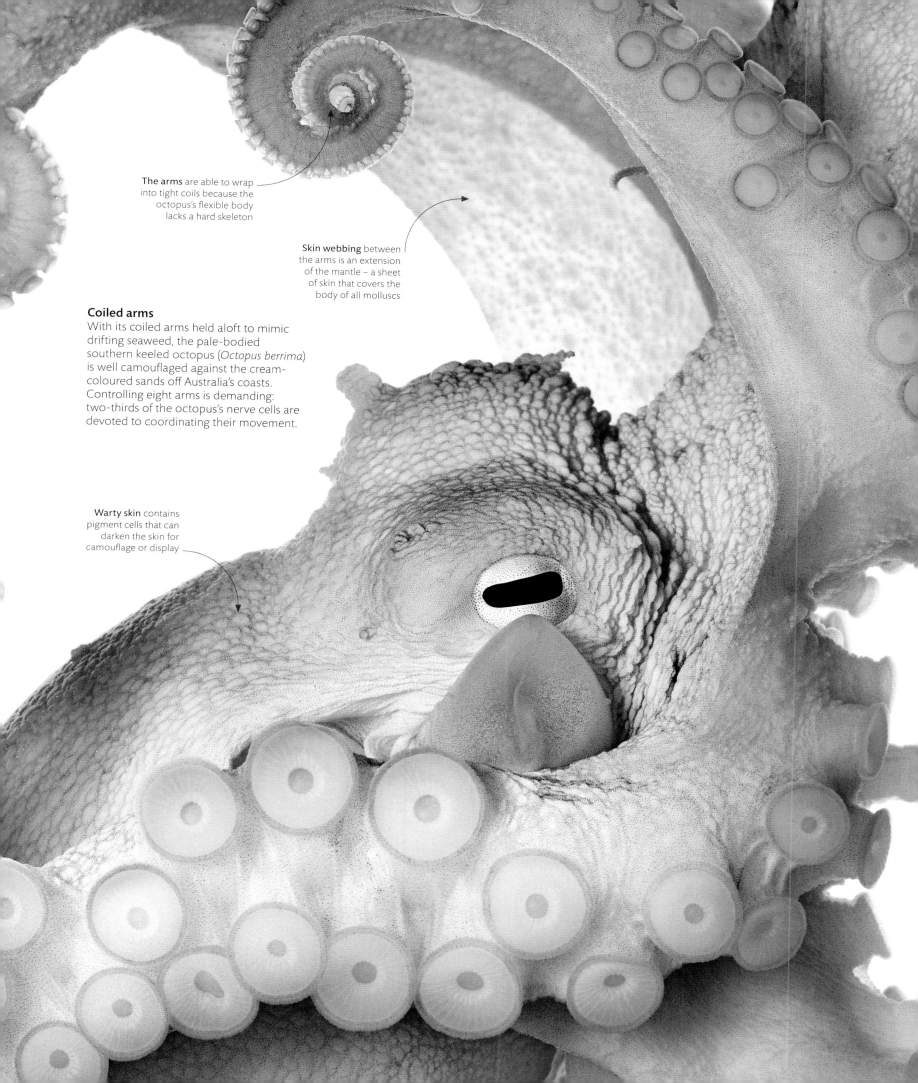

The arms are able to wrap into tight coils because the octopus's flexible body lacks a hard skeleton

Skin webbing between the arms is an extension of the mantle – a sheet of skin that covers the body of all molluscs

Coiled arms

With its coiled arms held aloft to mimic drifting seaweed, the pale-bodied southern keeled octopus (*Octopus berrima*) is well camouflaged against the cream-coloured sands off Australia's coasts. Controlling eight arms is demanding: two-thirds of the octopus's nerve cells are devoted to coordinating their movement.

Warty skin contains pigment cells that can darken the skin for camouflage or display

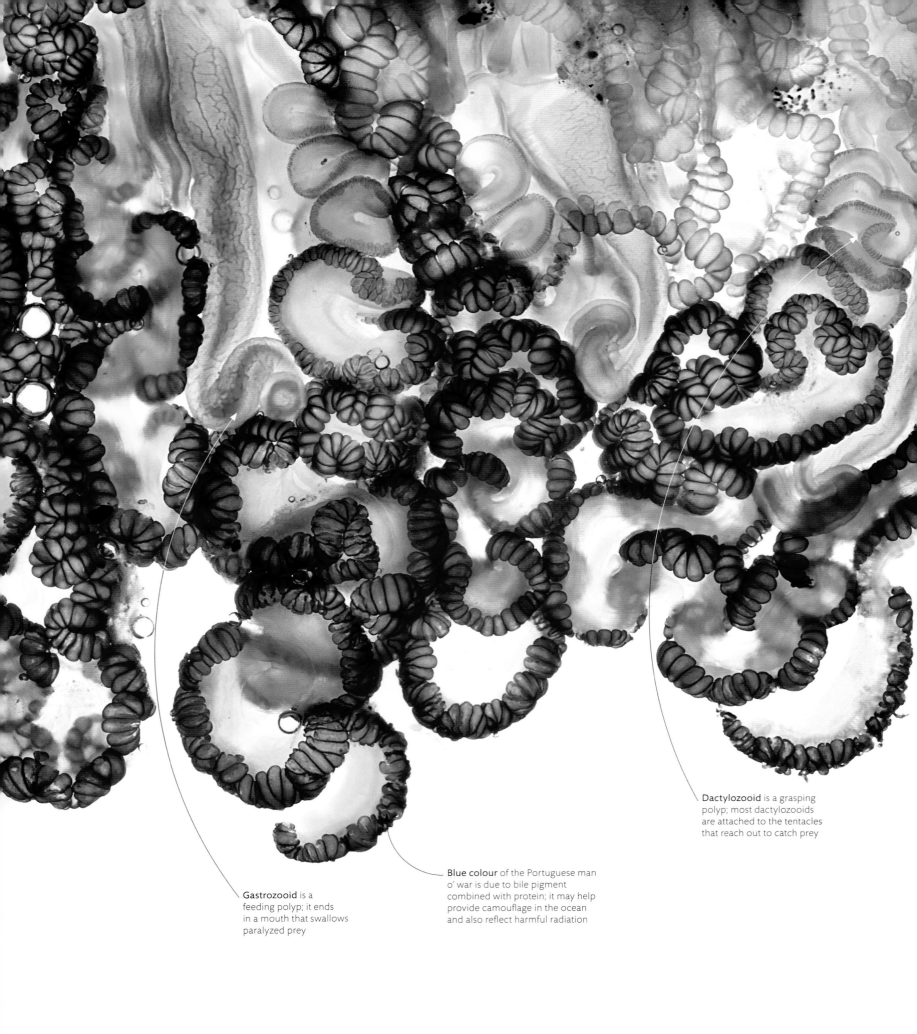

Dactylozooid is a grasping polyp; most dactylozooids are attached to the tentacles that reach out to catch prey

Blue colour of the Portuguese man o' war is due to bile pigment combined with protein; it may help provide camouflage in the ocean and also reflect harmful radiation

Gastrozooid is a feeding polyp; it ends in a mouth that swallows paralyzed prey

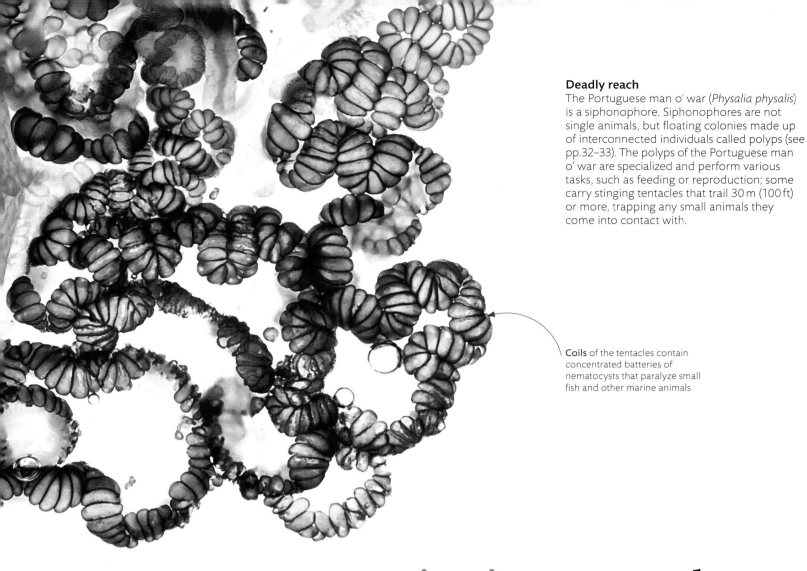

Deadly reach

The Portuguese man o' war (*Physalia physalis*) is a siphonophore. Siphonophores are not single animals, but floating colonies made up of interconnected individuals called polyps (see pp.32–33). The polyps of the Portuguese man o' war are specialized and perform various tasks, such as feeding or reproduction; some carry stinging tentacles that trail 30 m (100 ft) or more, trapping any small animals they come into contact with.

Coils of the tentacles contain concentrated batteries of nematocysts that paralyze small fish and other marine animals

stinging tentacles

Jellyfish and their relatives, such as the Portuguese man o' war, are predators, but they do not overcome prey by swift pursuit or muscular power. Instead, they rely on paralyzing venom to subdue their victims. The skin of their tentacles is peppered with specialized cells called nematocysts, each harbouring an elaborate apparatus that fires a microscopic, venom-laced harpoon into the prey's flesh.

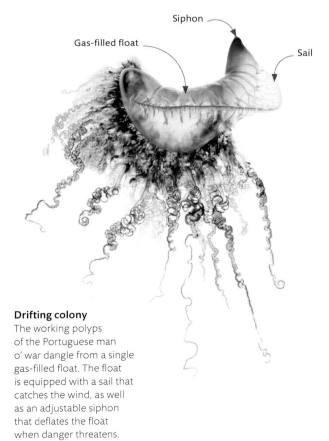

Siphon

Gas-filled float

Sail

Drifting colony

The working polyps of the Portuguese man o' war dangle from a single gas-filled float. The float is equipped with a sail that catches the wind, as well as an adjustable siphon that deflates the float when danger threatens.

DISCHARGING VENOM

Nematocysts contain a coiled inverted tube – like the finger of a rubber glove turned inside out – steeped in venom. Once triggered by contact with prey, the lid of the nematocyst snaps open and the tube springs out. Spines break the prey's skin and allow venom to flow into the wound.

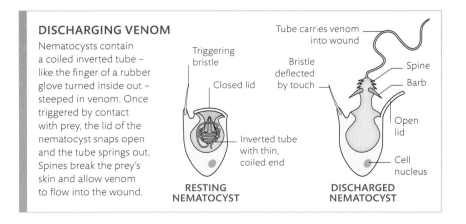

Triggering bristle

Closed lid

Inverted tube with thin, coiled end

RESTING NEMATOCYST

Tube carries venom into wound

Bristle deflected by touch

Spine

Barb

Open lid

Cell nucleus

DISCHARGED NEMATOCYST

prehensile tails

A "prehensile" part of the body is used for grasping. Jaws, hands, and feet are typically prehensile, but many animals can use their tail, too. A fully prehensile tail has the strength and flexibility to support the entire weight of the body. Spider monkeys use their tail as a fifth limb to suspend from branches, while seahorses use their tail to cling to waterweed so they are not swept away by currents (see pp.262–63). Many animals have tails that are prehensile to a lesser degree, but can still assist in climbing or in support.

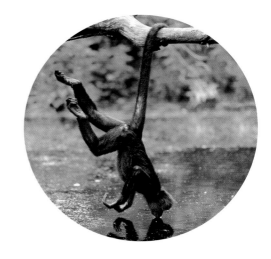

Sensitive tail
The tail of a brown-headed spider monkey (*Ateles fusciceps*) has a naked patch of skin near its tip, with ridges that resemble the sensitive pads on primate fingers (see p.238). It uses its tail to grip branches for swinging, feeding, or drinking.

Large wedge-shaped head has wide jaws and can feed on leaves, fruit, and flowers

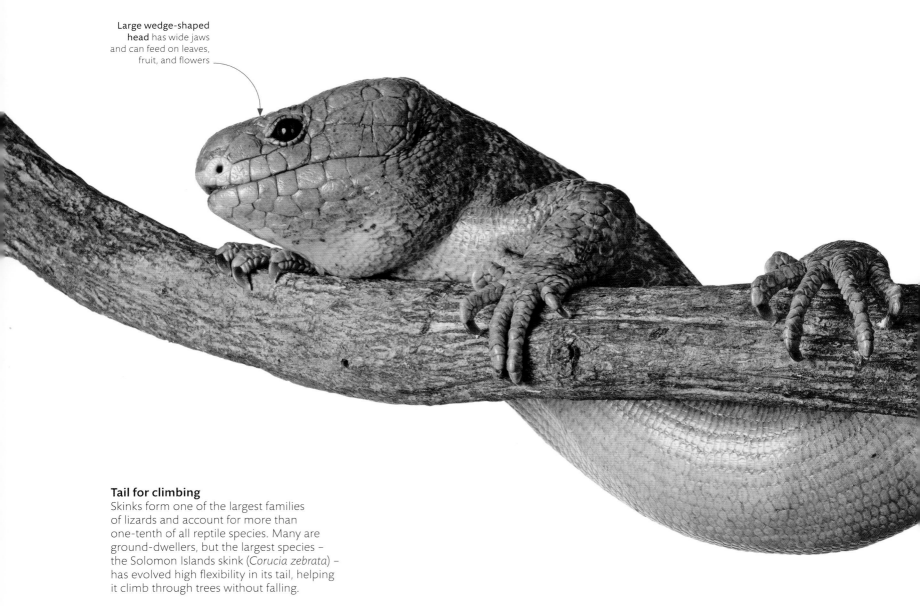

Tail for climbing
Skinks form one of the largest families of lizards and account for more than one-tenth of all reptile species. Many are ground-dwellers, but the largest species – the Solomon Islands skink (*Corucia zebrata*) – has evolved high flexibility in its tail, helping it climb through trees without falling.

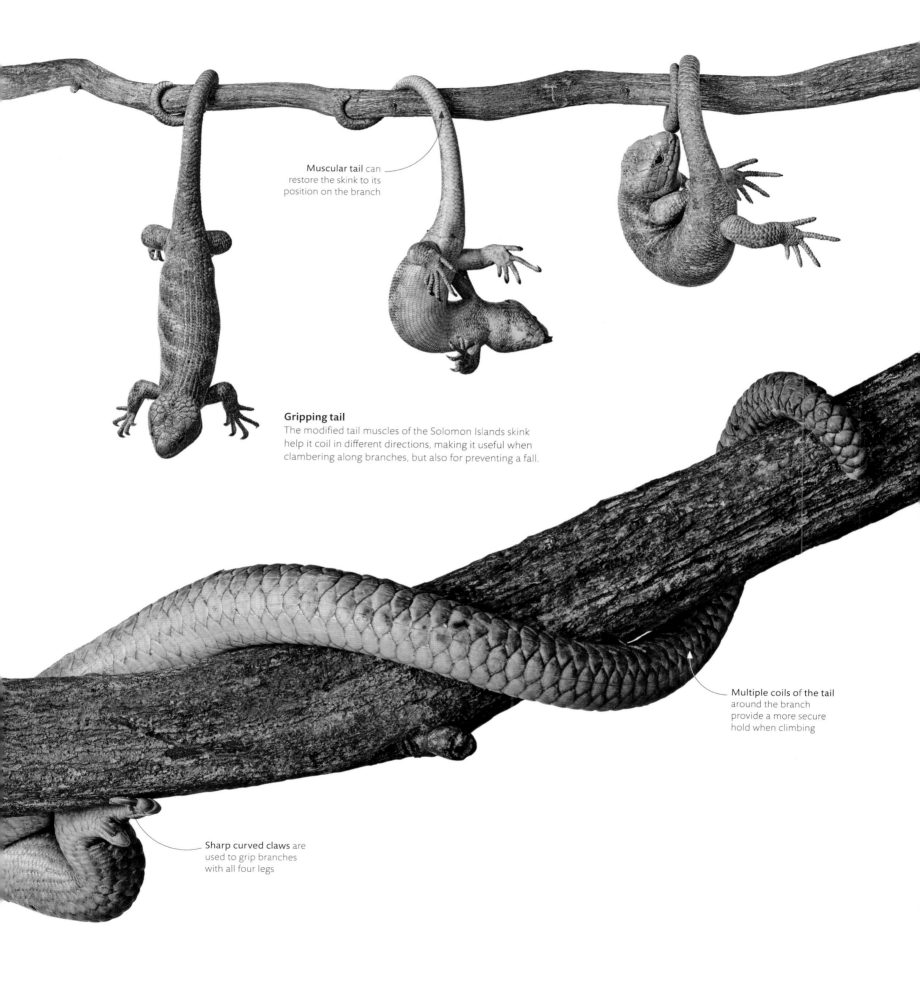

Muscular tail can restore the skink to its position on the branch

Gripping tail
The modified tail muscles of the Solomon Islands skink help it coil in different directions, making it useful when clambering along branches, but also for preventing a fall.

Multiple coils of the tail around the branch provide a more secure hold when climbing

Sharp curved claws are used to grip branches with all four legs

pintailed whydah

The remarkable finchlike pintailed whydah (*Vidua macroura*) is ecologically fascinating. The noisy song, spectacular display flights, and show-stopping, ribbonlike tail plumes are all male characteristics, yet their development is driven by the critical eye of the drab, inconspicuous female.

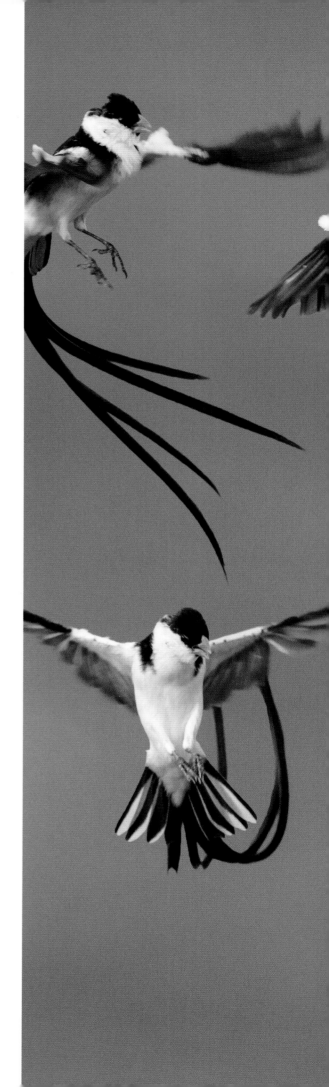

The pintailed whydah is widely distributed across sub-Saharan Africa, and non-native populations are also found in Portugal, Puerto Rico, California, and Singapore. Males and females are small – their bodies are about 12–13 cm (4¾–5 in) long. While the female is unremarkable to look at, she has a significant influence on the males of her species. A quirk of evolution resulted in female whydahs developing a preference for males with striking black-and-white plumage, a scarlet bill, and long tail feathers. A process called runaway sexual selection is probably responsible for then driving the last feature to preposterous proportions – at around 20 cm (8 in), the tail is almost twice as long as the body.

The bold colours and conspicuous tail were probably considered more desirable since they are a reliable indication of vitality, good nutritional status, and low parasite load. Over time, sexiness became what mattered most and the choices of generations of females has driven male tail length to the point where it is a survival disadvantage. The plumes take energy to grow and maintain, they make flight harder, and they must increase the chances of being caught by a predator. But biologically, longer tails carry the irresistible benefit of an opportunity to breed. In principle, shorter-tailed males may live longer, but on average father fewer offspring. However, tails cannot grow infinitely long, and natural selection picks off males with the longest tails, no matter how attractive they are.

Courtship display

This multiple-exposure photograph captures a single male bird using rhythmic wingbeats to emphasize his tail streamers, while he bobs up and down attempting to stay in one place in front of the female's perch.

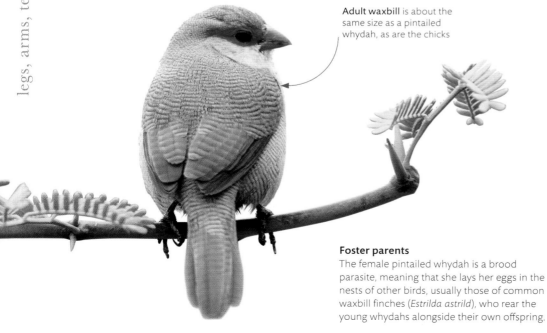

Adult waxbill is about the same size as a pintailed whydah, as are the chicks

Foster parents

The female pintailed whydah is a brood parasite, meaning that she lays her eggs in the nests of other birds, usually those of common waxbill finches (*Estrilda astrild*), who rear the young whydahs alongside their own offspring.

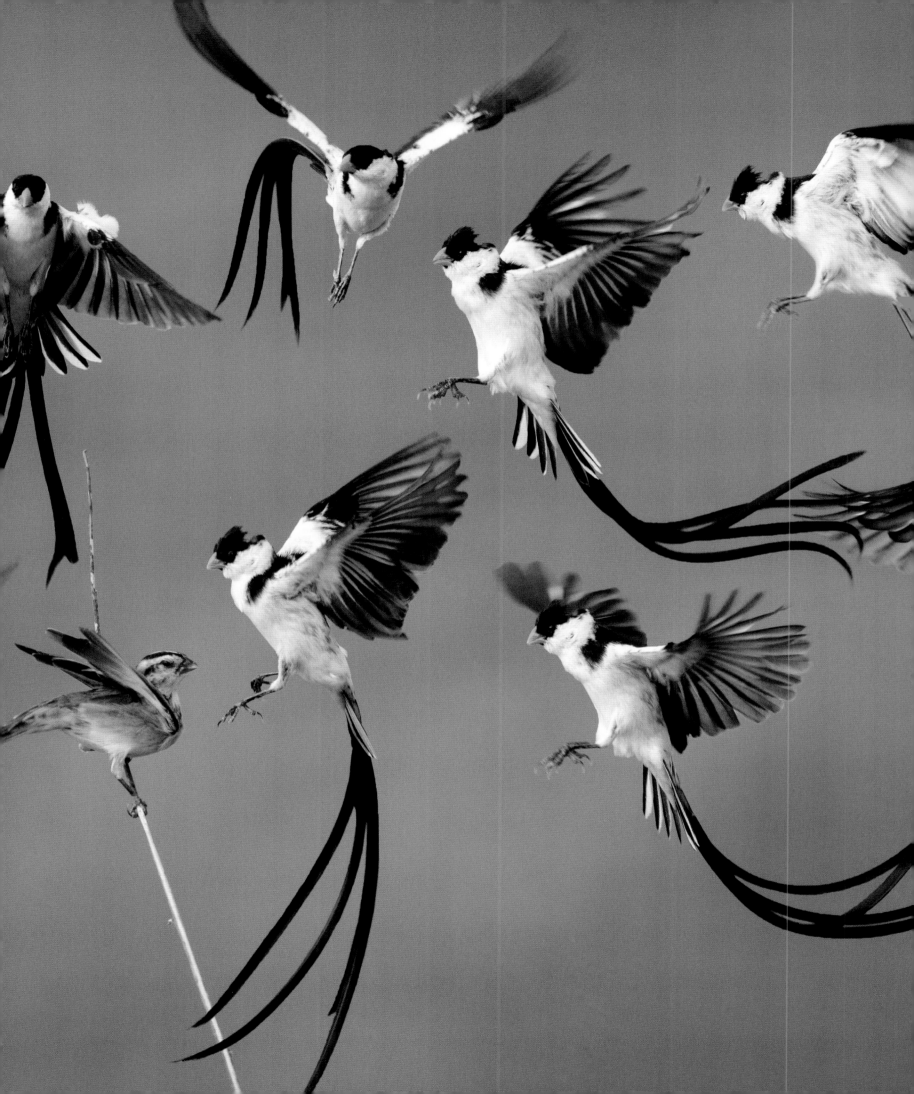

fins, flippers and paddles

fin. a thin, membranous appendage used for propulsion, steering, and balance.
flipper. a broad, flat limb, such as those found on seals, whales, and penguins, especially suited to swimming.
paddle. a fin or flipper of an aquatic animal.

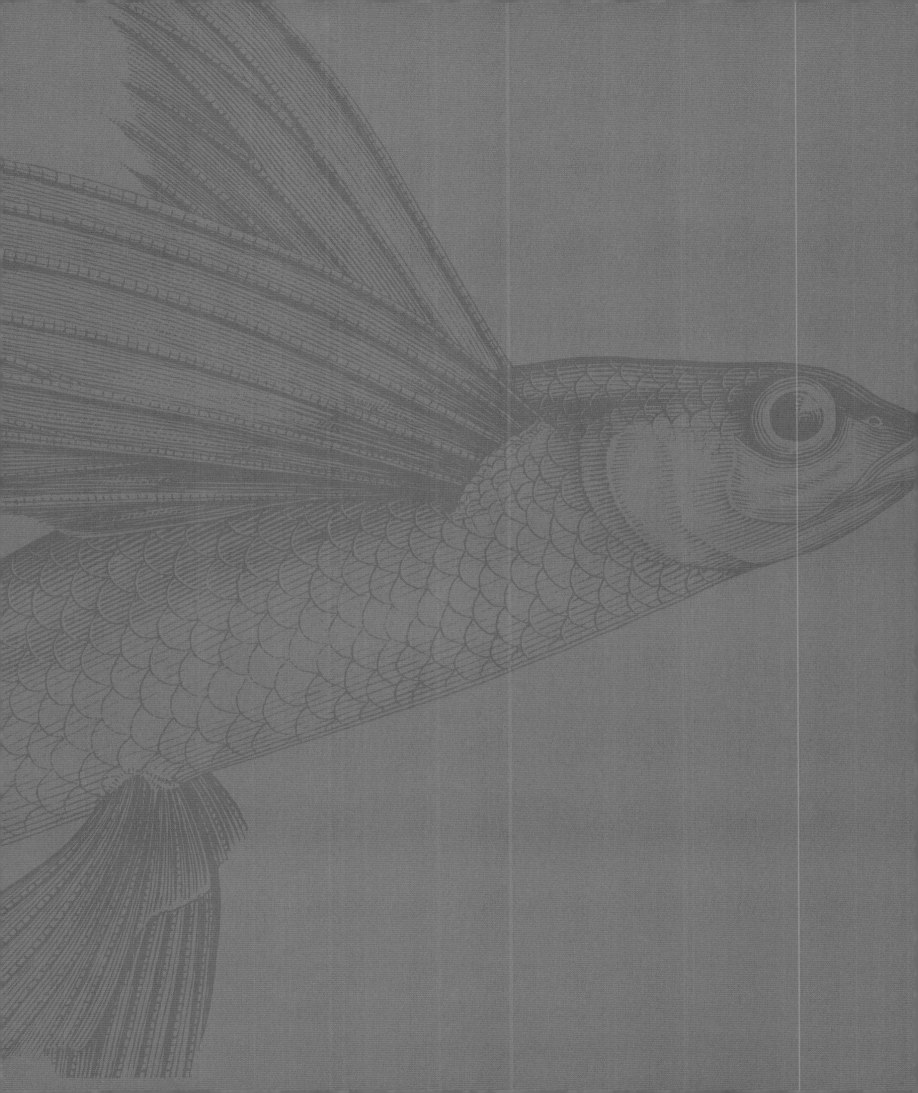

Strong swimmers
Animals with the size and strength to swim against currents, such as this courting pair of Caribbean reef squid (*Sepioteuthis sepiodea*), make up the ocean's nekton population.

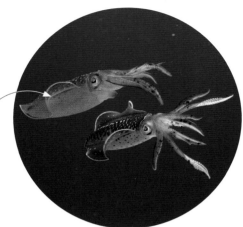

Undulating fins and a siphon jet, along either side of the body, propel this animal through water

nekton and
plankton

Many aquatic animals – so-called nektonic fauna – are strong swimmers and can battle against currents. But others, plankton, drift with currents. The tiniest animals are likely to be planktonic: water's viscosity makes progress for them equivalent to a human moving through syrup so currents, rather than muscle power, carry them from place to place. Some, such as fish and crabs, are larvae of animals while others stay as drifting animals their entire lives.

PLANKTONIC LIFE
Copepods (right) – crustaceans mostly less than 1 mm long – are components of zooplankton found in aquatic habitats from oceans to ponds. Many of these animals rely on long, spindly, antennae-like appendages that twitch backwards so they appear to jump through the water. These darting movements enable them to overcome water's viscosity without fatiguing, making them among the fastest and strongest of all animals for their size.

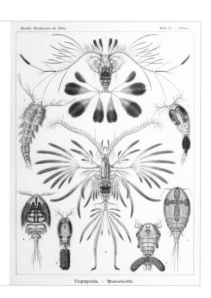

Planktonic slug
Sea angels, sea butterflies, and other sea slugs of the open ocean use fleshy, winglike flaps, or parapodia, to propel themselves. The largest sea angel, the naked sea butterfly (*Clione limacina*), is no more than 3 cm (1¹/₅ in) long. For all its flapping, it is at the mercy of ocean currents and classed as plankton.

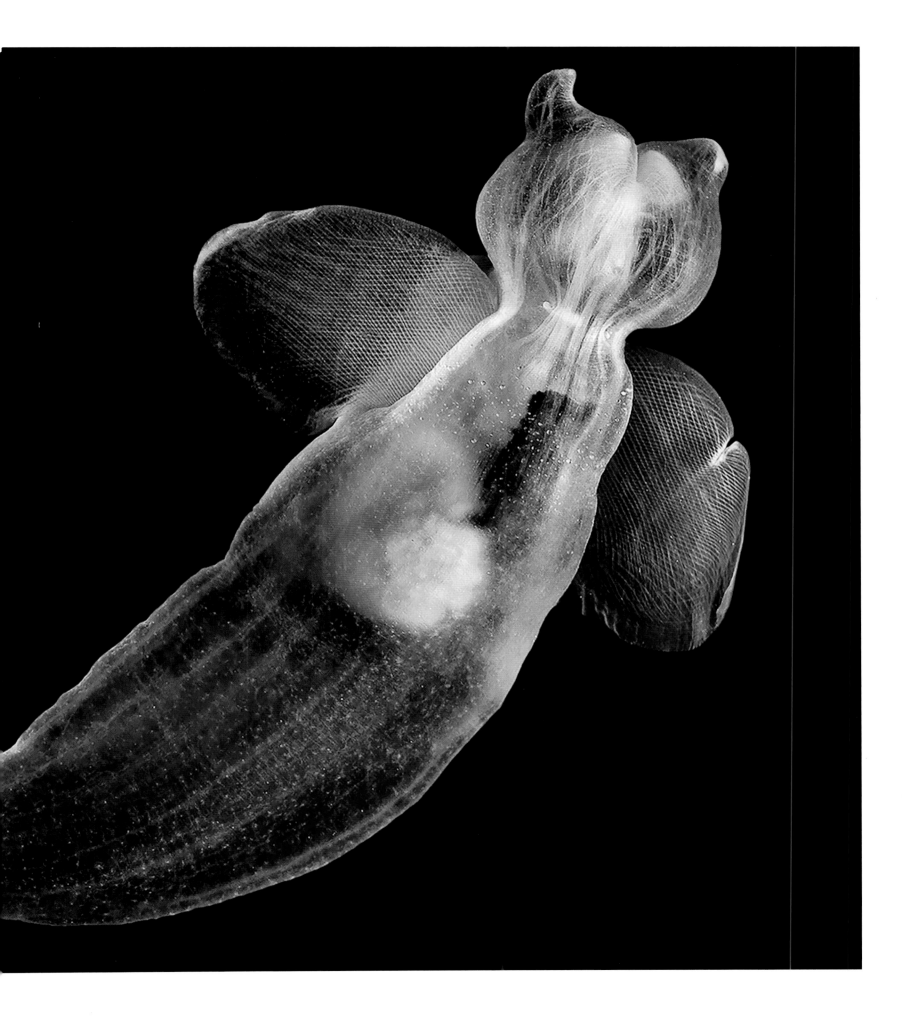

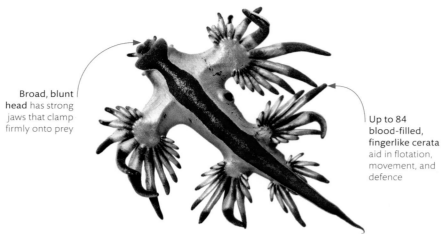

Broad, blunt head has strong jaws that clamp firmly onto prey

Up to 84 blood-filled, fingerlike cerata aid in flotation, movement, and defence

Drifting predator
The blue dragon's ability to float on its back, riding the surface tension, brings it into contact with the larger blue cnidarians on which it feeds. The dark blue at the tips of its cerata probably comes from the blue pigment of its prey.

spotlight species

blue dragon sea slug

A weak swimmer, the blue dragon sea slug (*Glaucus atlanticus*) spends most of its life floating upside down in the open ocean. This sea slug feeds mainly on the highly venomous Portuguese man o' war. The sting of this tiny predator, which reaches about 3 cm (1 in) long, can be more powerful than that of its much larger prey.

The blue dragon, also known as the blue angel or sea swallow, is a type of sea slug called a nudibranch. Like other sea slugs, blue dragons are shell-less marine snails, have soft bodies, and use a toothlike scraper called a radula to tear chunks off prey when feeding. However, unlike its benthic relatives that inhabit the ocean floor, the blue dragon is pelagic: it is found throughout the sea's upper levels and has been identified in temperate and tropical oceans worldwide.

The "wings" that give the animal its common name are fanlike projections called cerata, which protrude from the blue dragon's sides. Cerata usually grow on a sea slug's back, but in this species they sprout from body parts resembling stubby forearms and leg stumps. The blue dragon can move these appendages and their cerata, so they function rather like limbs and fingers, effectively enabling the sea slug to swim – albeit weakly – towards prey. Floating, however, is its chief mode of locomotion, accomplished by means of an air-filled capsule in its stomach. This capsule, coupled with the cerata's large surface area, allows the blue dragon to float easily, drifting wherever wind and wave take it. Countershading offers the blue dragon some protection: when lying on its back, its dark, silver-blue belly blends in with the water's surface, hiding it from the eyes of airborne predators. Its paler back is camouflaged against the sky when viewed by predators from below.

Dragon with tentacles
A Portuguese man o' war's tentacles are the blue dragon's favourite food. The sea slug consumes the man o' war's stinging cells, or nematocysts, intact. It concentrates them in special sacs at the tips of its cerata to deploy as a defence.

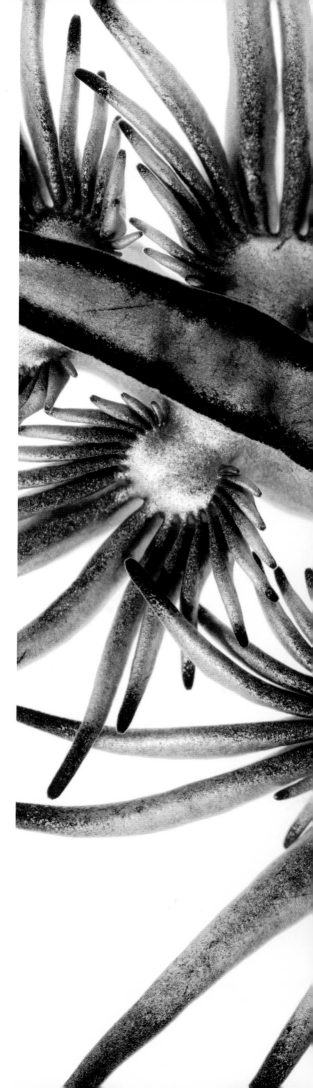

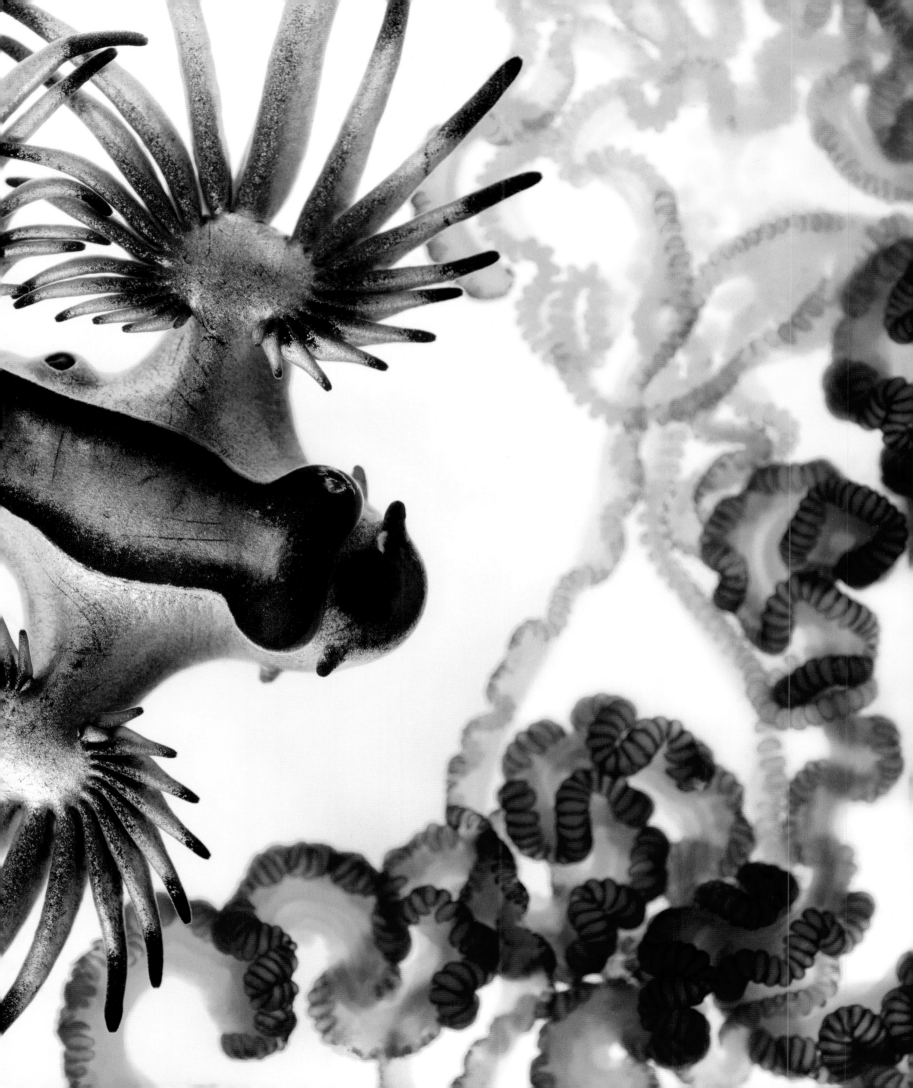

how fish swim

Fish swim forwards by undulating the body or oscillating appendages (see p.50 and p.259). But the challenge of moving in the three dimensions of open water involves more than propulsion: fish must control the way they move vertically and horizontally, as well as stay upright. Their body must also be buoyant to prevent them from sinking. Sharks achieve buoyancy with oily tissues, while most bony fish are kept buoyant by a gas-filled swim bladder.

Life on the bottom

Eel-shaped fish like the burbot (*Lota lota*), depicted here in a stylized 16th-century illustration, avoid the problems of mid-water swimming by living on the bottom. Multiple waves of undulation passing down the elongated body generate more drag than experienced by short-bodied fish. But a slower swimming speed is well-suited to a life spent in burrows and crevices.

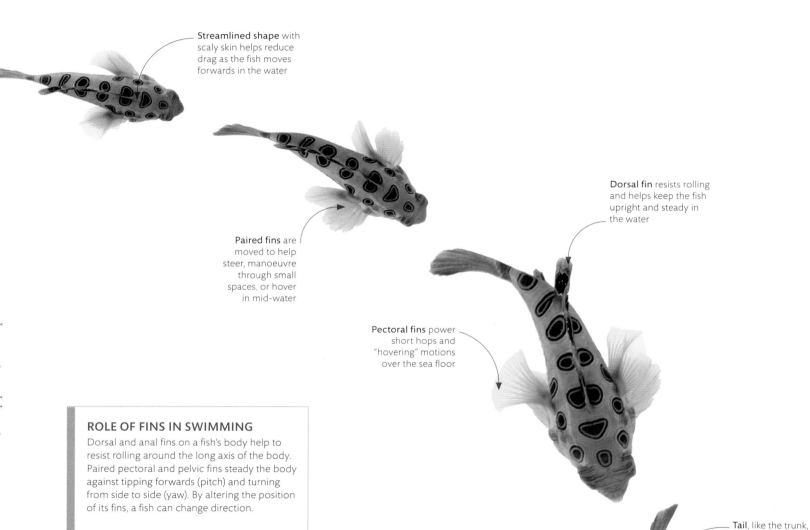

Streamlined shape with scaly skin helps reduce drag as the fish moves forwards in the water

Paired fins are moved to help steer, manoeuvre through small spaces, or hover in mid-water

Dorsal fin resists rolling and helps keep the fish upright and steady in the water

Pectoral fins power short hops and "hovering" motions over the sea floor

Tail, like the trunk, plays a much reduced role in swimming than in open- and mid-water species

ROLE OF FINS IN SWIMMING

Dorsal and anal fins on a fish's body help to resist rolling around the long axis of the body. Paired pectoral and pelvic fins steady the body against tipping forwards (pitch) and turning from side to side (yaw). By altering the position of its fins, a fish can change direction.

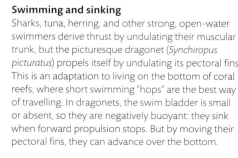

Anal fin: roll

Dorsal fin: roll

PITCH

Pelvic fins: yaw

ROLL

YAW

Pectoral fins: pitch and yaw

HOW SHARK FINS CONTROL MOVEMENT

Swimming and sinking

Sharks, tuna, herring, and other strong, open-water swimmers derive thrust by undulating their muscular trunk, but the picturesque dragonet (*Synchiropus picturatus*) propels itself by undulating its pectoral fins. This is an adaptation to living on the bottom of coral reefs, where short swimming "hops" are the best way of travelling. In dragonets, the swim bladder is small or absent, so they are negatively buoyant: they sink when forward propulsion stops. But by moving their pectoral fins, they can advance over the bottom.

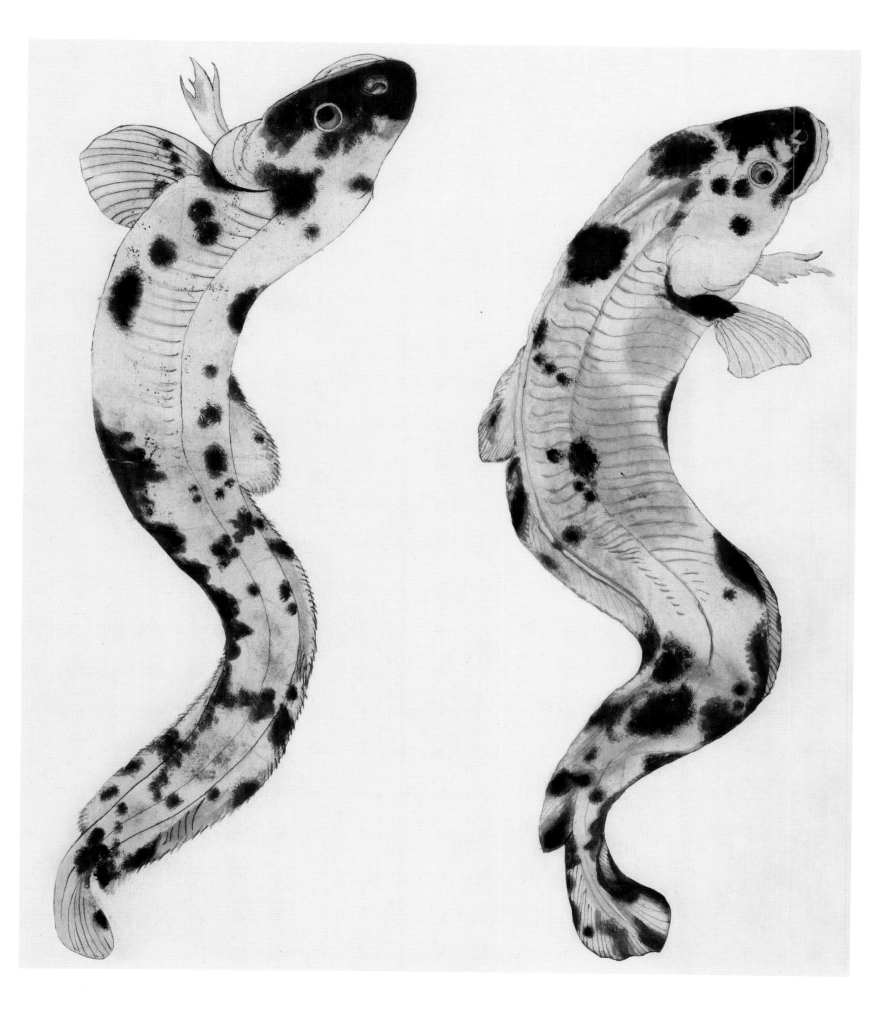

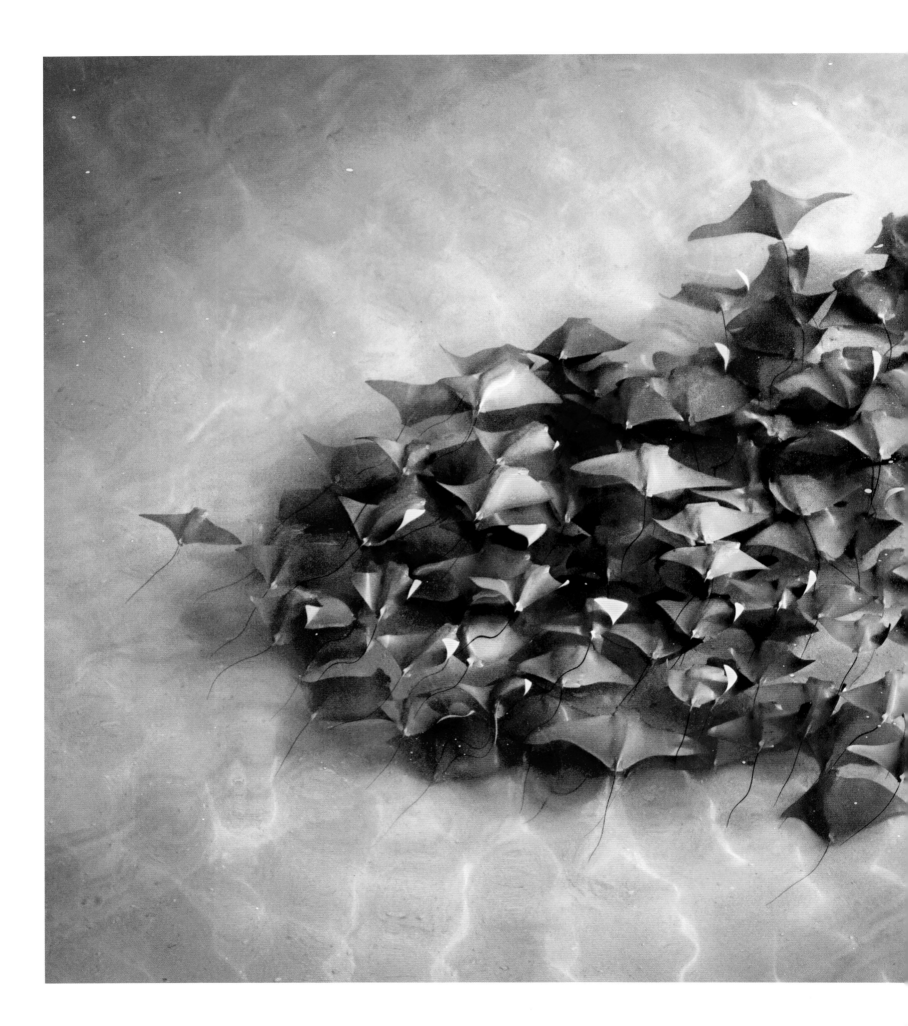

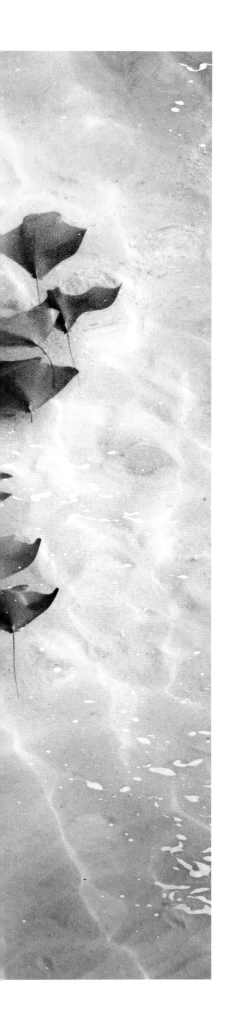

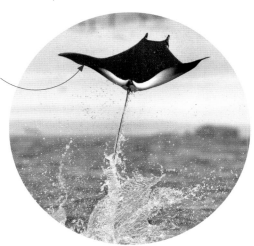

Pectoral fins
continue to flap up
and down when the
ray leaves the water

Giant ray
Cownose rays, with their rhomboid shape
and bulbous, protruding head, are among the
largest rays. They cruise in warm waters above
continental shelves or around estuaries and
bays. The short-tailed cownose ray (*Rhinoptera
jayakari*), reaching a "fin-span" of 90 cm (35 in),
has a shorter whiplike tail than most. It often
gathers into huge shoals, as seen here.

Mobula ray leaping
A few large-finned rays, such as mobulas (*Mobula* sp.),
leap from the water in spectacular displays. Whether
this dislodges parasites or is a social signal is not clear.

underwater wings

All fish need a source of propulsion. In rays, it comes from large, projecting
pectoral fins that connect to the flattened body along a wide base stretching
from head to tail. The smallest rays swim using delicate undulations that ripple
back along the margins of these fins – they can even hover in mid-water – but
larger species flap them like wings to "fly" through the ocean.

FORWARD PROPULSION

Some fish propel themselves by undulating body
parts (shown in orange). Jacks and trevallies, which use
a wavelike body motion, are faster swimmers than
those, like rays, with rippling fins. Other fish use
oscillation, where a structure is moved back and forth
(shown in blue). Box fish have a rigid, armoured body,
so they flap their tails to generate force, while wrasses
row through water with their pectoral fins.

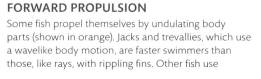

Undulating
pectoral fins

RETICULATED FRESHWATER STINGRAY
Potamotrygon orbignyi

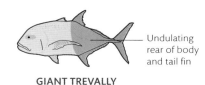

Undulating
rear of body
and tail fin

GIANT TREVALLY
Caranx ignobilis

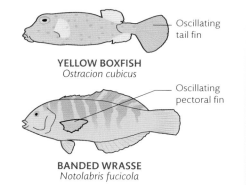

Oscillating
tail fin

YELLOW BOXFISH
Ostracion cubicus

Oscillating
pectoral fin

BANDED WRASSE
Notolabris fucicola

fish fins

Fins provide directional control as fish propel their bodies through the water (see p.259), and they contribute to the propulsion itself – the caudal, or tail, fin increases the force of an undulating fish body. Many fish, such as seahorses (see pp.262–63), rely entirely on fin movement for swimming. The most prominent types of fin are the caudal and dorsal fins, and the paired pectoral and pelvic fins.

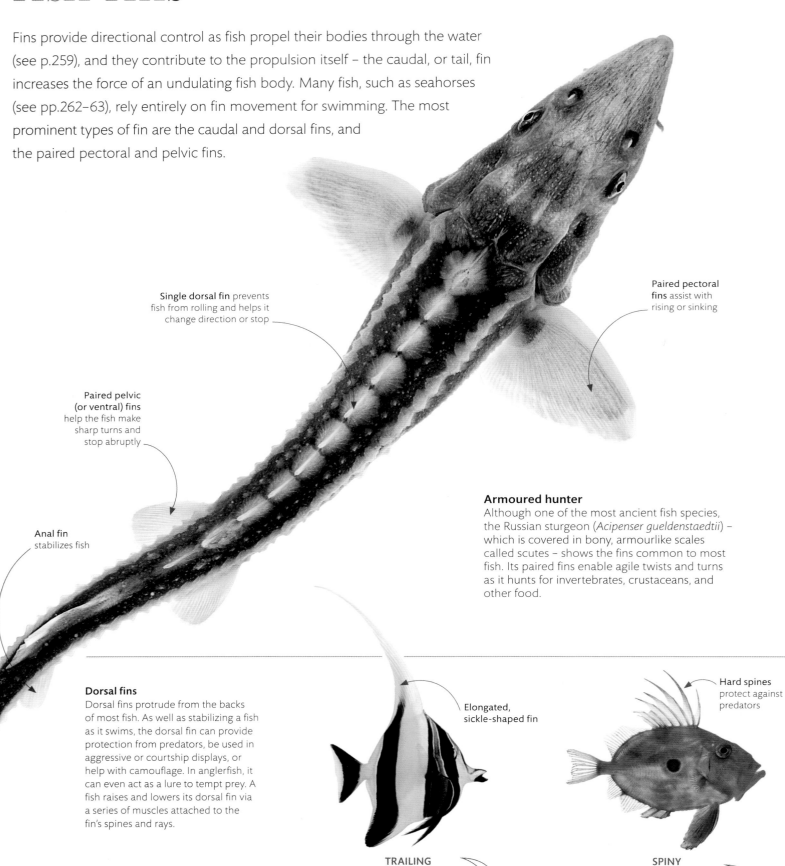

Single dorsal fin prevents fish from rolling and helps it change direction or stop

Paired pectoral fins assist with rising or sinking

Paired pelvic (or ventral) fins help the fish make sharp turns and stop abruptly

Anal fin stabilizes fish

Armoured hunter
Although one of the most ancient fish species, the Russian sturgeon (*Acipenser gueldenstaedtii*) – which is covered in bony, armourlike scales called scutes – shows the fins common to most fish. Its paired fins enable agile twists and turns as it hunts for invertebrates, crustaceans, and other food.

Dorsal fins
Dorsal fins protrude from the backs of most fish. As well as stabilizing a fish as it swims, the dorsal fin can provide protection from predators, be used in aggressive or courtship displays, or help with camouflage. In anglerfish, it can even act as a lure to tempt prey. A fish raises and lowers its dorsal fin via a series of muscles attached to the fin's spines and rays.

Elongated, sickle-shaped fin

TRAILING
Moorish idol
Zanclus cornutus

Hard spines protect against predators

SPINY
John Dory
Zeus faber

Caudal fins

Most fish propel themselves using their caudal, or tail, fin. The shape of a fish's tail provides clues as to how and where it lives. Unlobed, rounded, or straight caudal fins are found mainly on slower-moving, shallow-water dwellers, while lunate and forked caudal fins are typical of fast-moving or long-distance swimmers found in open or deeper waters.

Broad tail is good for agile manoeuvres, but creates high drag

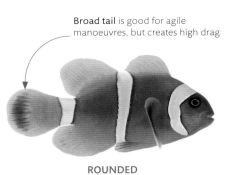

ROUNDED
Maroon clownfish
Premnas biaculeatus

Thin, rigid tail enables high-speed swimming

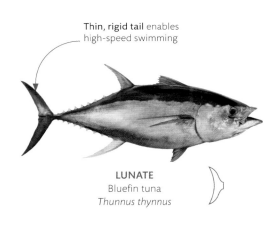

LUNATE
Bluefin tuna
Thunnus thynnus

Narrow tail base reduces drag

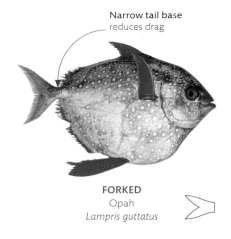

FORKED
Opah
Lampris guttatus

Flattened shape provides good acceleration and manoeuvrability

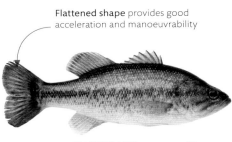
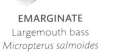

EMARGINATE
Largemouth bass
Micropterus salmoides

Tail point formed by fused anal and caudal fins

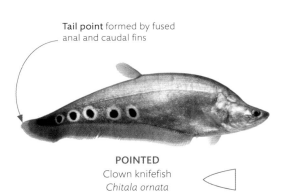

POINTED
Clown knifefish
Chitala ornata

Large surface area helps manoeuvrability

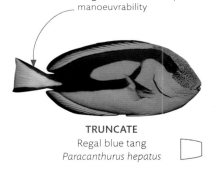

TRUNCATE
Regal blue tang
Paracanthurus hepatus

Spine can be used to lock the fish into crevices at night for protection

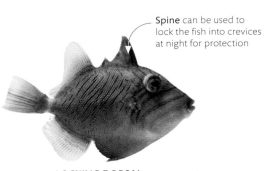

LOCKING DORSAL
Queen triggerfish
Balistes vetula

Long dorsal fin stops fish from rolling

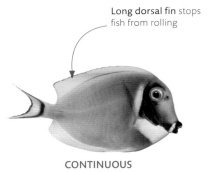

CONTINUOUS
Powder blue tang
Acanthurus leucosternon

Second dorsal fin made of soft rays

First dorsal fin has hard spines

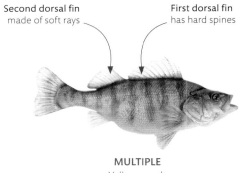

MULTIPLE
Yellow perch
Perca flavescens

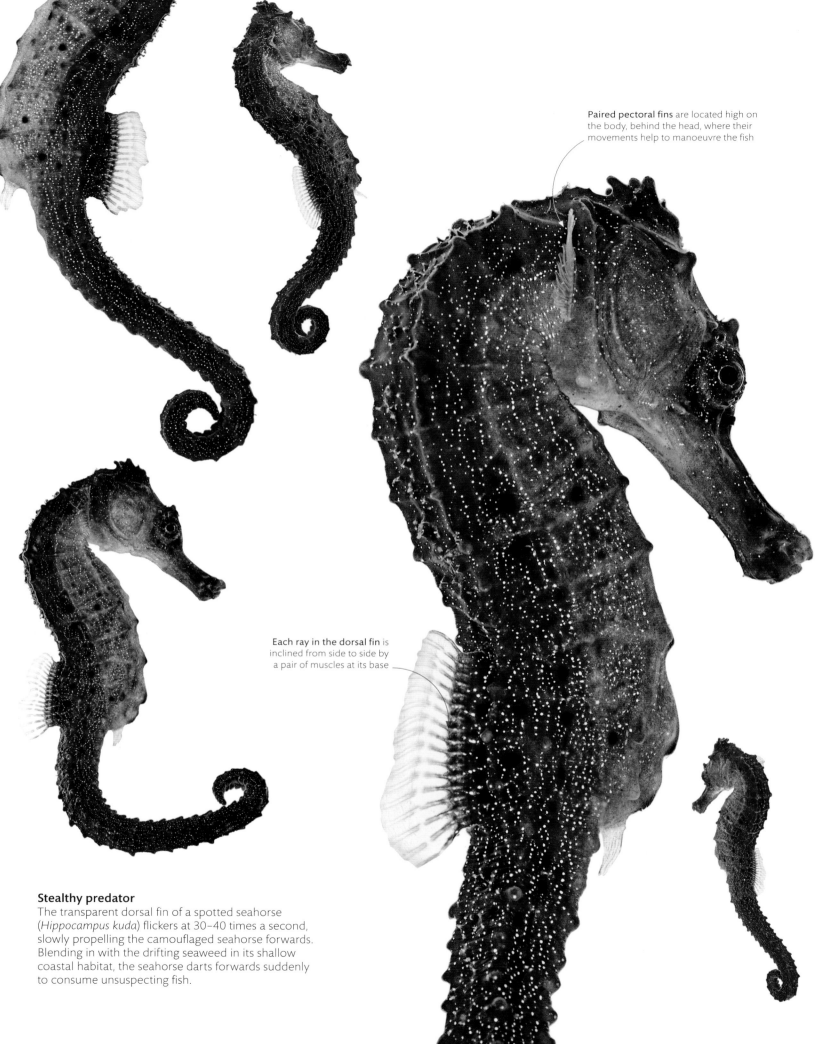

Paired pectoral fins are located high on the body, behind the head, where their movements help to manoeuvre the fish

Each ray in the dorsal fin is inclined from side to side by a pair of muscles at its base

Stealthy predator
The transparent dorsal fin of a spotted seahorse (*Hippocampus kuda*) flickers at 30–40 times a second, slowly propelling the camouflaged seahorse forwards. Blending in with the drifting seaweed in its shallow coastal habitat, the seahorse darts forwards suddenly to consume unsuspecting fish.

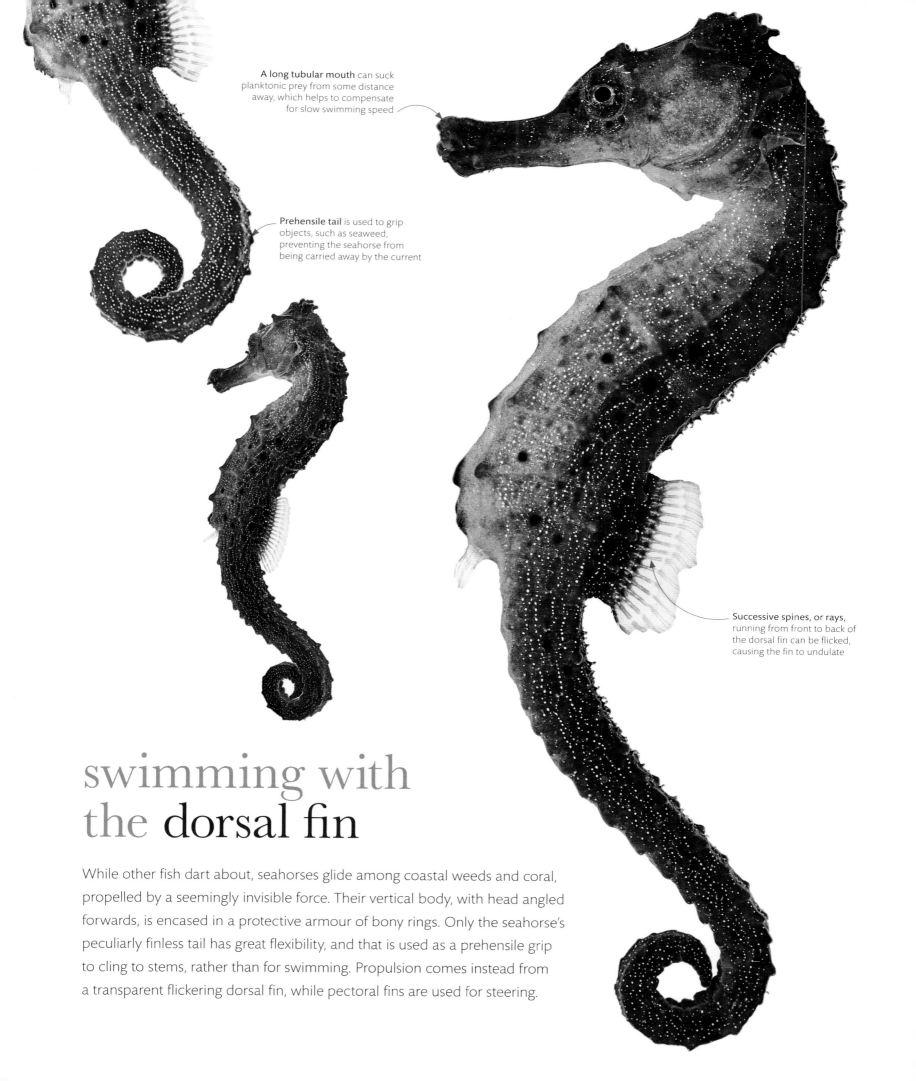

A long tubular mouth can suck planktonic prey from some distance away, which helps to compensate for slow swimming speed

Prehensile tail is used to grip objects, such as seaweed, preventing the seahorse from being carried away by the current

Successive spines, or rays, running from front to back of the dorsal fin can be flicked, causing the fin to undulate

swimming with the dorsal fin

While other fish dart about, seahorses glide among coastal weeds and coral, propelled by a seemingly invisible force. Their vertical body, with head angled forwards, is encased in a protective armour of bony rings. Only the seahorse's peculiarly finless tail has great flexibility, and that is used as a prehensile grip to cling to stems, rather than for swimming. Propulsion comes instead from a transparent flickering dorsal fin, while pectoral fins are used for steering.

four-winged flying fish

When it comes to escaping marine predators, the four-winged flying fish (*Hirundichthys affinis*) has a distinct advantage. It can propel itself out of the water to reach an airborne speed of up to 72 kph (44 mph), gliding 400 m (1,312 ft) at a single stretch, and also turn and alter its altitude in mid-glide.

An estimated 65 species of flying fish inhabit tropical and temperate oceans, divided into two categories based on their "wing" numbers. All have two greatly enlarged pectoral fins that allow them to glide over water, a streamlined body, and an unevenly forked tail with a much larger lower lobe. Four-winged species such as *Hirundichthys affinis*, which is found in the eastern and northwestern Atlantic, the Gulf of Mexico, and the Caribbean, also have enlarged pelvic fins; this second set of wings gives them an aerodynamic edge over their two-winged cousins.

Unlike birds and bats, flying fish do not actively flap but glide, relying on the surface area of their extended pectoral fins to stay airborne. The pelvic fins of *H. affinis* and other four-wingers likely evolved as a stabilizing aid that controls pitch, functioning in a similar way to an aeroplane's horizontal tail fins. Pelvic wings may also be a response to size:

flying fish range in length from around 15 to 50 cm (6 to 20 in), and four-winged species like *H. affinis* tend to be longer than two-wingers. Only four-wingers are able to change direction mid-air.

To get airborne, the four-winged flying fish swims towards the surface at up to 36 kph (22 mph). As the tail, beating at 50 times per second, launches the fish clear of the water, the pectoral fins unfurl and the fish begins to glide. Flying fish are found in waters warmer than 20–23°C (68–73°F); experts believe that their muscles could not contract rapidly enough to achieve takeoff in cooler temperatures.

Extended air time
Flying fish re-enter the water tail-first, but by rapidly beating their tails whenever the larger lobe touches the water – behaviour called taxiing – they can prolong flights several times. In the air, tails are held high for stability.

Flying fish fresco
People have been fascinated and inspired by flying fish since ancient times. This Minoan fresco, from the site of Phylakopi on the Mediterranean island of Milos, dates from around 2500 BCE.

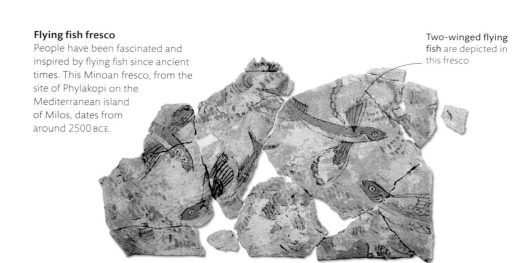

Two-winged flying fish are depicted in this fresco

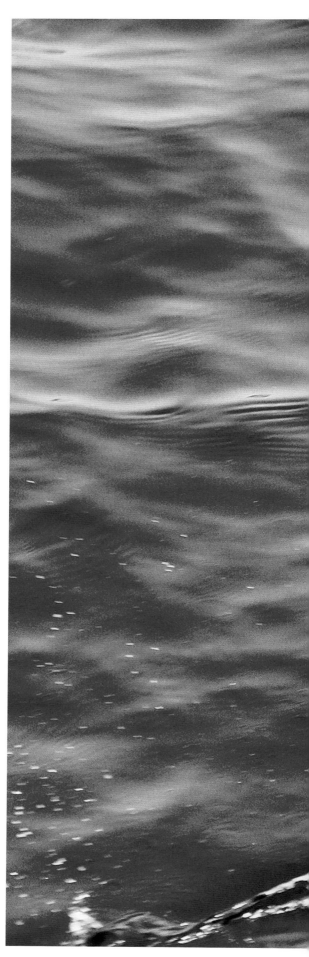

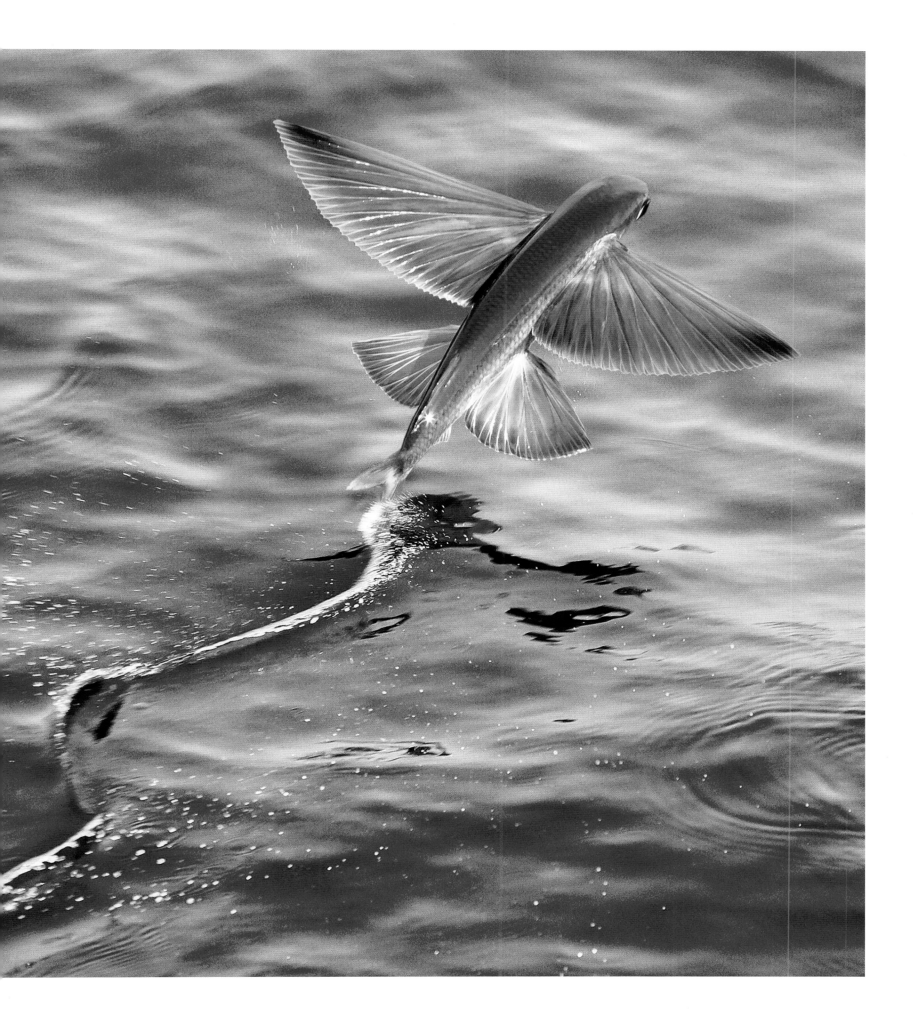

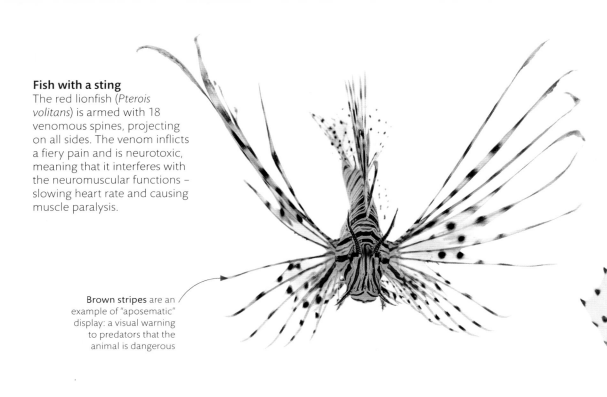

Fish with a sting
The red lionfish (*Pterois volitans*) is armed with 18 venomous spines, projecting on all sides. The venom inflicts a fiery pain and is neurotoxic, meaning that it interferes with the neuromuscular functions – slowing heart rate and causing muscle paralysis.

Brown stripes are an example of "aposematic" display: a visual warning to predators that the animal is dangerous

venomous spines

Nearly all fish alive today have fins that are supported by stiff, flexible rays that grow within the surface of the skin, then embed deeper and fan out to provide structural support in the adult fish. In some fish, however, the fronts of the fins have reinforcement in the form of harder, bony spines. Spines alone offer some physical protection from predators, but other fish, such as those in the scorpionfish family, go further and use their spines as weapons for injecting venom. These stingers produce some of the most lethal cocktails of toxins found in the animal kingdom.

The anal fin has three venomous spines on the leading edge

DELIVERING VENOM

A cross-section through a venomous spine of a lionfish reveals a length of solid bone – deeply grooved on either side to carry a pair of long glands – beneath a spongy coat. As the spine punctures the flesh, the skin rolls back and the pressure of the impact squeezes the glands to release venom into the wound.

Venom gland

Bony core

Skin seals in venom

SPINE BEFORE PENETRATING VICTIM

Surface epithelium rubbed away

Venom released

SPINE AFTER PENETRATING VICTIM

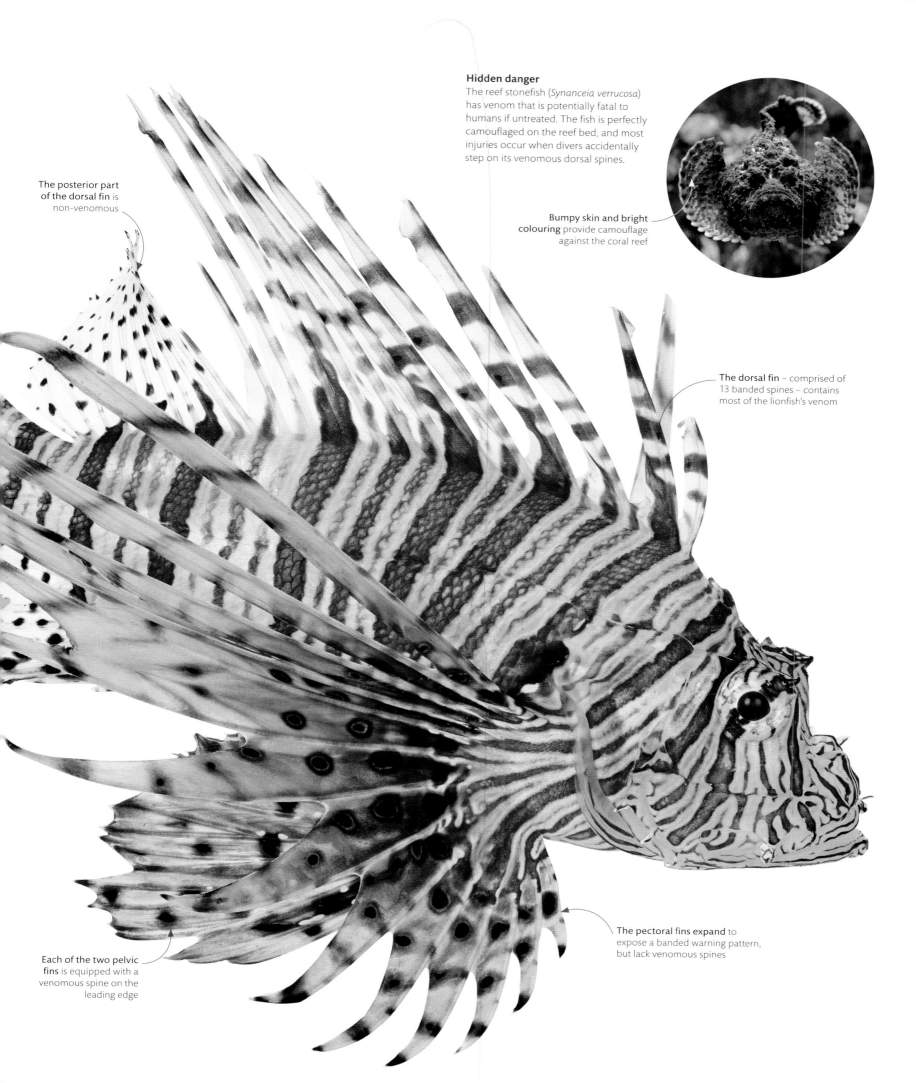

Hidden danger
The reef stonefish (*Synanceia verrucosa*) has venom that is potentially fatal to humans if untreated. The fish is perfectly camouflaged on the reef bed, and most injuries occur when divers accidentally step on its venomous dorsal spines.

Bumpy skin and bright colouring provide camouflage against the coral reef

The posterior part of the dorsal fin is non-venomous

The dorsal fin – comprised of 13 banded spines – contains most of the lionfish's venom

The pectoral fins expand to expose a banded warning pattern, but lack venomous spines

Each of the two pelvic fins is equipped with a venomous spine on the leading edge

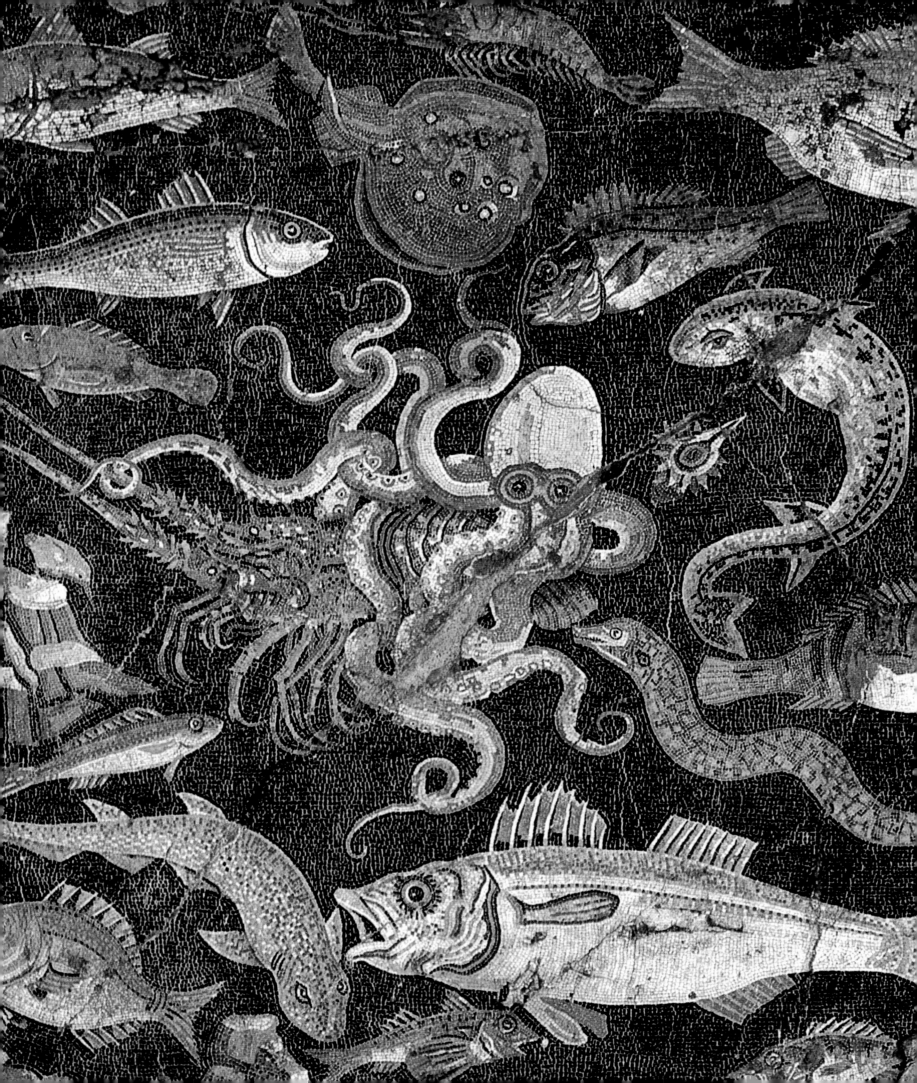

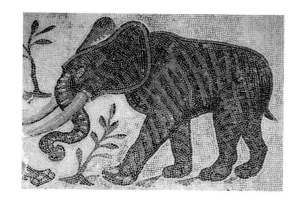

Elephant mosaic
Romans prized elephants as workhorses, for combat, and for their status as exotic animals. This elephant, part of a trio with a horse and bear, is from a 2nd- or 3rd-century floor mosaic from the "House of the Laberii" in Oudhna, Tunisia.

animals in art

empire of abundance

The vibrant frescoes and mosaics that survive in every corner of the old Roman empire reveal an easy acceptance of nature's bounty. Images of fish and animals for the table, exotic pets, sacred creatures, and hunting and circus scenes celebrate the abundance of the natural world. Yet in reality, many Roman pursuits showed a total disregard for animal wellbeing.

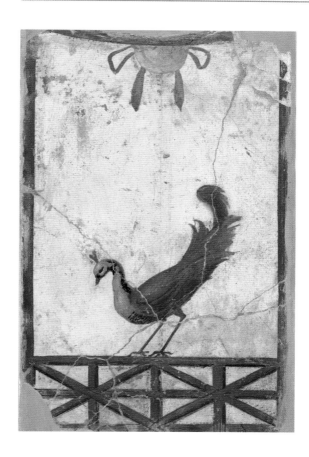

The interior designers of the Roman world were unnamed craftsmen and painters commissioned to adorn walls and floors with frescoes and mosaics that revealed the wealth and status of noble households. Large sea-life mosaics were the preferred decoration for public baths and private houses, where fresh and saltwater oyster and fish ponds provided a ready source of food. In the finest works, artisans created identifiable species, from dogfish and rays to sea bream and bass. Some villas had gullies bringing sea water from the bay of Naples to their ponds.

The majority of elephants in Roman art are thought to be a small subspecies of North African elephant that is now extinct. Larger Indian elephants are shown carrying soldiers into battle. Mosaics in Sicily show how they were captured in their hundreds in India and Africa along with leopards, lions, tigers, rhinoceroses, and bears, among many other animals and birds. Transported on ships, they were kept for public exhibition and *venatios* – staged hunts in arenas.

Peacock fresco
Romans imported peacocks from India to be bred as creatures sacred to the Goddess Juno, exotic pets for the wealthy, and even as delicacies for banquets. This bright peacock on a fence is on a recovered fragment of a wall fresco (63BCE-79CE), probably from Pompeii, Italy.

Marine Life
The rich harvest of the sea around Naples is depicted with a gourmet's eye in this 1st-century CE mosaic recovered from the ruins of Pompeii, Italy. The tableau of plump fish, shellfish, and eels, with a life and death struggle between a lobster and an octopus as its centre, reads like a menu-board for the "House of the Faun" where it was found.

❝ What pleasure can a cultivated man find in seeing a noble beast run through with a hunting spear. ❞

CICERO, *AD FAMILIARES (LETTERS TO FRIENDS)*, 62–43 BCE

walking on the sea bed

Some ocean fish have given up swimming in open water, instead settling for life on the sea bed. For many, their fins have evolved accordingly, and rather than helping to control the body in mid-water, they are used for walking on the ocean floor. The paired pectoral and pelvic fins are more robust and help to support the fish's weight. They have become broader at the end so that they work more like feet than fins. In frogfish and related cousins of deep-sea anglerfish, the pectoral fins are angled like elbows to give extra flexibility.

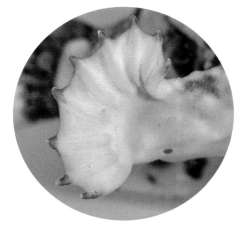

Handlike fin
The fingerlike bony rays of the frogfish's pectoral fin project beyond the webbing and help to improve traction on the sea bed.

Clown among the rocks
Perfectly camouflaged among the sponges at the base of a rocky coral reef, a clown frogfish (*Antennarius maculatus*) lacks the speed and agility of many open-water fish, but has the ability to clamber over the underwater terrain. It uses a flaglike lure to draw smaller fish towards its waiting jaws.

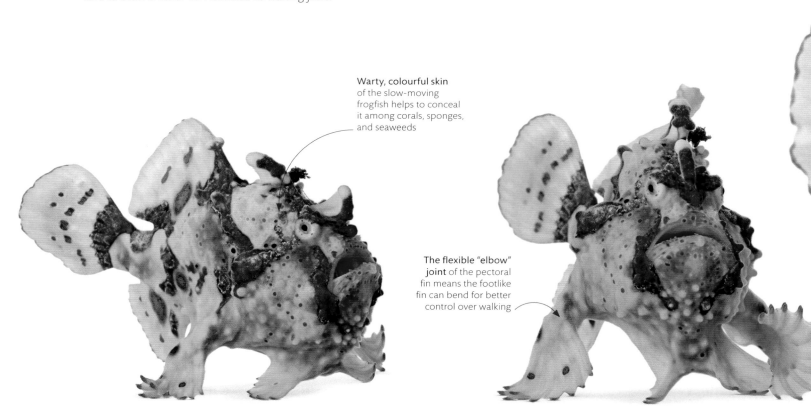

Warty, colourful skin of the slow-moving frogfish helps to conceal it among corals, sponges, and seaweeds

The flexible "elbow" joint of the pectoral fin means the footlike fin can bend for better control over walking

FINS AS LEGS

The fins of several groups of fish have evolved in a number of ways to help them to walk. The pelvic fins are located closer to the front of the body to stabilize the fish, while the pectoral fins have become elongated to become more leglike. The mudskipper, which spends much of its time out of the water, relies mainly on pectoral fins for thrust, and uses its suckerlike pelvic fins for stability. In frogfish, both pairs of fins are leglike, and it can use them together to maximize thrust when walking, more like a four-legged land animal.

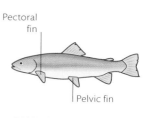

Pectoral fin

Pelvic fin

CONVENTIONAL FISH

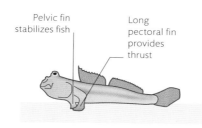

Pelvic fin stabilizes fish

Long pectoral fin provides thrust

MUDSKIPPER ON LAND

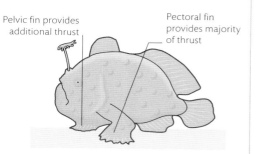

Pelvic fin provides additional thrust

Pectoral fin provides majority of thrust

FROGFISH WALKING UNDER WATER

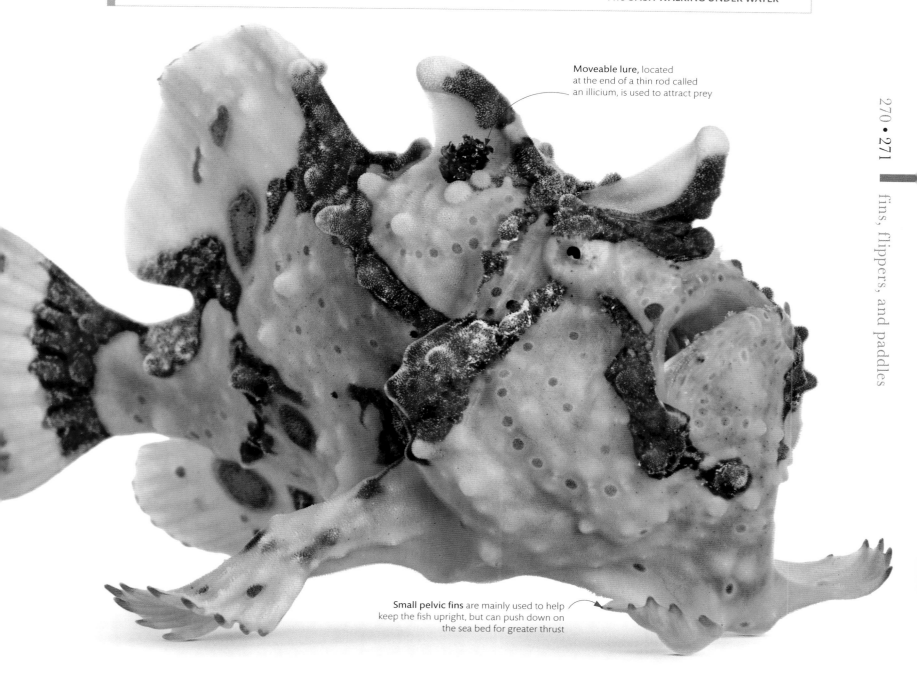

Moveable lure, located at the end of a thin rod called an illicium, is used to attract prey

Small pelvic fins are mainly used to help keep the fish upright, but can push down on the sea bed for greater thrust

returning to water

Reptiles – with hard, water-resistant skin – evolved on land, but many of their kind have returned to the underwater habitat of their ancestors. For turtles that live in the oceans, the transition is virtually complete: together with one species of freshwater river turtle, they are now the only reptiles with feet perfectly modified into flippers. They only venture onto land to lay their eggs.

Cruising the ocean
Like all saltwater turtles, the green turtle (*Chelonia mydas*) swims by flapping its flippers, which push against the water to propel the body forwards. Both upstroke and downstroke give propulsion, leaving the webbed hindfeet to work like a rudder for steering.

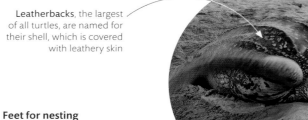

Leatherbacks, the largest of all turtles, are named for their shell, which is covered with leathery skin

Feet for nesting
Weighing more than half a tonne, a leatherback turtle (*Dermochelys coriacea*) uses its flippers to heave itself ashore and its back feet to dig a nest for its air-breathing eggs.

CONVERGENT EVOLUTION
Even though they are only distantly related, different kinds of swimming vertebrates have evolved flippers that match the hydrodynamic shapes of the fins of sharks and other fish. Turtles and dolphins evolved from walking ancestors, while penguin flippers are modified wings.

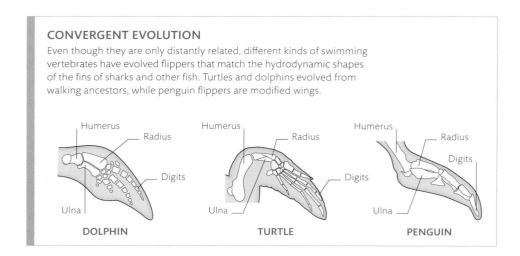

DOLPHIN   TURTLE   PENGUIN

(Humerus, Radius, Digits, Ulna)

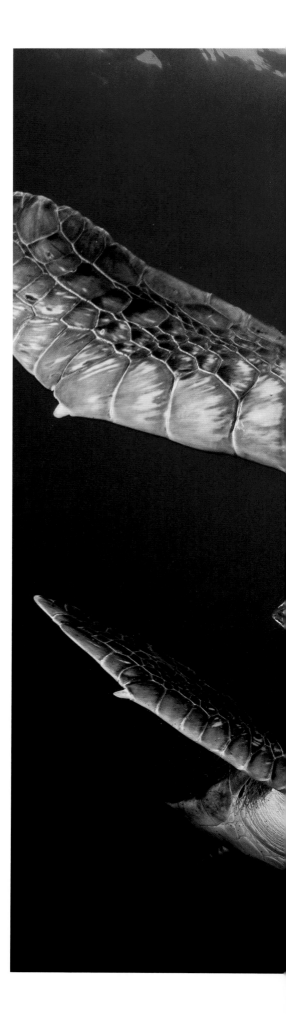

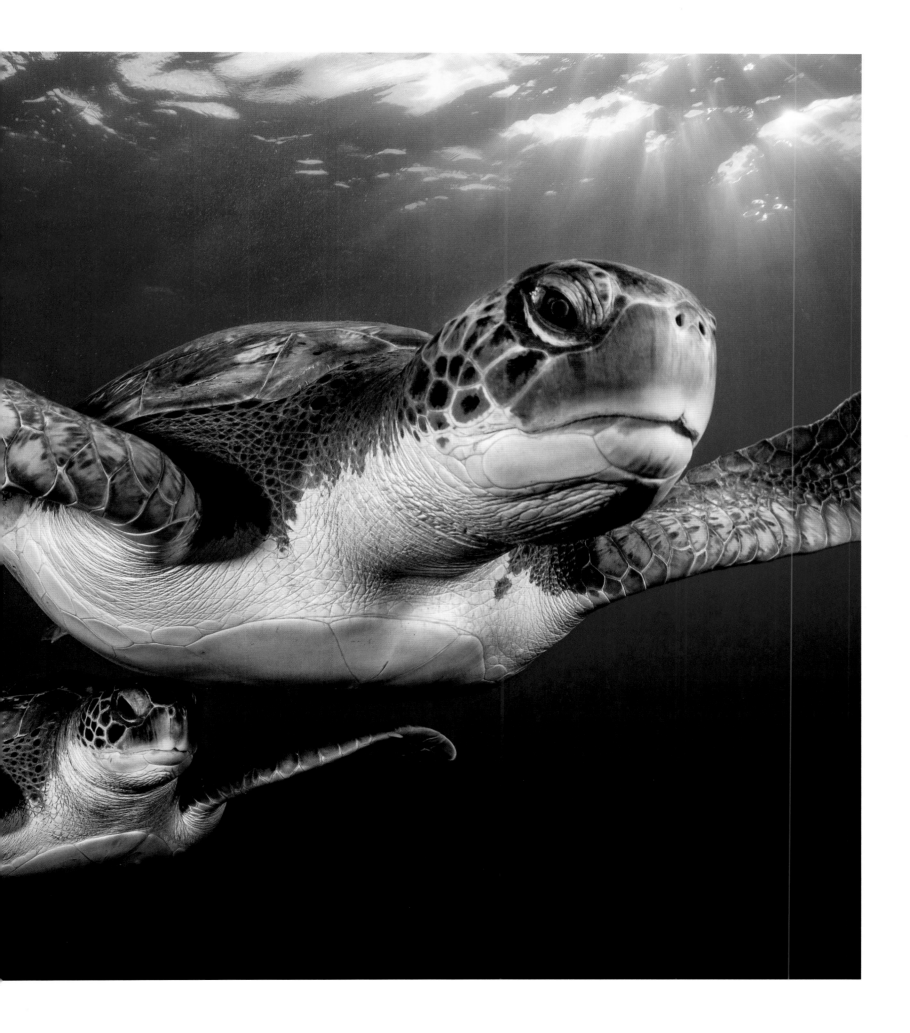

Extraordinary power
A cetacean's tail fluke contains no skeletal elements – the animal's vertebral column ends at its base. This massive boneless blade generates a powerful up-down sweep that pushes against the water to drive the whale's enormous body forwards.

tail flukes

No mammals are better adapted for life in water than dolphins and whales, together known as cetaceans. Their torpedo-shaped bodies are streamlined, which minimizes drag, and their front limbs, modified as flippers, help with stability. But it is their huge tail blade, or fluke, that provides their forward thrust. This blade is made from a solid mass of connective tissue, strengthened by parallel bundles of the tough protein, collagen.

HORIZONTAL TAILS

The manatee and dugong, together known as sea cows, or sirenians, are aquatic mammals that have evolved similar adaptations to those of cetaceans. They use the same horizontal, up-down tail propulsion, which contrasts with the side-to-side tail movement of fish. The dugong has a notched tail fluke similar to that of a cetacean, but the manatee has evolved a spatula-shaped tail paddle.

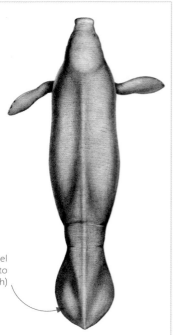

Flattened tail can propel the manatee at up to 15 mph (24 kph)

MANATEE *Trichechus* species

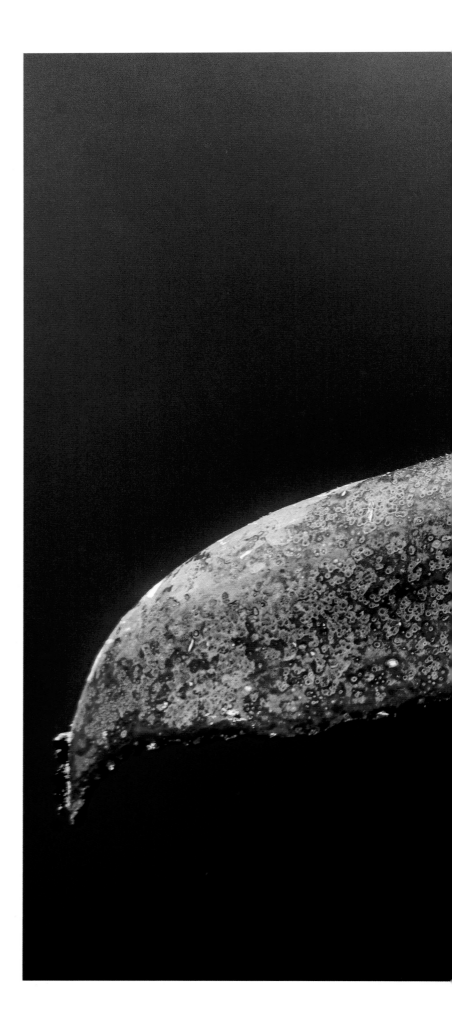

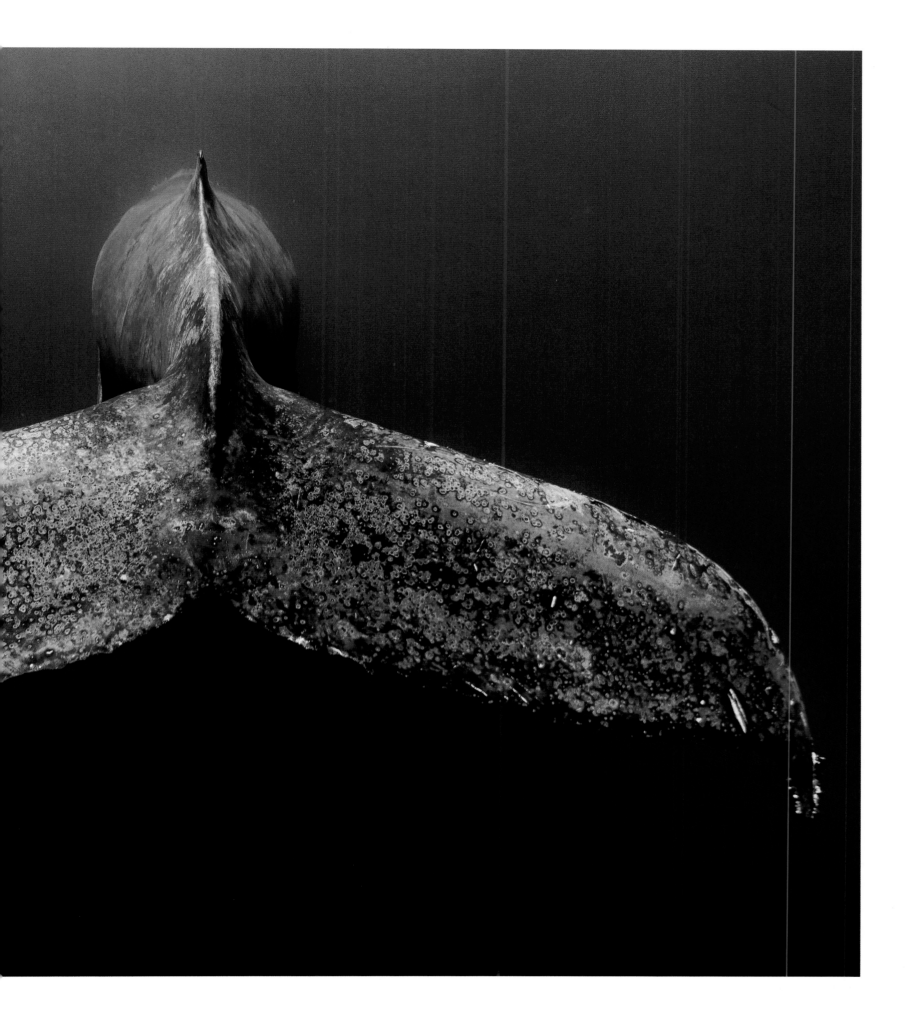

wings and parachutes

wing. any structure that generates lift as an aerofoil, including either of the two modified forelimbs in birds or bats, or one of the extensions of the thoracic cuticle in an insect, or one of the skin flaps of a colugo or flying squirrel.

parachute. any structure – including membranes and folds of skin between digits and limbs – that maximizes air resistance to slow an animal's fall.

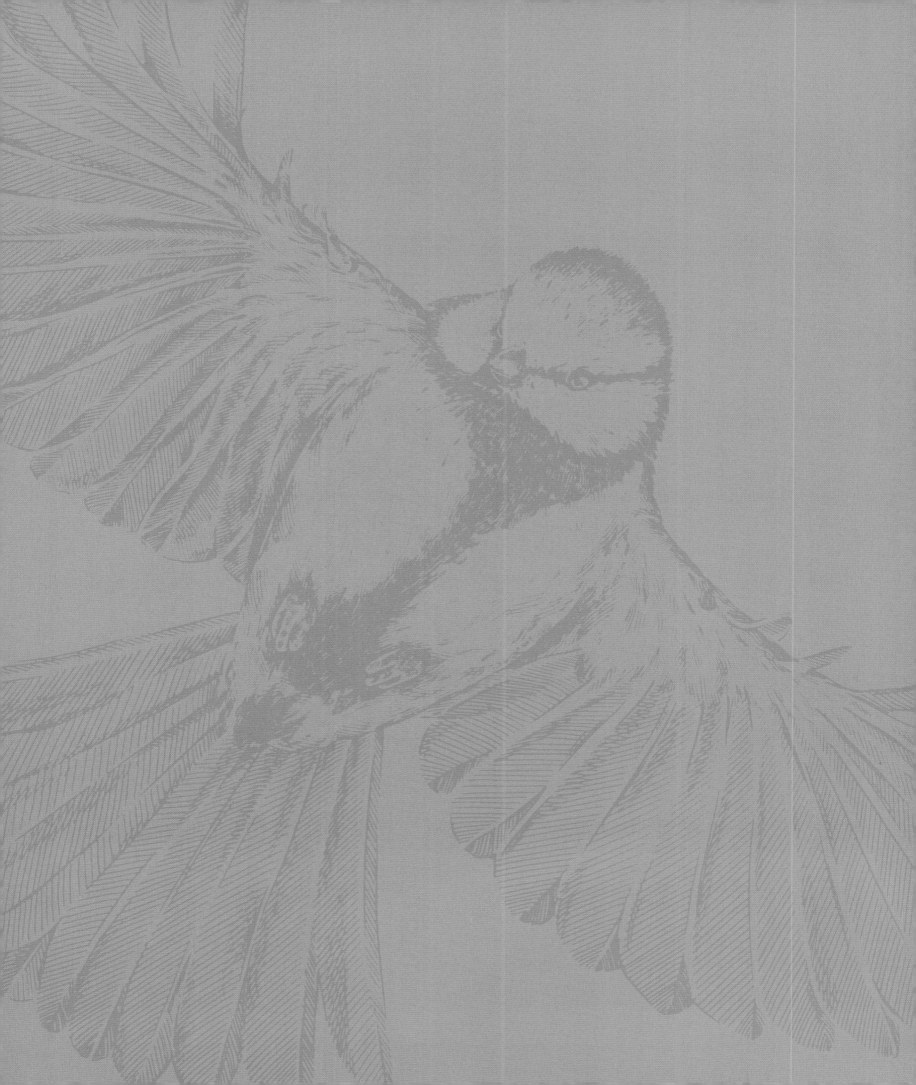

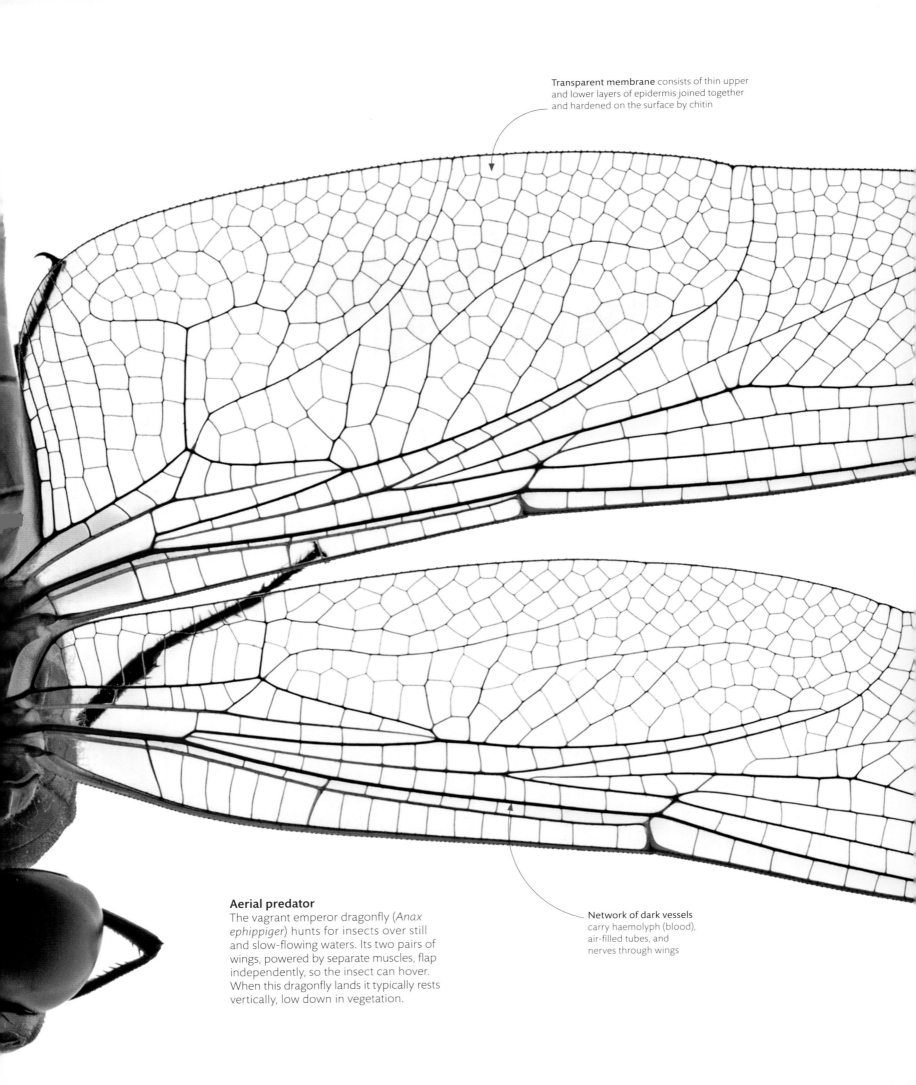

Transparent membrane consists of thin upper and lower layers of epidermis joined together and hardened on the surface by chitin

Aerial predator
The vagrant emperor dragonfly (*Anax ephippiger*) hunts for insects over still and slow-flowing waters. Its two pairs of wings, powered by separate muscles, flap independently, so the insect can hover. When this dragonfly lands it typically rests vertically, low down in vegetation.

Network of dark vessels carry haemolyph (blood), air-filled tubes, and nerves through wings

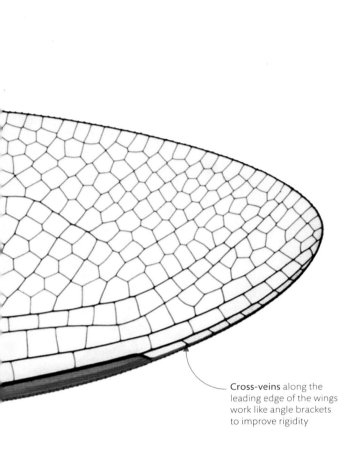

Cross-veins along the leading edge of the wings work like angle brackets to improve rigidity

Aerodynamics in miniature
When insect wings flap they create tiny vortices, or swirling eddies of air. These provide the lift that counteracts the gravitational pull of the animal's weight as seen in this sombre goldenring (*Cordulegaster bidentata*) dragonfly.

Dragonflies clamp their legs beneath their bodies when in flight

insect flight

Four hundred million years ago, insects became the first animals ever to take to the skies. Their flight was powered by flapping wings, and today they are still the only invertebrates capable of this. The challenges of becoming airborne and controlling mid-air manoeuvres were met by the emergence of wings that evolved from flaps of hard exoskeleton. Their wings are strong and lightweight, and worked by combinations of powerful muscles.

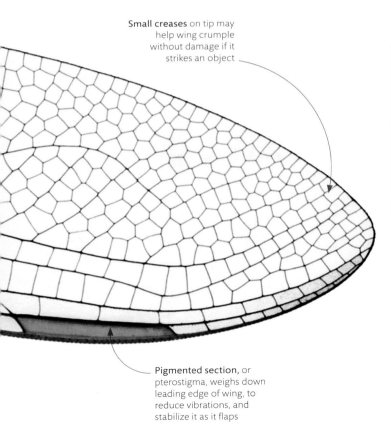

Small creases on tip may help wing crumple without damage if it strikes an object

Pigmented section, or pterostigma, weighs down leading edge of wing, to reduce vibrations, and stabilize it as it flaps

MUSCLES IN THE THORAX

The first insects moved their wings with one set of muscles attached directly to the wings that pulled them down, and another set of indirect muscles that pulled on the roof of the thorax to "pivot" the wings back up. Dragonflies and mayflies still do this. Later insects rely more completely on indirect muscles that deform the shape of the thorax to move wings up and down, which increases the speed of the wingbeat to hundreds of times per second in true flies and bees.

Thorax

Indirect muscle relaxes, so thorax can spring upward

Base of wing swings up and pivots, as blade is pulled down

Wing blade

DOWNSTROKE

Direct muscle contracts to pull wing down

Thorax

Indirect muscle contracts, pulling top of thorax downward

Base of wing swings down and pivots, so blade moves up

UPSTROKE

Direct muscle relaxes, allowing wing to rise

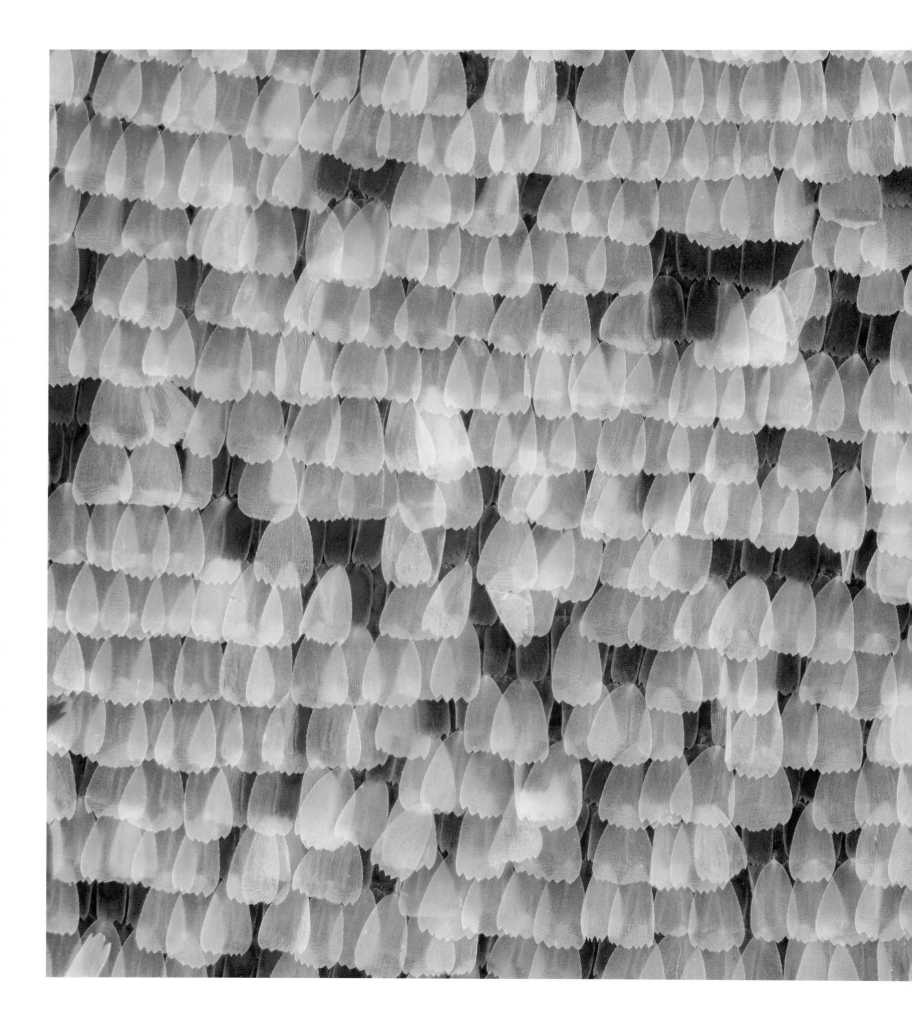

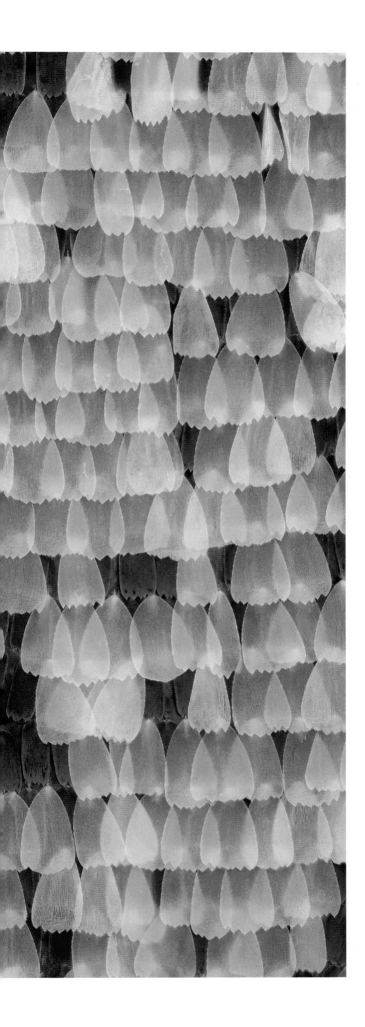

Trick of the light

The brilliant electric blue of the common blue morpho (*Morpho peleides*) butterfly is due to structure, not pigment. Tiny ridges on each scale – scarcely a thousandth of a millimetre apart – reflect light in such a way that interference between light rays cancels out all colours except blue, which is intensified.

COMMON BLUE MORPHO
Morpho peleides

Black edge of wing is due to a dark pigment, melanin, rather than light interference

scaly wings

Insects of the order Lepidoptera – butterflies and moths – are unmistakable in that their wings and bodies are covered in tiny scales so delicately attached that they rub off like dust. Under the microscope, these overlapping scales resemble miniature roofing tiles. They may trap air to improve lift or even help the insects slip from the webs of predatory spiders. But they are also responsible for stunning colour – whatever purpose this might serve.

MULTI-PURPOSE COLOUR

Butterflies, such as *Consul fabius* (top) and *Myscelia orsis* (centre left), and even some day-flying moths, such as *Ceretes thais* (centre right), are among the most colourful of lepidopterans and may flaunt their patterns in courtship or to repel predators. But colours can be also used in camouflage: the underwing of *C. fabius* (bottom) mimics a dead leaf, making the insect almost invisible when it closes its wings.

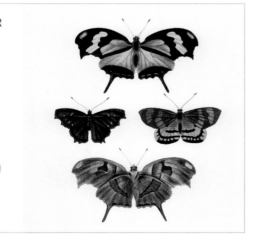

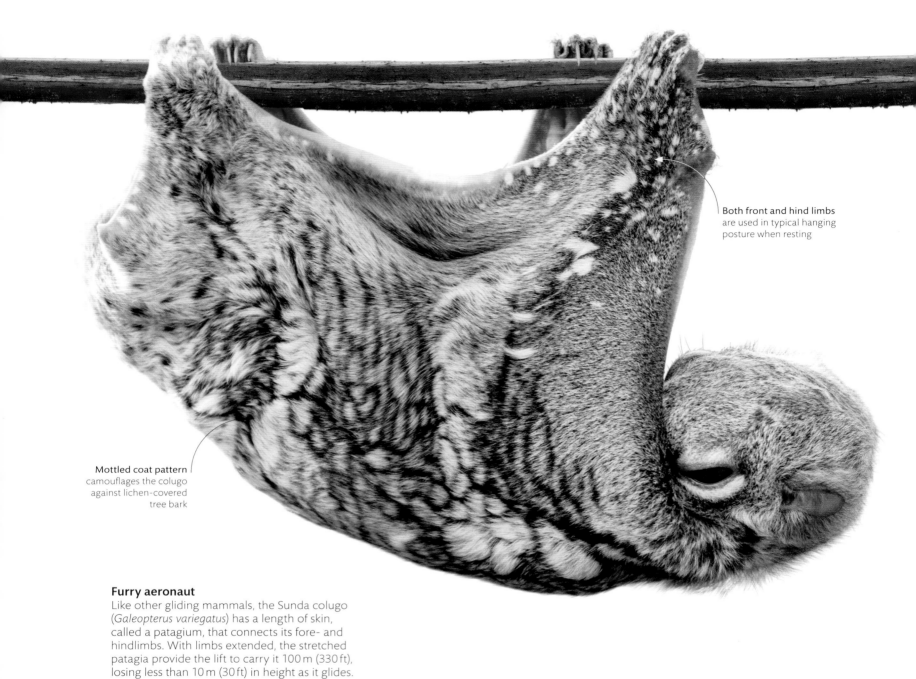

Both front and hind limbs are used in typical hanging posture when resting

Mottled coat pattern camouflages the colugo against lichen-covered tree bark

Furry aeronaut
Like other gliding mammals, the Sunda colugo (*Galeopterus variegatus*) has a length of skin, called a patagium, that connects its fore- and hindlimbs. With limbs extended, the stretched patagia provide the lift to carry it 100 m (330 ft), losing less than 10 m (30 ft) in height as it glides.

gliding and parachuting

Every major group of vertebrates – from fish to mammals – has some species that can glide through the air. This is unsurprising because it is a remarkably efficient way to move: as long as the animal can launch itself, it relies only on lift provided by its aerodynamic shape (rather than muscle-powered thrust) to cover a distance. Gliding minimizes drag during flight, but in contrast, parachuting maximizes drag. By turning a wing into a parachute, animals can reduce impact speed for a safe landing.

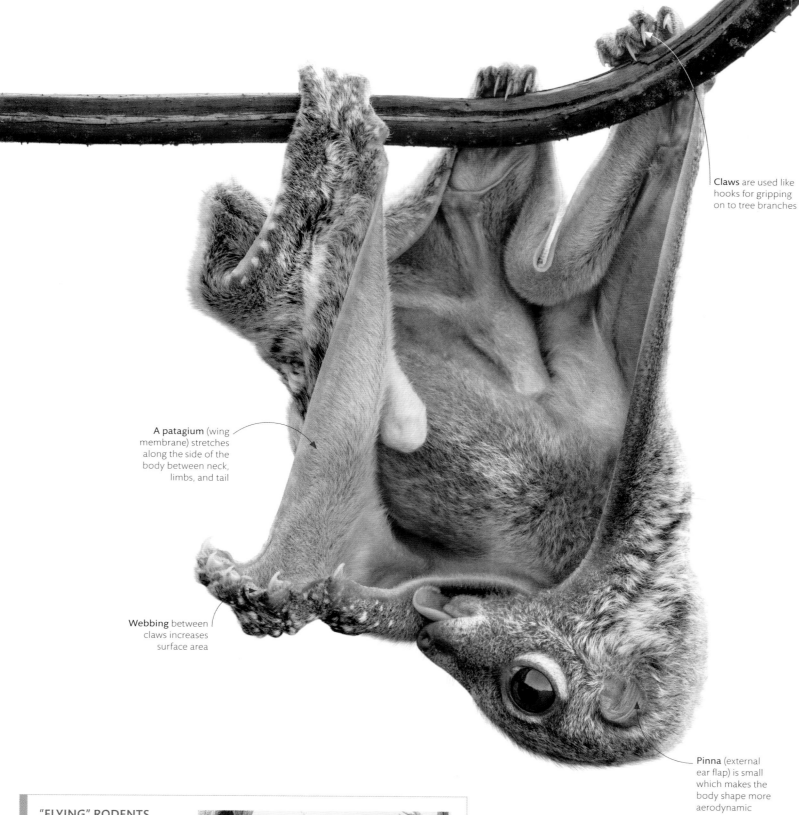

Claws are used like hooks for gripping on to tree branches

A patagium (wing membrane) stretches along the side of the body between neck, limbs, and tail

Webbing between claws increases surface area

Pinna (external ear flap) is small which makes the body shape more aerodynamic

"FLYING" RODENTS

Gliding with patagia has evolved independently in several groups of mammals, including among marsupial possums and squirrels. The patagia of flying squirrels stretch from wrist to ankle. They can change direction mid-flight where necessary.

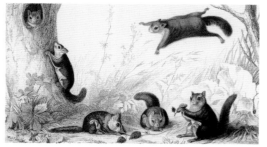

FLYING SQUIRRELS

how birds fly

For an animal to fly, it needs to overcome the pull of gravity and lift itself into the air, as well as propel itself forwards. No vertebrate group has as many flying species as birds. When birds evolved from upright, feathered dinosaurs, their muscle-powered forelimbs became modified into wings that gave both the lift and the propulsion necessary to conquer the skies. With fingers reduced, their arm bones provide the wings' frames, while stiff-bladed flight feathers – rooted firmly in the bones – form the aerodynamic surfaces.

Aerial predator
In normal flight, the common kestrel (*Falco tinnunculus*) has a cruising speed of about 32 kph (20 mph). Like other falcons, the kestrel has long, pointed wings that allow high-speed flight, occasional soaring, and also the ability to hover against the wind while scanning the ground below for small mammal prey.

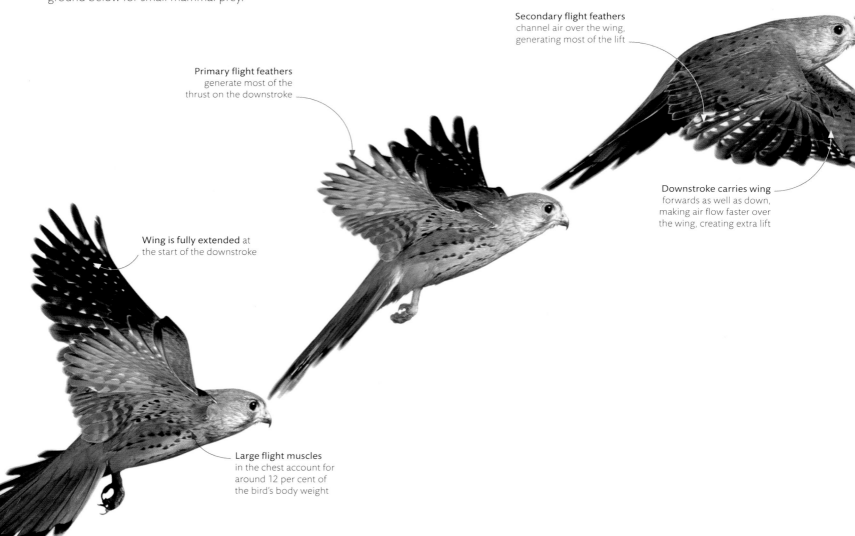

Secondary flight feathers channel air over the wing, generating most of the lift

Primary flight feathers generate most of the thrust on the downstroke

Downstroke carries wing forwards as well as down, making air flow faster over the wing, creating extra lift

Wing is fully extended at the start of the downstroke

Large flight muscles in the chest account for around 12 per cent of the bird's body weight

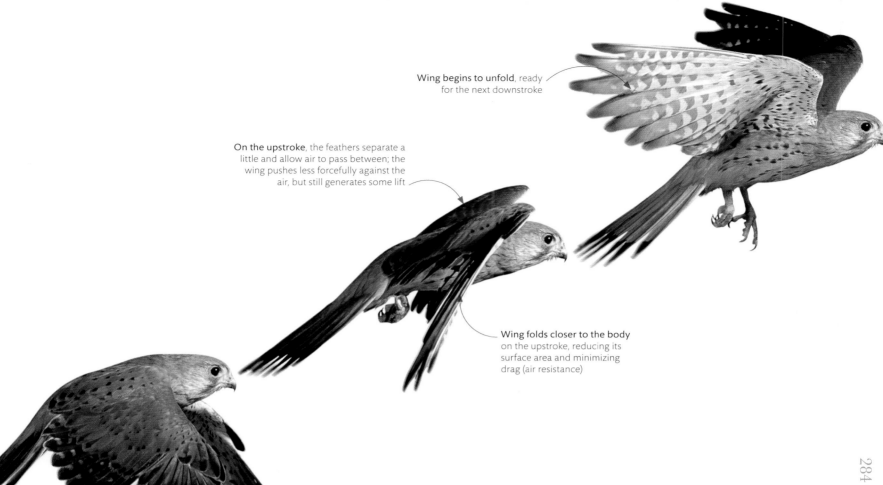

Wing begins to unfold, ready for the next downstroke

On the upstroke, the feathers separate a little and allow air to pass between; the wing pushes less forcefully against the air, but still generates some lift

Wing folds closer to the body on the upstroke, reducing its surface area and minimizing drag (air resistance)

Outer wing is stiff on the downstroke and pushes against the air, due to the way the feathers overlap, with the inner edge of each underneath the next feather

WINGS IN ACTION

Powerful chest muscles, connected to a strongly keeled sternum (breastbone), contract to flap the wings. This action gives the thrust needed to move forwards. In most birds, the downstroke provides the propulsion. Lift is generated by the wings' shape – their convex upper surface causes air to flow faster and with lower pressure over the top of the wings than beneath, pulling the body upwards.

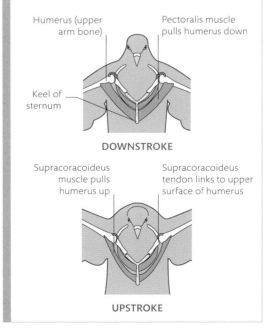

Humerus (upper arm bone)

Pectoralis muscle pulls humerus down

Keel of sternum

DOWNSTROKE

Supracoracoideus muscle pulls humerus up

Supracoracoideus tendon links to upper surface of humerus

UPSTROKE

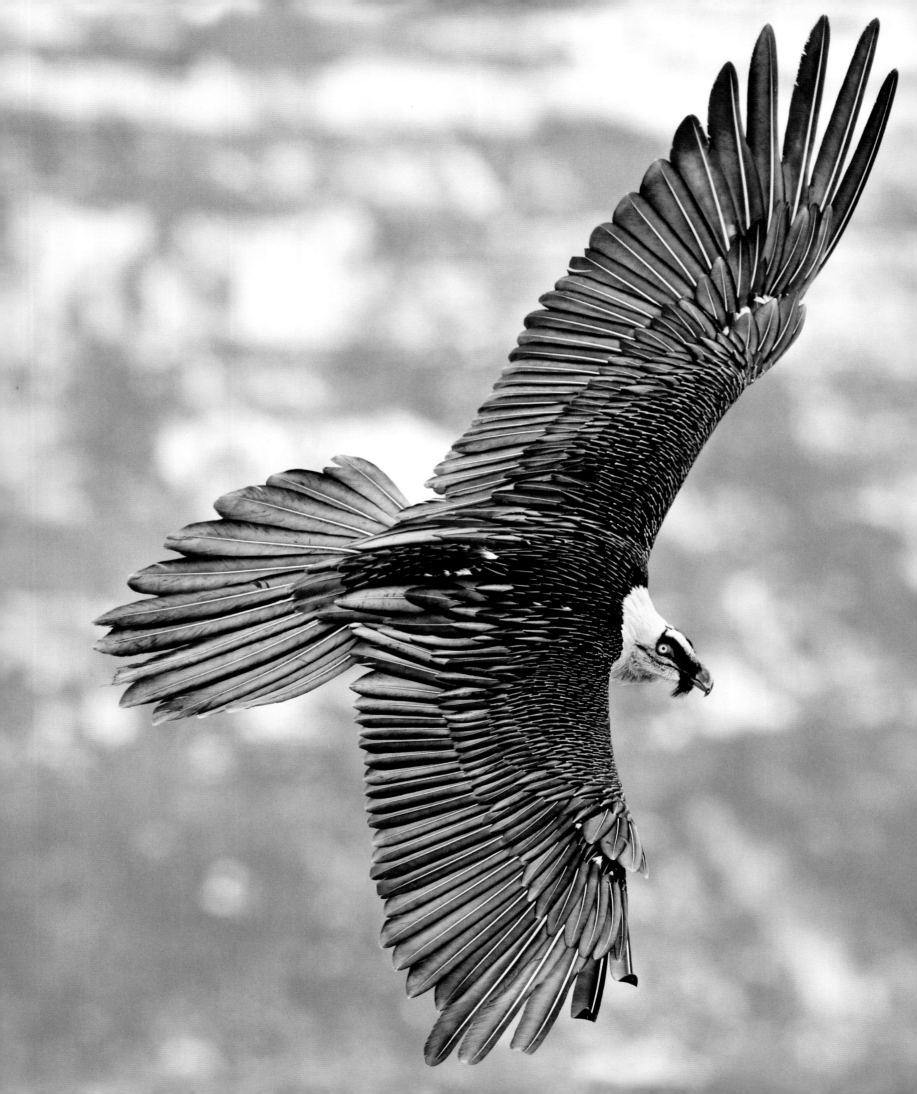

lammergeier

As long as a five-year-old child is tall, and weighing twice as much as many a domestic cat, the lammergeier (*Gypaetus barbatus*) is impressive when seen on the ground. In the air, this vulture is even more striking: a superb glider, it soars for hours over high, rugged terrain in search of bones to eat.

A mountain dweller of Europe, Asia, and Africa, the lammergeier, or bearded vulture, roosts on cliffs, usually above 1,000 m (3,300 ft) and possibly even higher than 5,000 m (16,400 ft) in Nepal.

Lammergeiers are the only vertebrates whose diet consists almost exclusively of bones (up to 85 per cent). While facing little competition for this unusual dietary staple, they must range far and wide to find enough bones to sustain themselves; some birds have been known to travel as far as 700 km (435 miles) in a single day. For this reason, lammergeiers tend to be

Life on the wing
Adult lammergeiers spend up to 80 per cent of daylight hours in soaring flight, scouring the land below for food. Once airborne, they glide on mountain winds at speeds of 20–77 kph (12–48 mph).

sparsely distributed and patrol vast territories that in some locations cover hundreds of square kilometres.

Their huge wingspan – up to 3 m (10 ft) in the slightly larger females – enables the birds to soar with barely a wingbeat, hitching rides on rising air currents. By scanning the terrain from on high or skimming over the ground, they can spot the carcasses of mountain goats or wild sheep on clifftops and in remote canyons.

While other vultures will gorge in a feast-or-famine strategy, lammergeiers are steady consumers, swallowing about 8 per cent of their body weight – roughly 465 g (1 lb) – each day in bones. Small bones are swallowed whole; larger ones are carried aloft and dropped from 50 to 80 m (165 to 260 ft) onto rocks to shatter them into manageable pieces. Acidic gastric fluids dissolve these skeletal meals in 24 hours.

Red-necked raptor
A lammergeier habitually stains its plumage a rusty red by bathing in soil or water rich in iron oxide. A greater degree of staining between individuals of a similar age may signifiy more dominant birds.

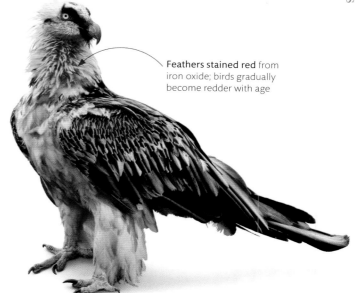

Feathers stained red from iron oxide; birds gradually become redder with age

Slotted high-lift wings

Deep notches between primary feathers on a broad wing shape provides greater lift, not only allowing birds to gain altitude with less effort but also to land in more confined spaces. Typical of raptors such as hawks and eagles, and soaring birds such as vultures, this shape is also found in swans and larger waders.

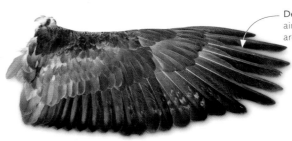

Deep slots reduce air turbulence around wingtips

GOLDEN EAGLE
Aquila chrysaetos

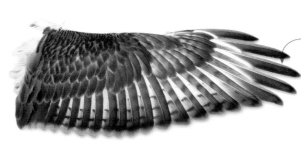

Long wing remains broad even at the tip

FERRUGINOUS HAWK
Buteo regalis

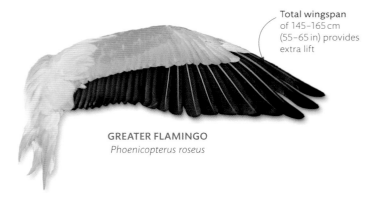

Total wingspan of 145–165 cm (55–65 in) provides extra lift

GREATER FLAMINGO
Phoenicopterus roseus

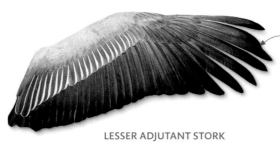

Feathers at tip become fingerlike extensions

LESSER ADJUTANT STORK
Leptoptilos javanicus

High-speed wings

Thin wings, tapering to a point produce the high-speed, ultra-manoeuvrable flights of aerial feeders such as swallows, swifts, and martins, and stealth hunters such as falcons. They are also seen in ducks and shorebird species, which, while not known for aerodynamic displays, use rapid wingbeats to move at high speed in level flight.

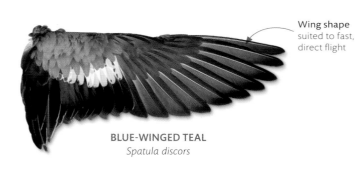

Wing shape suited to fast, direct flight

BLUE-WINGED TEAL
Spatula discors

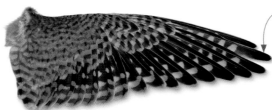

Pointed wingtip enables the high-speed "stoop", or dive, of this falcon

AMERICAN KESTREL
Falco sparverius

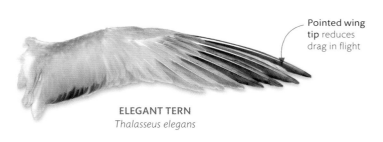

Pointed wing tip reduces drag in flight

ELEGANT TERN
Thalasseus elegans

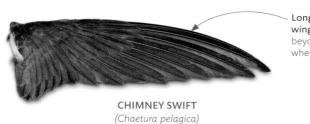

Long, curved wing extends beyond tail when folded

CHIMNEY SWIFT
(Chaetura pelagica)

Elliptical wings

Good for quick takeoffs and short bursts of speed, the elliptical wing shape of passerines (perching birds such as sparrows and thrushes) and game birds provides excellent manoeuvrability in dense, brushy habitats, but at a high energy cost. Although many passerines still achieve long-distance migrations, game birds cannot sustain long flights.

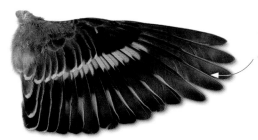

Classic elliptical shape aids manoeuvrability in thrush's dense forest habitat

TOWNSEND'S SOLITAIRE
Myadestes townsendi

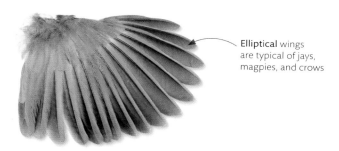

Elliptical wings are typical of jays, magpies, and crows

GREEN JAY
Cyanocorax luxuosus

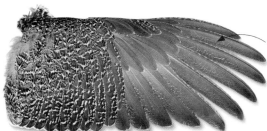

Short, rounded wing shape enables quick takeoff

GREATER SAGE GROUSE
Centrocercus urophasianus

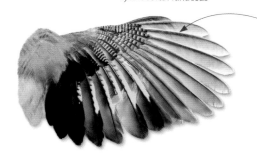

Distinctive wing pattern makes bird conspicuous in flight

EURASIAN JAY
Garrulus glandarius

High-aspect ratio wings

Just like the pole used by a human tightrope walker, long, narrow wings give a bird more stability. This wing shape also produces less drag, which means birds can fly for longer using less energy. High-aspect ratio wings are responsible for the long, soaring flights of sea birds such as albatrosses, gulls, gannets, and petrels.

Wing tips are black above and white below

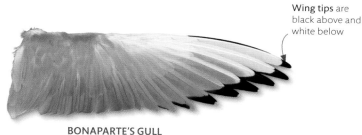

BONAPARTE'S GULL
Chroicocephalus philadelphia

Huge 2.1 m (7 ft) wing span allows bird to soar for many hours without flapping

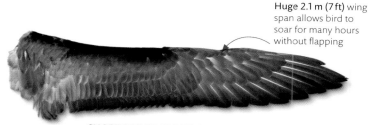

SHORT-TAILED ALBATROSS
Phoebastria albatrus

bird wings

Wing shape is an indicator of how a bird flies, and whether or not it is a predator or prey species. Apart from specialized wings, such as those of hummingbirds (see pp.292–93), whose shifting feathers provide the control needed to hover, most birds' wings can be grouped into four basic shapes, each related to a range of flight speeds, styles, and distances.

emperor penguin

At up to 46 kg (101 lb) in weight and 1.35 m (4.4 ft) tall, Antarctica's emperor penguin (*Aptenodytes forsteri*) is the world's largest penguin. Although flightless like other penguins, a streamlined body and wings modified into flippers gives this species diving abilities that are unparalleled among birds.

In abandoning flight, emperor penguins have also lost buoyancy. Their bones are denser than those of flying birds, making their bodies slightly heavier than water. Their stiffened wings, originally used to stay airborne, now provide thrust in the denser medium of water. The only flexible wing joint connects the humerus (upper arm) to the shoulder. But the birds' wing muscles retain similar power to those of flying birds, so they can flap vigorously to dive deep in search of food. The feet, used for steering, are set so far back that on land the penguin must stand vertically.

With a heavier body, less energy is needed to stay submerged, allowing dives to be longer and deeper. As a result, each foraging expedition results in a bigger net energy gain. Emperor penguins catch shrimplike krill under the ice floes, but the bulk of their diet is fish and squid, often caught at depths of 200 m (660 ft) or more. A typical dives lasts 3 to 6 minutes, but dive times of 20 to 30 minutes have been recorded, and emperors are capable of reaching as deep as 565 m (1,850 ft).

Body fat built up by the summer catch sustains the emperors through the harsh Antarctic winter, when the birds breed. Having laid a single egg, the female treks across the ice to feed out at sea. The male rests the egg on his feet and incubates it under a fold of warm abdominal skin, in temperatures as low as −62°C (−80°F); he does not eat at all during this time. When the chick hatches, he feeds it on nutritious "milk" secreted by his oesophagus. After the female returns, the roles reverse: she cares for the chick, while he sets off for the ocean – and his first meal in four months.

Flying with flippers
Bubbles stream off the waterproof coats of these diving emperors. Once under water, they flap their wings through such a wide arc that the tips nearly touch on the upstroke.

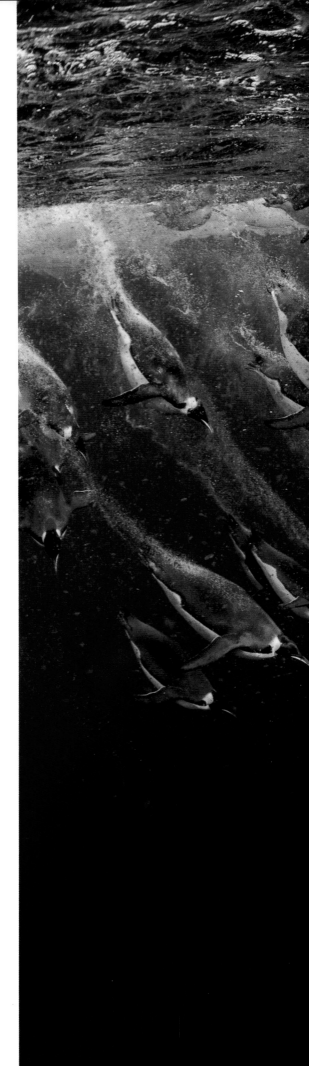

DIVING WINGS

Like penguins, auks (such as razorbills, puffins, and guillemots) use wing-propulsion to help them dive. Penguins' short, rigid wings only move at the base, but auks have larger, more flexible wings that limit their diving depth but permit them to get airborne.

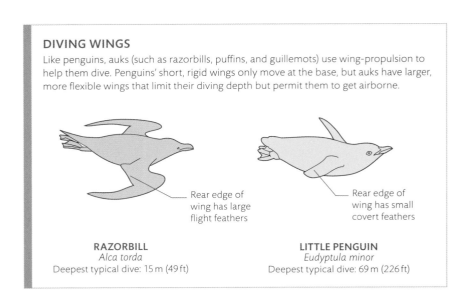

Rear edge of wing has large flight feathers

Rear edge of wing has small covert feathers

RAZORBILL
Alca torda
Deepest typical dive: 15 m (49 ft)

LITTLE PENGUIN
Eudyptula minor
Deepest typical dive: 69 m (226 ft)

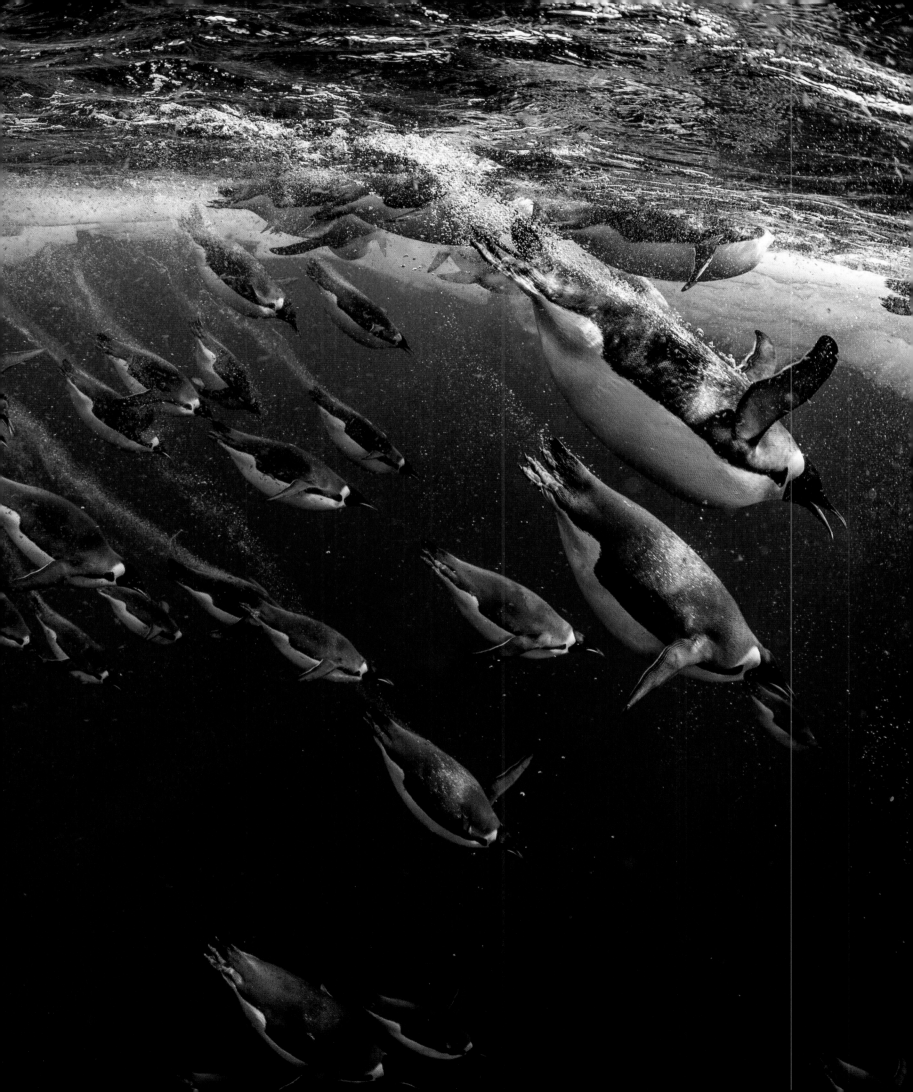

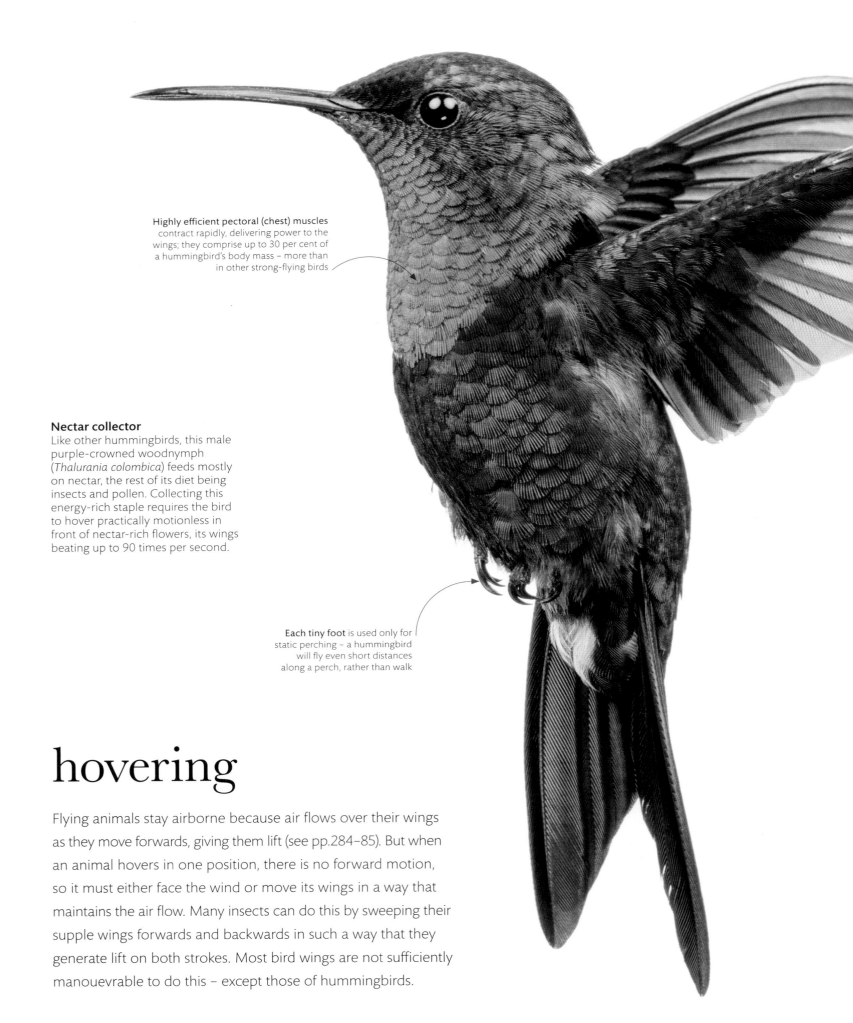

Highly efficient pectoral (chest) muscles contract rapidly, delivering power to the wings; they comprise up to 30 per cent of a hummingbird's body mass – more than in other strong-flying birds

Nectar collector
Like other hummingbirds, this male purple-crowned woodnymph (*Thalurania colombica*) feeds mostly on nectar, the rest of its diet being insects and pollen. Collecting this energy-rich staple requires the bird to hover practically motionless in front of nectar-rich flowers, its wings beating up to 90 times per second.

Each tiny foot is used only for static perching – a hummingbird will fly even short distances along a perch, rather than walk

hovering

Flying animals stay airborne because air flows over their wings as they move forwards, giving them lift (see pp.284–85). But when an animal hovers in one position, there is no forward motion, so it must either face the wind or move its wings in a way that maintains the air flow. Many insects can do this by sweeping their supple wings forwards and backwards in such a way that they generate lift on both strokes. Most bird wings are not sufficiently manouevrable to do this – except those of hummingbirds.

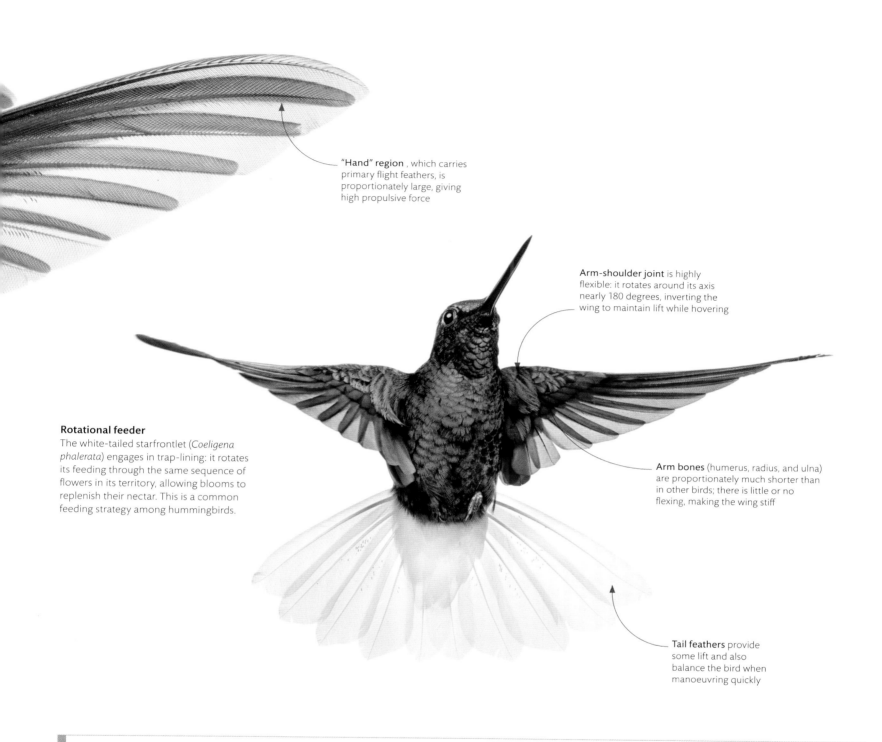

"Hand" region , which carries primary flight feathers, is proportionately large, giving high propulsive force

Arm-shoulder joint is highly flexible: it rotates around its axis nearly 180 degrees, inverting the wing to maintain lift while hovering

Rotational feeder
The white-tailed starfrontlet (*Coeligena phalerata*) engages in trap-lining: it rotates its feeding through the same sequence of flowers in its territory, allowing blooms to replenish their nectar. This is a common feeding strategy among hummingbirds.

Arm bones (humerus, radius, and ulna) are proportionately much shorter than in other birds; there is little or no flexing, making the wing stiff

Tail feathers provide some lift and also balance the bird when manoeuvring quickly

---

HOW HOVERING WORKS

Hummingbird wings do not fold on the upstroke, like those of other birds, but stay open and flip upside-down. As a result, air flowing over the surface of the wings produces lift on both the forward "downstroke" and the backward "upstroke". Thrust is vertical, not horizontal, so it steadies the bird against gravity. The wings' figure-of-eight path helps overcome the momentum of the downstroke and swiftly reverses the wings' direction.

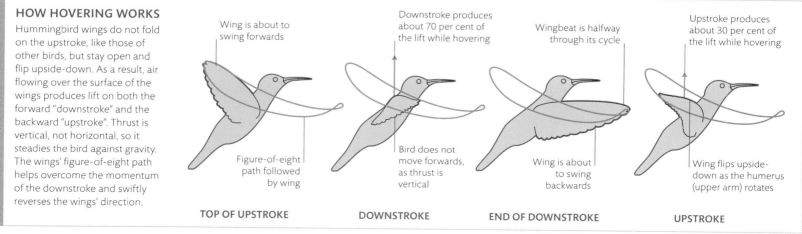

Wing is about to swing forwards

Figure-of-eight path followed by wing

TOP OF UPSTROKE

Downstroke produces about 70 per cent of the lift while hovering

Bird does not move forwards, as thrust is vertical

DOWNSTROKE

Wingbeat is halfway through its cycle

Wing is about to swing backwards

END OF DOWNSTROKE

Upstroke produces about 30 per cent of the lift while hovering

Wing flips upside-down as the humerus (upper arm) rotates

UPSTROKE

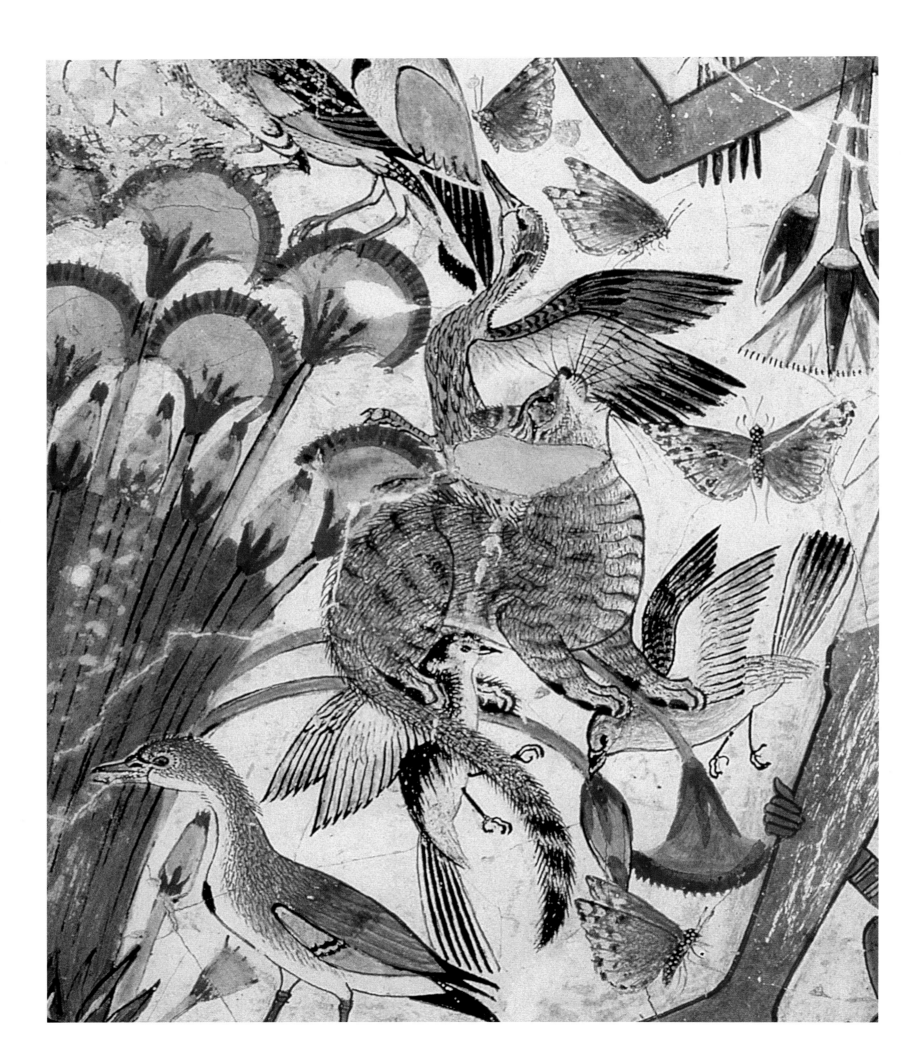

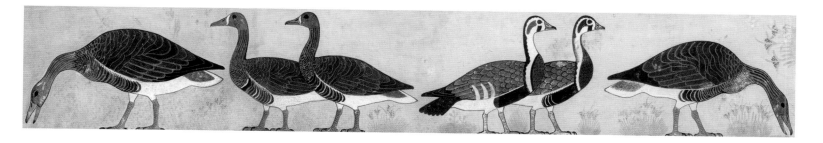

Meidum geese (4th Dynasty)
An Old Kingdom masterpiece in inscribed plaster and paint depicts three species: white-fronted, bean, and red-breasted geese. It was found in the tomb-chapel of Atet , wife of Prince Nefermaat, alongside Pharaoh Sneferu's pyramid at Meidum.

animals in art

egyptian birdlife

In ancient Egypt, the banks of the Nile teemed with bird life. Homes were built close to the water, and residents were faithful observers, replicating bird shapes in religious icons and hieroglyphs and endowing their gods with avian powers and characteristics. The value of birds as a source of food and religious inspiration continued into the afterlife, where the deceased could choose to take on one of their many forms.

Egyptian art forms abound with reflections of the natural world, and birds enjoyed a particular reverence. The rich variety from sky-borne falcons, swallows, kites, and owls to waterside herons, cranes, and ibis is reflected in Egyptian script: up to 70 different species are incorporated in hieroglyphs.

Gods were apportioned with bird traits that enhanced their powers; Horus, ruler of the sky, is depicted with a falcon head because of the dizzying height of his flight. With one eye representing the sun and the other the moon, his journey across the sky mimicks the path of the sun from dawn to dusk. Thoth, the god of magic, wisdom, and the moon, bears an Ibis head with a curved beak like a crescent moon.

Tomb paintings reveal the afterlife as a land of plenty. Using funerary spells, the deceased could transform into a chosen avian or take the form of the Ba, a human-headed bird that flees the tomb each night. In the 18th-dynasty tomb of Nebamum, the scribe is shown restored to human form to hunt game in fertile marshes and take stock of the domestic fowl provided for his eternity.

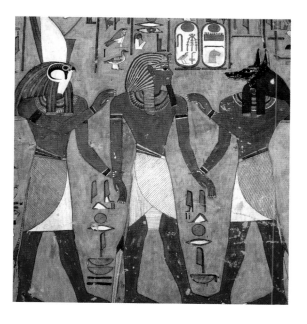

Animal deities (19th Dynasty 1306–1304 BCE)
In a tomb painting, falcon-headed Horus, son of Isis, god of the sky, welcomes Pharaoh Ramses I to the afterlife, assisted by the jackal-headed god of funerals and death, Anubis.

Nebamum's cat (18th dynasty c.1350 BCE)
An ecstatic tawny cat hunting birds is captured in a detail from the fresco *Nebamum Hunting in Marshes* found in the tomb of an Egyptian scribe. Cats were often symbols of Bastet, the goddess of fertility and pregnancy. The gilded eye on Nebamum's cat hints at religious significance.

❝ As a falcon, I live in the light. My crown and my radiance have given me power. ❞

CHAPTER 78, *THE BOOK OF THE DEAD*

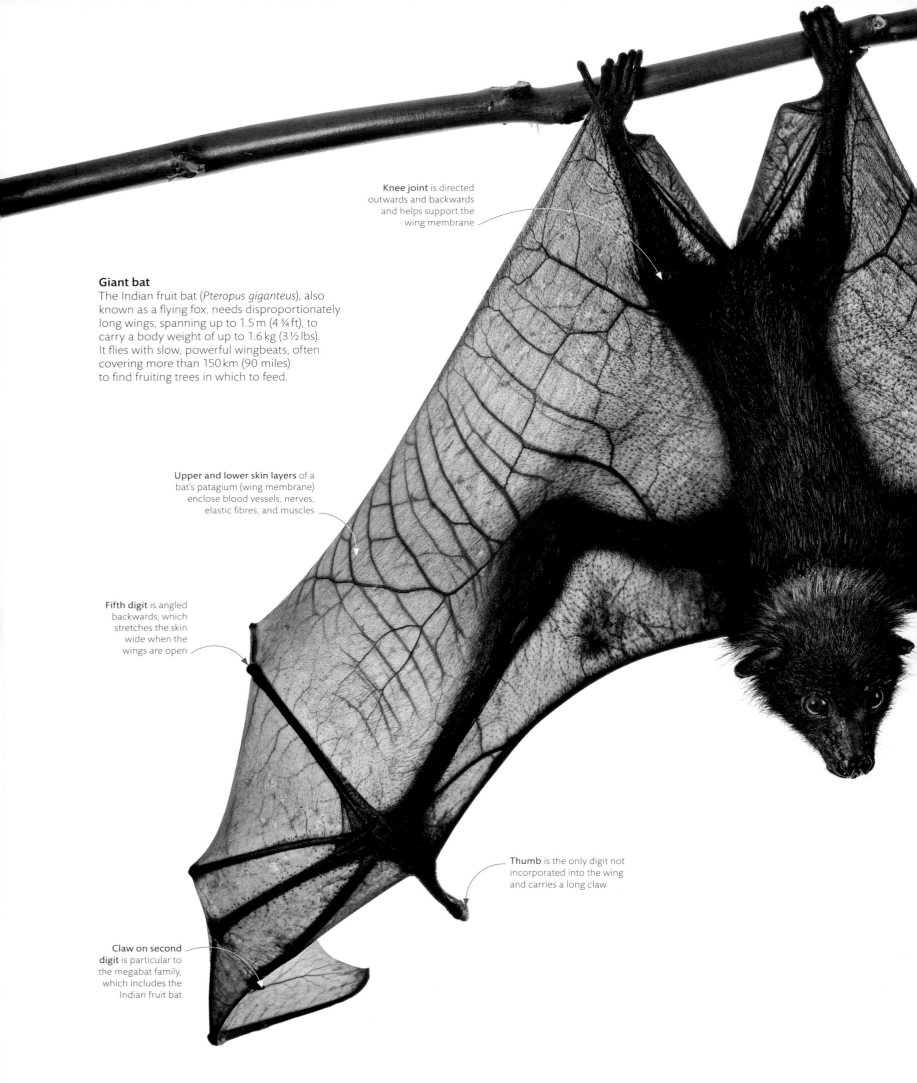

Giant bat

The Indian fruit bat (*Pteropus giganteus*), also known as a flying fox, needs disproportionately long wings, spanning up to 1.5 m (4 ¾ ft), to carry a body weight of up to 1.6 kg (3 ½ lbs). It flies with slow, powerful wingbeats, often covering more than 150 km (90 miles) to find fruiting trees in which to feed.

Knee joint is directed outwards and backwards and helps support the wing membrane

Upper and lower skin layers of a bat's patagium (wing membrane) enclose blood vessels, nerves, elastic fibres, and muscles

Fifth digit is angled backwards, which stretches the skin wide when the wings are open

Thumb is the only digit not incorporated into the wing and carries a long claw

Claw on second digit is particular to the megabat family, which includes the Indian fruit bat

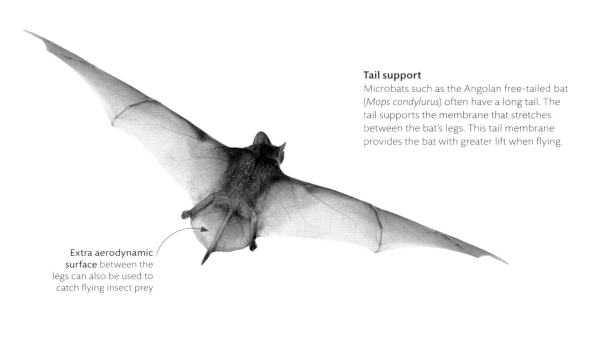

Tail support
Microbats such as the Angolan free-tailed bat (*Mops condylurus*) often have a long tail. The tail supports the membrane that stretches between the bat's legs. This tail membrane provides the bat with greater lift when flying.

Extra aerodynamic surface between the legs can also be used to catch flying insect prey

wings of skin

Like birds, bats have powered flight; they beat their wings to generate the thrust necessary to move them through the air. However, instead of stiff flight feathers (see pp.122–23) providing the aerodynamic surface, bats have a skin membrane, stretched between elongated finger bones, which reaches back to their feet. Wings of living skin are more adjustable and sensitive to air around them, and mean bats are able to manoeuvre tightly while catching flying insects or foraging for fruit and nectar.

Third digit is usually the longest and reaches to the wingtip

WING SHAPES

The shape of a wing is described by its aspect ratio: the proportion of its length to its width. Short, wide wings (low aspect ratio) occur in bats whose manoeuvrability and precision are important – such as in dense forest – whereas long, narrow wings (high aspect ratio) provide faster sustained flight high above the ground.

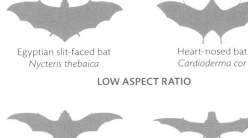

Egyptian slit-faced bat
Nycteris thebaica

Heart-nosed bat
Cardioderma cor

LOW ASPECT RATIO

Pel's pouched bat
Saccolaimus peli

Midas' free-tailed bat
Mops midas

HIGH ASPECT RATIO

eggs and offspring

egg. (1) the sex cell of a female animal before it is fertilized to become an embryo; (2) a protective package laid by a female animal, containing an embryo, along with a food supply and environment that supports its development.

offspring. the young of an animal.

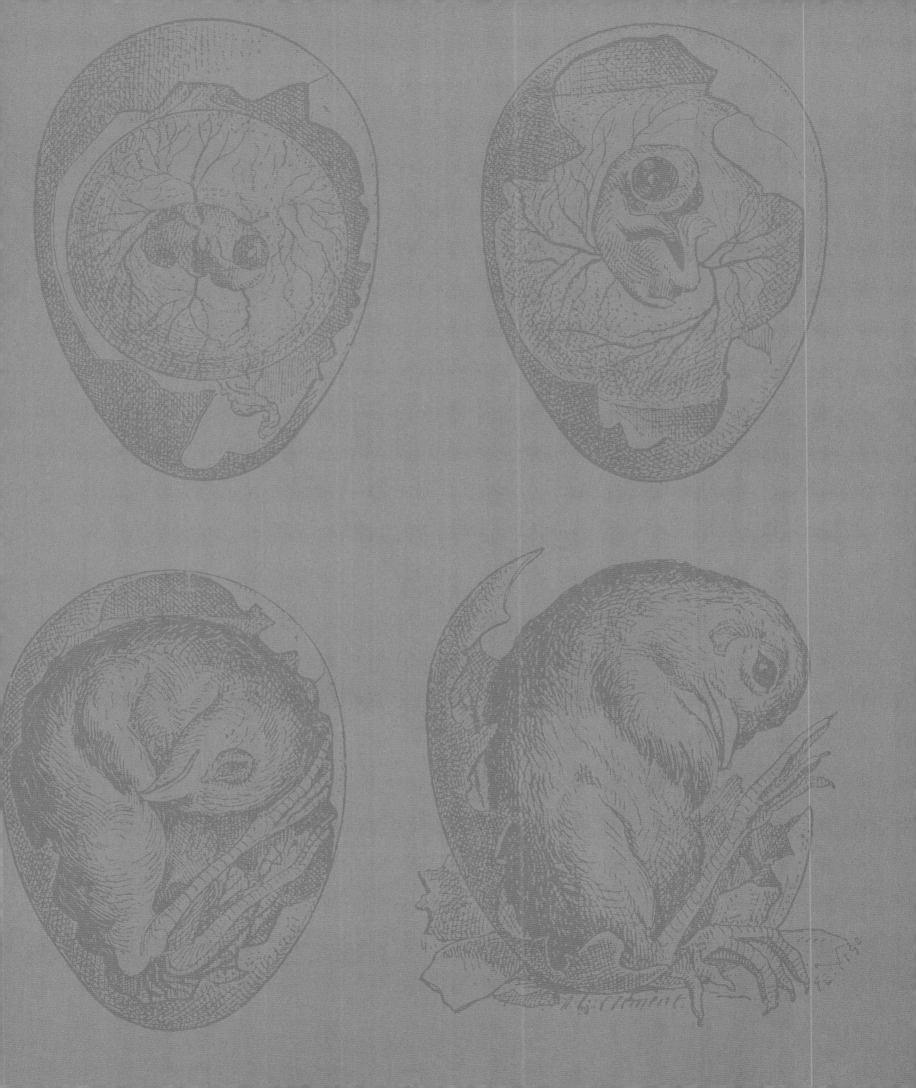

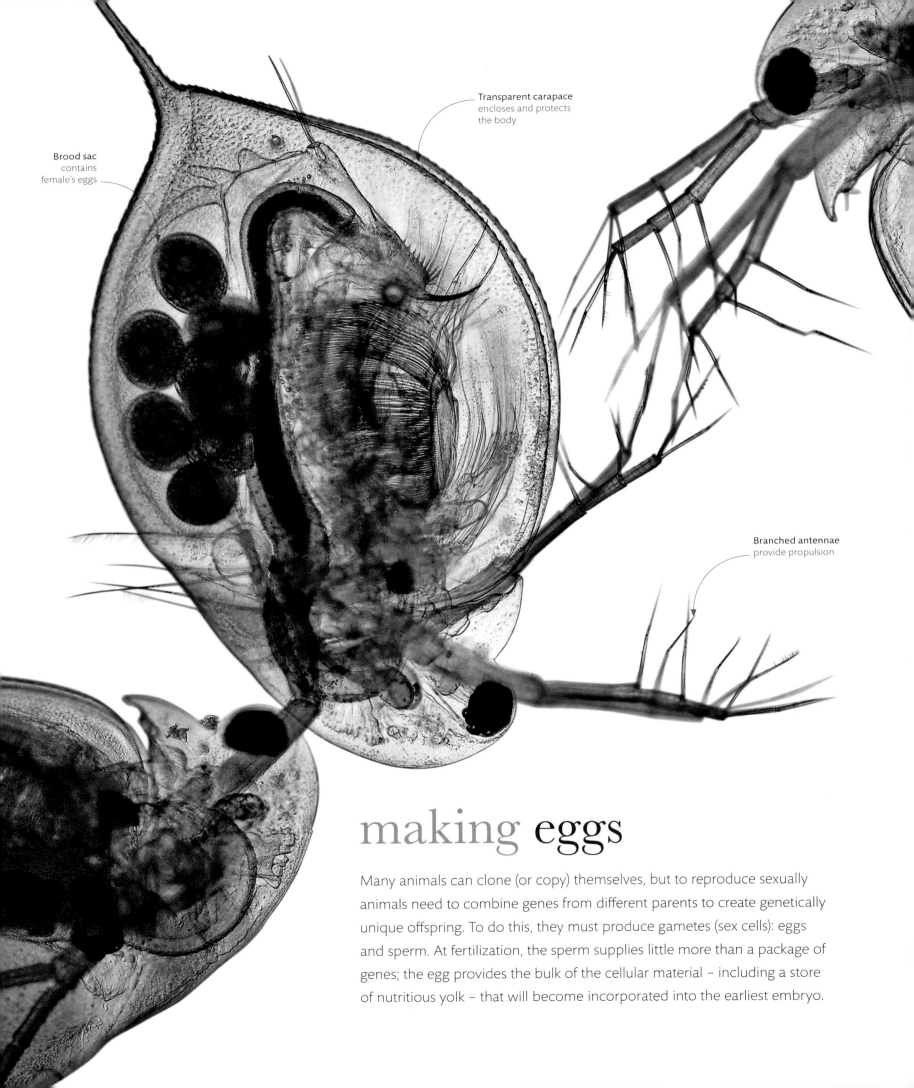

Brood sac
contains
female's eggs

Transparent carapace
encloses and protects
the body

Branched antennae
provide propulsion

making eggs

Many animals can clone (or copy) themselves, but to reproduce sexually
animals need to combine genes from different parents to create genetically
unique offspring. To do this, they must produce gametes (sex cells): eggs
and sperm. At fertilization, the sperm supplies little more than a package of
genes; the egg provides the bulk of the cellular material – including a store
of nutritious yolk – that will become incorporated into the earliest embryo.

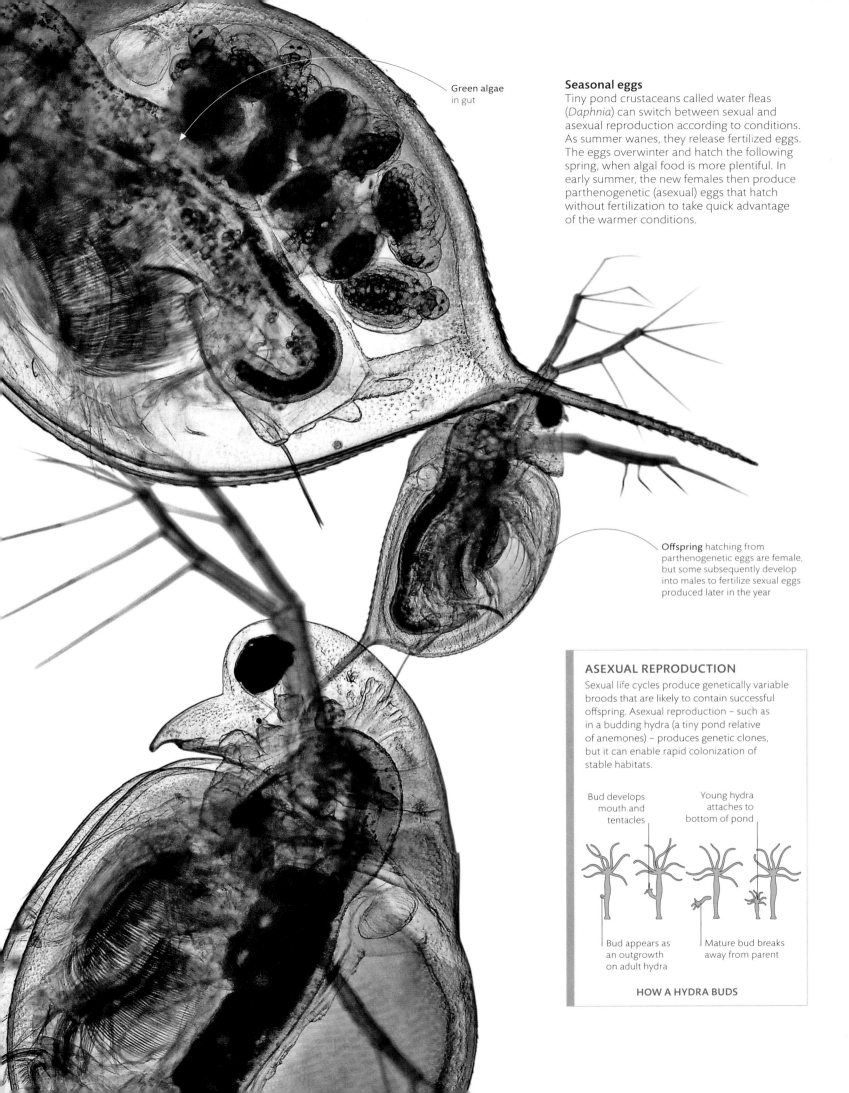

Green algae
in gut

Seasonal eggs
Tiny pond crustaceans called water fleas
(*Daphnia*) can switch between sexual and
asexual reproduction according to conditions.
As summer wanes, they release fertilized eggs.
The eggs overwinter and hatch the following
spring, when algal food is more plentiful. In
early summer, the new females then produce
parthenogenetic (asexual) eggs that hatch
without fertilization to take quick advantage
of the warmer conditions.

Offspring hatching from
parthenogenetic eggs are female,
but some subsequently develop
into males to fertilize sexual eggs
produced later in the year

ASEXUAL REPRODUCTION

Sexual life cycles produce genetically variable
broods that are likely to contain successful
offspring. Asexual reproduction – such as
in a budding hydra (a tiny pond relative
of anemones) – produces genetic clones,
but it can enable rapid colonization of
stable habitats.

Bud develops
mouth and
tentacles

Young hydra
attaches to
bottom of pond

Bud appears as
an outgrowth
on adult hydra

Mature bud breaks
away from parent

HOW A HYDRA BUDS

Spawning frogs
Almost all amphibians fertilize their eggs externally, and the majority of frogs maximize their chances of success by a behaviour called amplexus. The male grasps the female, usually from behind, to bring their reproductive openings close together. Sperm and eggs are then released simultaneously.

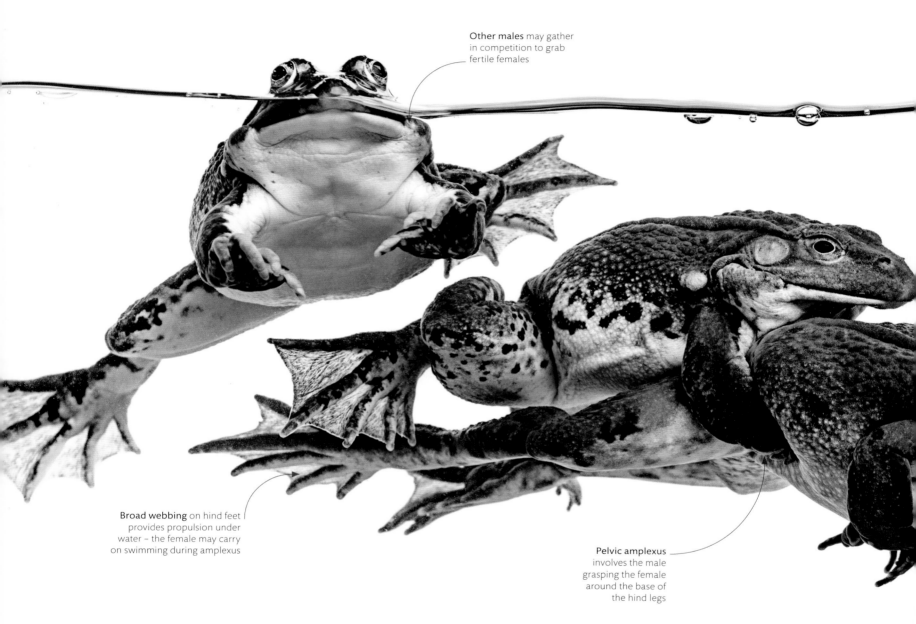

Other males may gather in competition to grab fertile females

Broad webbing on hind feet provides propulsion under water – the female may carry on swimming during amplexus

Pelvic amplexus involves the male grasping the female around the base of the hind legs

fertilization

In sexual reproduction, genetic variety is created when DNA from different individuals mixes as a sperm fertilizes an egg. Animals employ a variety of ways to ensure this happens as effectively as possible: many that spawn (release eggs and sperm) in water simply produce as many sex cells as possible to increase the chance of sperm meeting egg; in other animals, females retain fewer eggs inside the body and the act of copulation achieves fertilization internally.

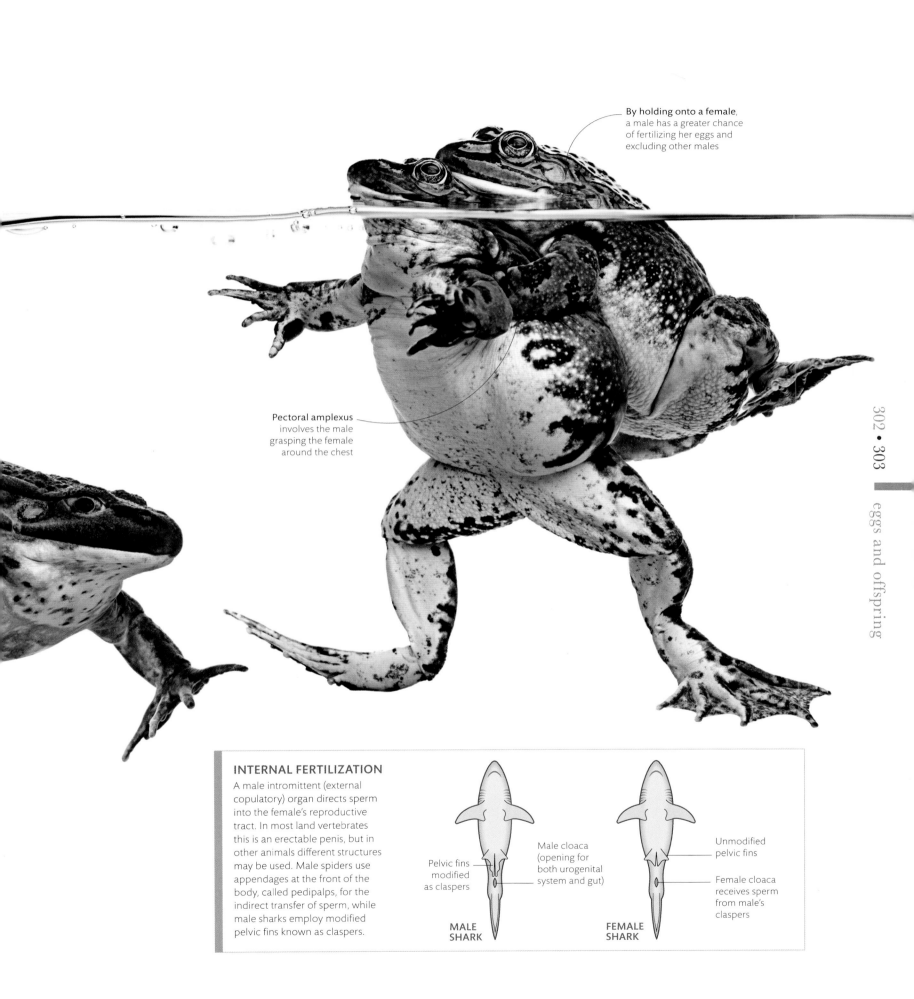

By holding onto a female, a male has a greater chance of fertilizing her eggs and excluding other males

Pectoral amplexus involves the male grasping the female around the chest

INTERNAL FERTILIZATION

A male intromittent (external copulatory) organ directs sperm into the female's reproductive tract. In most land vertebrates this is an erectable penis, but in other animals different structures may be used. Male spiders use appendages at the front of the body, called pedipalps, for the indirect transfer of sperm, while male sharks employ modified pelvic fins known as claspers.

Pelvic fins modified as claspers

Male cloaca (opening for both urogenital system and gut)

Unmodified pelvic fins

Female cloaca receives sperm from male's claspers

MALE SHARK

FEMALE SHARK

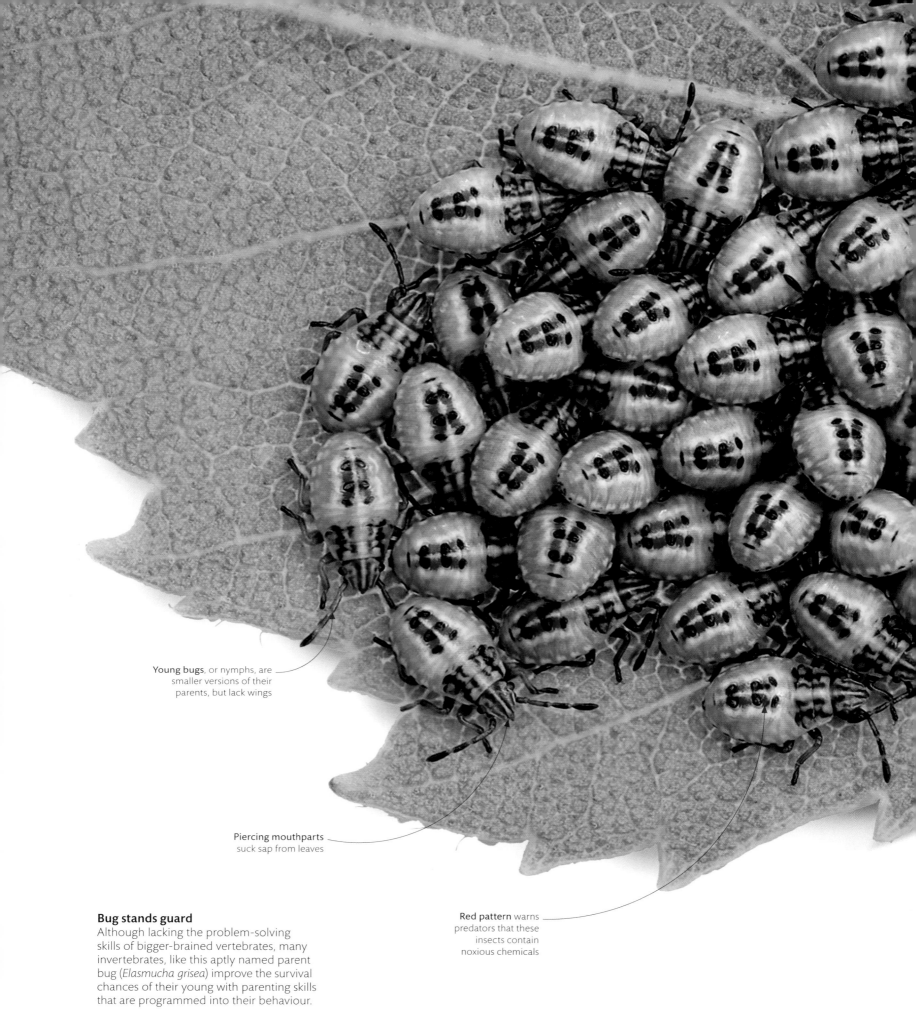

Young bugs, or nymphs, are smaller versions of their parents, but lack wings

Piercing mouthparts suck sap from leaves

Bug stands guard
Although lacking the problem-solving skills of bigger-brained vertebrates, many invertebrates, like this aptly named parent bug (*Elasmucha grisea*) improve the survival chances of their young with parenting skills that are programmed into their behaviour.

Red pattern warns predators that these insects contain noxious chemicals

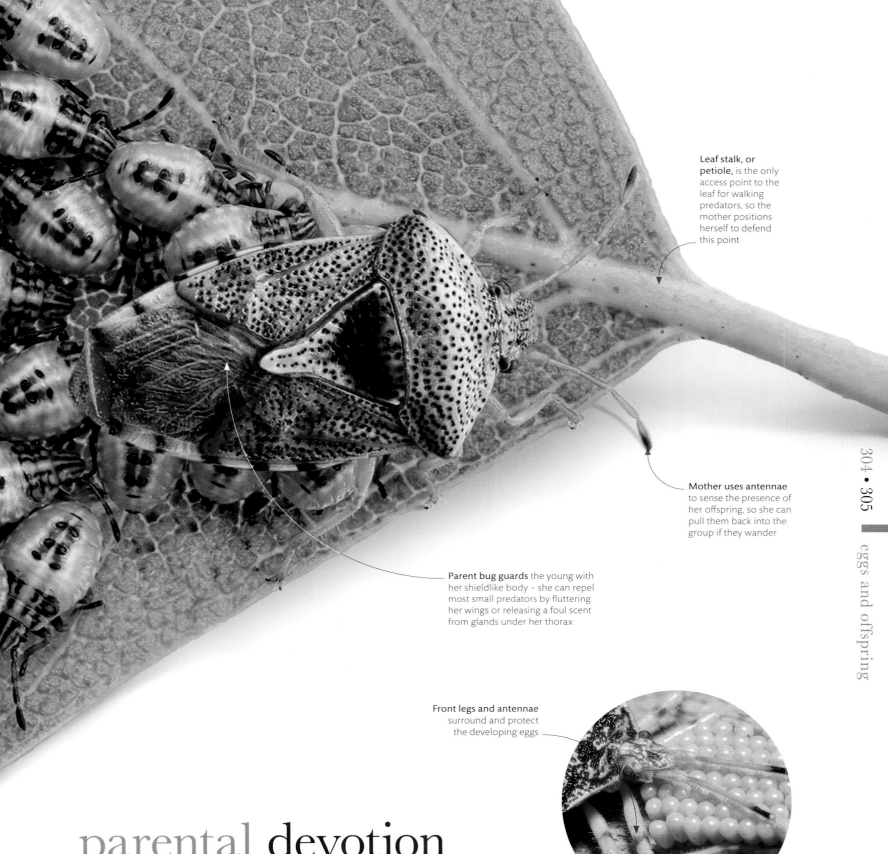

Leaf stalk, or petiole, is the only access point to the leaf for walking predators, so the mother positions herself to defend this point

Mother uses antennae to sense the presence of her offspring, so she can pull them back into the group if they wander

Parent bug guards the young with her shieldlike body – she can repel most small predators by fluttering her wings or releasing a foul scent from glands under her thorax

Front legs and antennae surround and protect the developing eggs

Waiting to hatch
Brazilian stink bug (*Antiteuchis* sp.) mothers - like the related parent bug - may guard over their eggs to prevent them being attacked by parasitic wasps.

parental devotion

All parents invest time and energy into reproduction – even if this is just in producing eggs and sperm. But some animals go further and take care of their offspring. This is not without risk as time spent parenting can also mean going without food or being exposed to danger, but their tended young are more likely to survive and grow into adulthood.

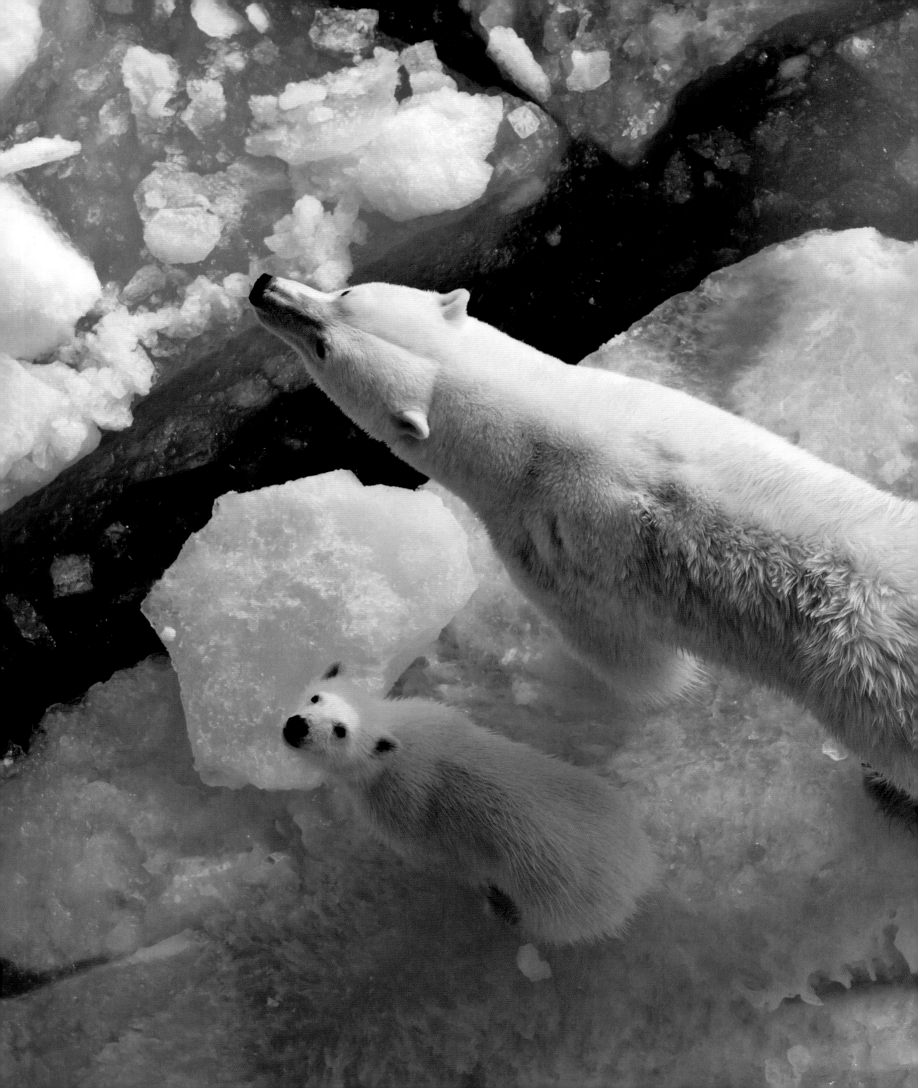

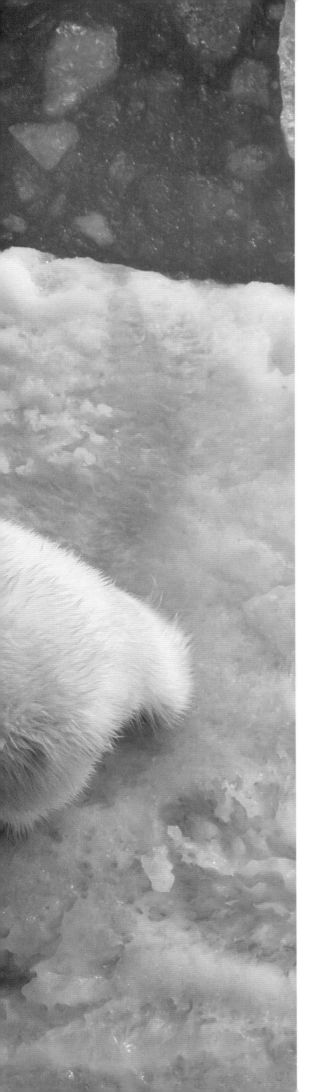

Twin cubs
Three-month-old cubs shelter in their maternity den. With sufficient food, the mother bear may successfully raise both of them, but it will be more than a year before they are independent.

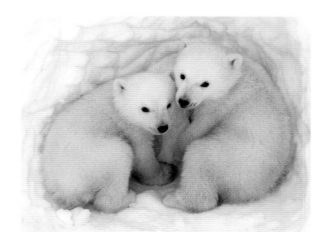

spotlight species

polar bear

The polar bear (*Ursus maritimus*) is the top predator in the frozen Arctic, where temperatures can dip below -50°C (-58°F), but the species is highly vulnerable in a modern age of global warming. To survive the winter, the cubs rely on a devoted mother, a maternity den, and plenty of calorific milk.

Polar bears are the largest of the bear species. Genetic evidence suggests it took just 200,000 years for them to evolve from their brown bear ancestors into bigger, whiter, more carnivorous animals, better attuned to the Arctic seasons. Fattened by gorging on fat-rich seals all summer, polar bears can generate more body heat. Their thick coat of translucent hairs, left hollow by lost pigment, traps extra warm air near the skin. This exceptional insulation is critical as most polar bears stay active even in the coldest months – it is only the pregnant females that hibernate in winter.

Survival on the ice
Fractured floating sea ice is the best hunting ground for a polar bear and her cub as, beneath their feet, is a good supply of food in the form of seal meat. The mother can ambush seals as they surface for air in the gaps in the ice.

Summer mating triggers ovulation, but as in other bears, the fertilized eggs do not implant in the womb until autumn, by which time the polar bear has settled down for the winter in a snow cave or underground peat bank that will serve as a maternity den. The tiny cubs – typically two, each scarcely bigger than a guinea pig – are born between November and January and do not emerge from the den until the harshest months are over.

The mother suckles her cubs with milk enriched with seal fat from the previous summer, but while they feed, she fasts. When the family emerges in spring, she may not have eaten for 8 months, so must refuel on seal meat. As the ice retreats under the summer sun, polar bears hunt closer to shore or migrate north. But each year rising temperatures shrink the icy platforms they depend on, so these bears face an uncertain future.

shelled eggs

The first backboned animals evolved on land more than 350 million years ago, but many remained tied to moist habitats because they laid soft eggs that dehydrated in air and that typically hatched into swimming larvae. Reptiles and birds escaped this restriction by producing hard-shelled eggs. Inside, the embryo is nurtured in a pool of fluid until it is ready to hatch into a miniature version of its air-breathing parents.

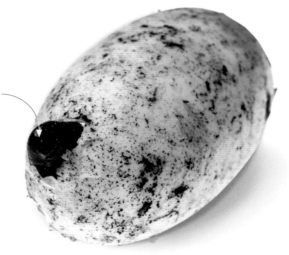

Eggshell splits open as the alligator forces its way through the egg membrane

Hatching out
After two months of incubation, warmed by the heat from a nest of decomposing vegetation, an American alligator (*Alligator mississippiensis*) is ready to hatch. Yelping while still inside the egg alerts the waiting mother, who will be ready to carry her newborn in her mouth from the nest to the water.

An egg-tooth (a horny patch of skin at the front of the upper jaw) is used to pierce the egg membrane under the shell

---

LIFE INSIDE THE EGG
A series of membrane-bound sacs provide the embryo's life-support system: an amnion that cushions its body, a yolk sac that supplies nourishment, and an allantois that absorbs oxygen seeping into the egg and stores waste.

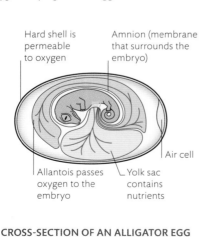

Hard shell is permeable to oxygen

Amnion (membrane that surrounds the embryo)

Air cell

Allantois passes oxygen to the embryo

Yolk sac contains nutrients

CROSS-SECTION OF AN ALLIGATOR EGG

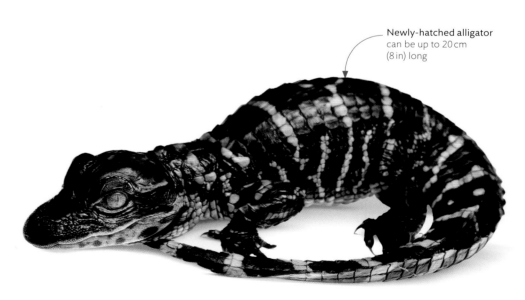

Newly-hatched alligator can be up to 20 cm (8 in) long

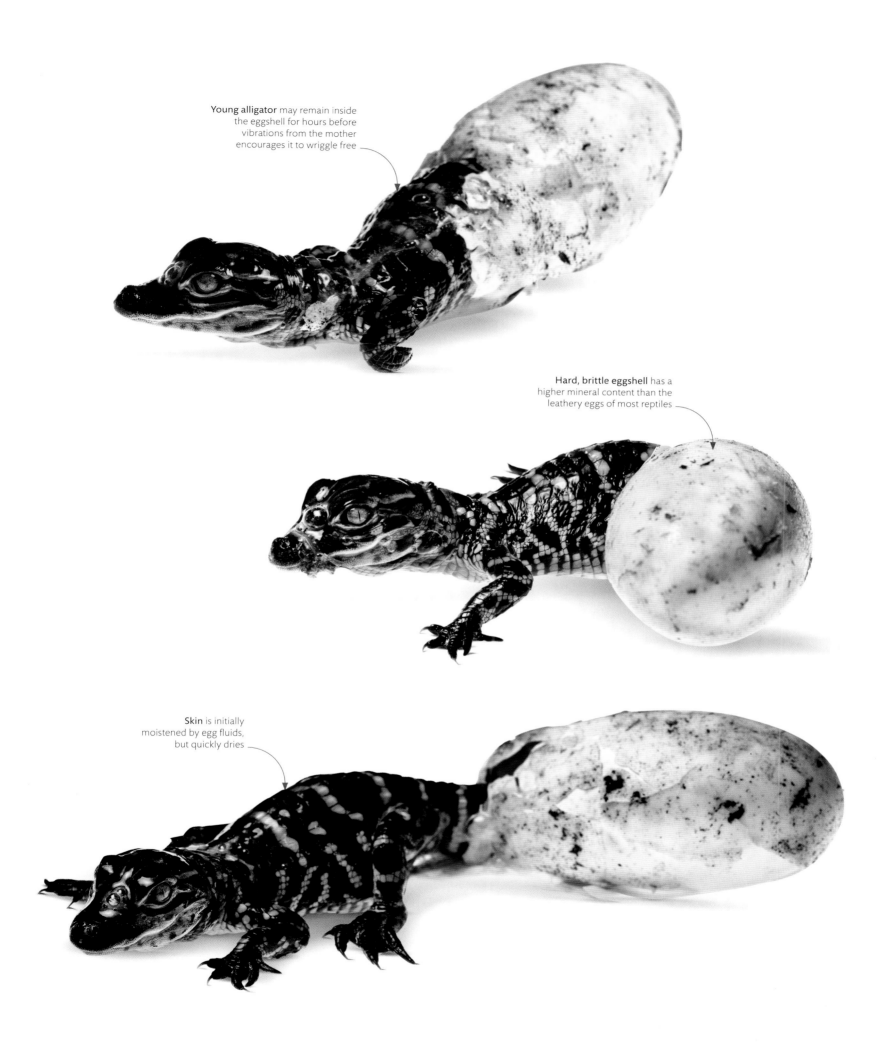

Young alligator may remain inside the eggshell for hours before vibrations from the mother encourages it to wriggle free

Hard, brittle eggshell has a higher mineral content than the leathery eggs of most reptiles

Skin is initially moistened by egg fluids, but quickly dries

bird eggs

Whatever their size or shape, birds' eggs consist of a membrane-covered embryo protected by an outer shell of calcium carbonate. The variety in eggshell colours, which may conceal the egg from predators, derives from just two pigments: protoporphyrin (reddish-brown) and biliverdin (bluish-green).

White eggs
Birds that lay eggs in nests hidden out of sight – in cavities such as tree holes or burrows, or in bowl-shaped nests – often lay white or pale eggs.

Tiny egg is plain as it is hidden in a cup-shaped nest

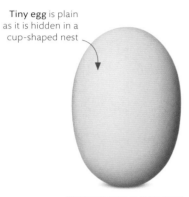

RUFOUS HUMMINGBIRD
Selasphorus rufus

Glossy white egg laid in nest burrow

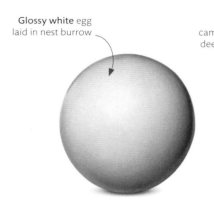

COMMON KINGFISHER
Alcedo atthis

Elliptical egg lacks camouflage as it is laid deep within tree trunk

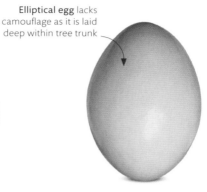

BLACK WOODPECKER
Dryocopus martius

Blue and green eggs
Tree- or shrub-nesting birds tend to lay blue or green eggs. The colour may act as a sunscreen, while the pattern often reflects the nest material, providing camouflage.

Plain blue and slightly glossy

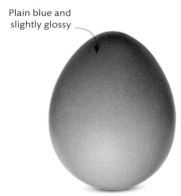

DUNNOCK
Prunella modularis

Blue with white speckles

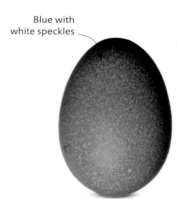

GREEN WOOD-HOOPOE
Phoeniculus purpureus

White layer flakes off during incubation, creating marbled pattern

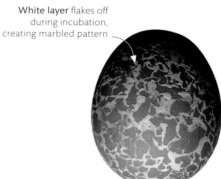

GUIRA CUCKOO
Guira guira

Earth-coloured eggs
Birds that nest on the ground rely on camouflage protecting their eggs. Plain brown or speckled eggs are hard to see in sandy, scrubby, or rocky habitats.

Light brown and glossy

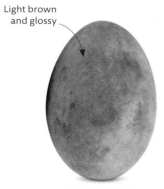

BLACK SWAN
Cygnus atratus

Reddish-brown with darker blotches

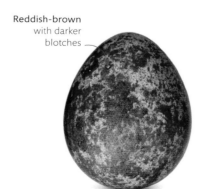

PEREGRINE FALCON
Falco peregrinus

Brown speckles blend into ground-nesting sites

WOOD WARBLER
Phylloscopus sibilatrix

Dull white and
oval-shaped

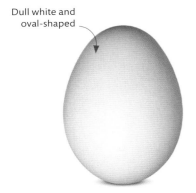

BARN OWL
Tyto alba

Sparsely dappled
and round

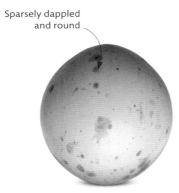

MASKED FINFOOT
Heliopais personatus

Reddish-brown
speckles may mimic
nesting site colours

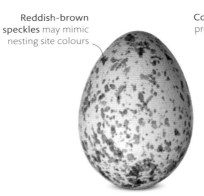

GREAT TIT
Parus major

Conical shape might help
prevent egg rolling off the
cliff-ledge nesting site

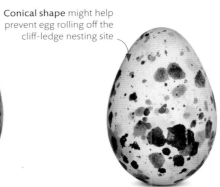

BLACK GUILLEMOT
Cepphus grylle

Blue colour
caused by
biliverdin
deposited
during laying

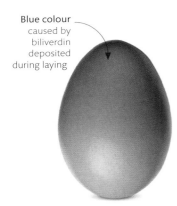

AMERICAN ROBIN
Turdus migratorius

Blotched
pattern

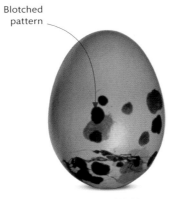

PLAIN PRINIA
Prinia inornata

Glossy, speckled,
and oval-shaped

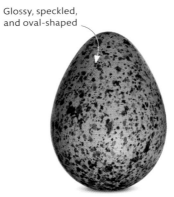

LARGE-BILLED CROW
Corvus macrorhynchos

Markings and
colour blend in with
nest-site materials

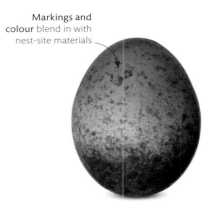

ANDEAN SPARROW
Zonotrichia capensis

Brown colour
mimics
rock-ledge
nesting sites

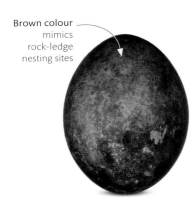

EGYPTIAN VULTURE
Neophron percnopterus

Colour and shape
blend in with
beach pebbles

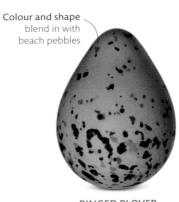

RINGED PLOVER
Charadrius hiaticula

Pebble shape
disguises egg in
coastal habitat

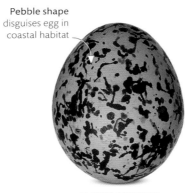

OYSTERCATCHER
Haematopus ostralegus

Almost black
colour – fresh
eggs are green

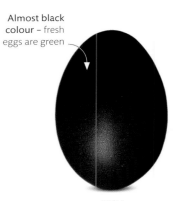

EMU
Dromaius novaehollandiae

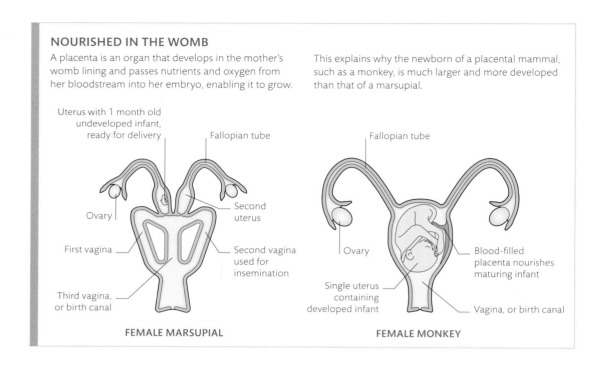

NOURISHED IN THE WOMB

A placenta is an organ that develops in the mother's womb lining and passes nutrients and oxygen from her bloodstream into her embryo, enabling it to grow.

This explains why the newborn of a placental mammal, such as a monkey, is much larger and more developed than that of a marsupial.

Uterus with 1 month old undeveloped infant, ready for delivery

Fallopian tube

Ovary

Second uterus

First vagina

Second vagina used for insemination

Third vagina, or birth canal

FEMALE MARSUPIAL

Fallopian tube

Ovary

Blood-filled placenta nourishes maturing infant

Single uterus containing developed infant

Vagina, or birth canal

FEMALE MONKEY

marsupial pouches

In most mammals, the unborn infant is nourished inside the mother's womb by a blood-filled organ called a placenta over a prolonged pregnancy. But in marsupials, the reproductive strategy is different; the pregnancy is short and the newborn completes most of its development outside the mother's body. Her pouch provides a warm shelter, and as with other mammals, the mother's milk is the baby's lifeline.

Out of pocket
At 8 months, too big to remain in the pouch, this juvenile swamp wallaby hides in vegetation for protection. It still returns to its mother to suckle milk – something that it will continue to do for another 6 months.

Pouched protection
This infant swamp wallaby (*Wallabia bicolor*) spends around 8 months in its mother's pouch. During this time, a new embryo may already be developing – ready to take possession of the pouch when it is their turn.

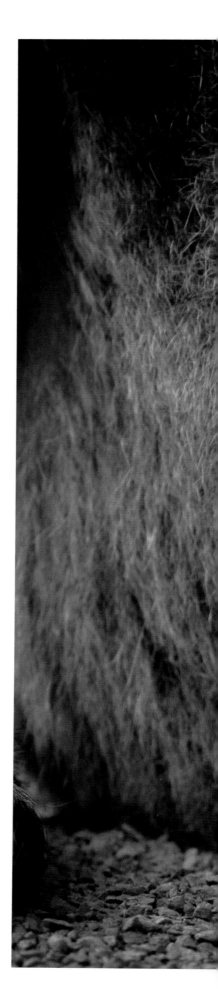

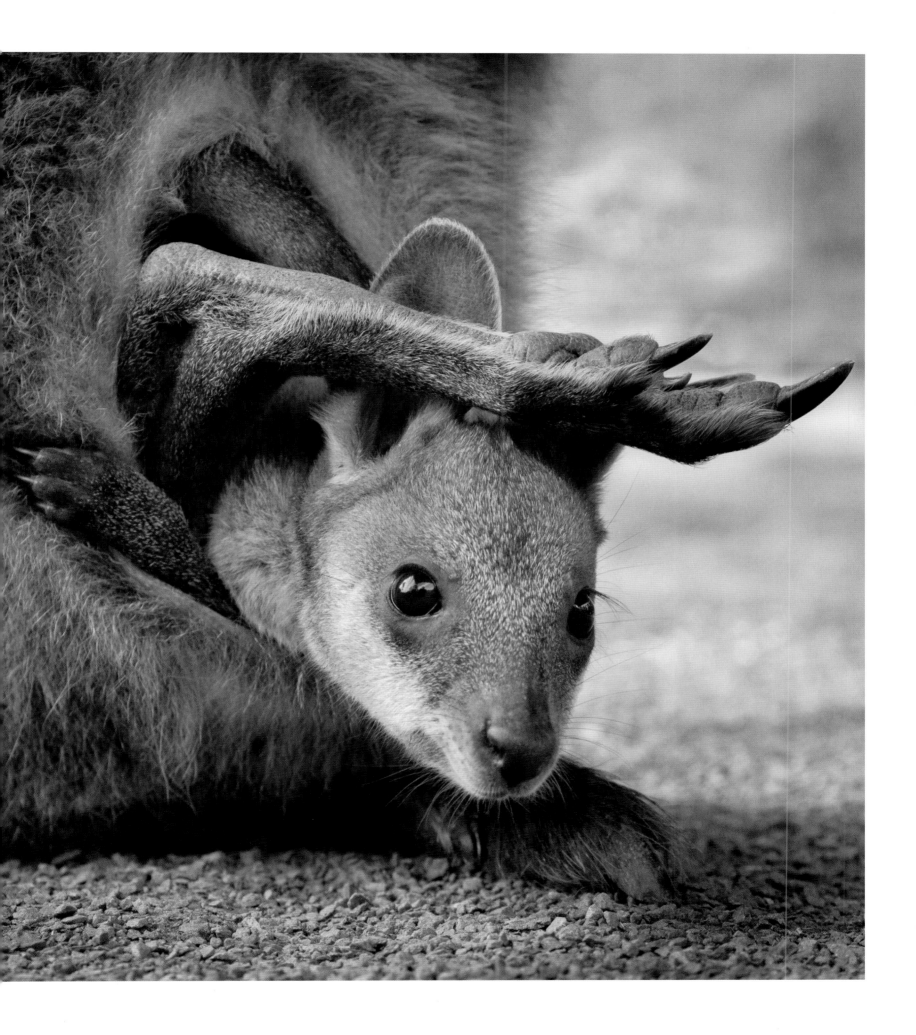

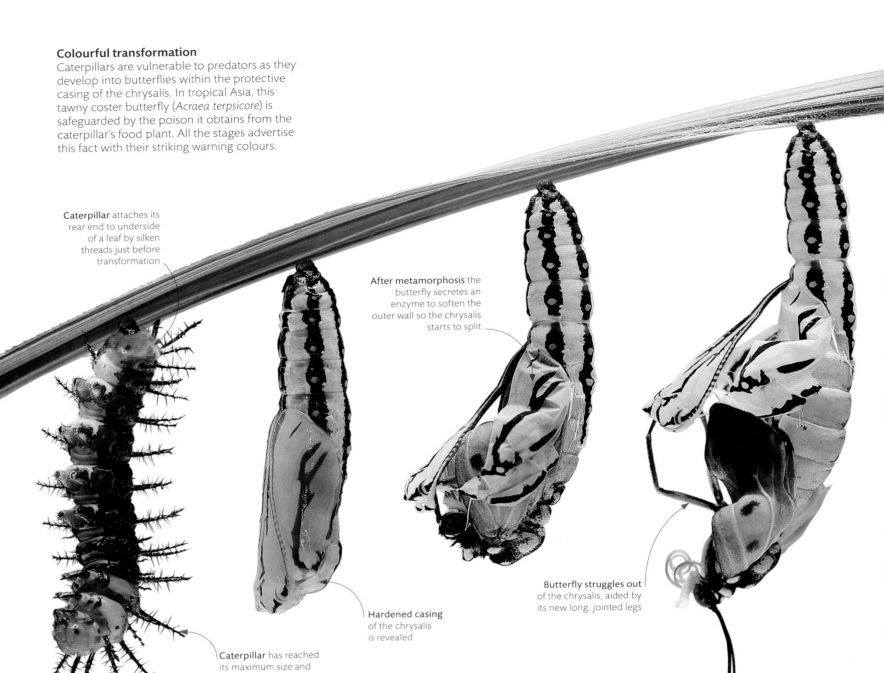

Colourful transformation

Caterpillars are vulnerable to predators as they develop into butterflies within the protective casing of the chrysalis. In tropical Asia, this tawny coster butterfly (*Acraea terpsicore*) is safeguarded by the poison it obtains from the caterpillar's food plant. All the stages advertise this fact with their striking warning colours.

Caterpillar attaches its rear end to underside of a leaf by silken threads just before transformation

After metamorphosis the butterfly secretes an enzyme to soften the outer wall so the chrysalis starts to split

Hardened casing of the chrysalis is revealed

Caterpillar has reached its maximum size and is ready to shed skin for final time

Butterfly struggles out of the chrysalis, aided by its new long, jointed legs

Butterfly antennae are revealed for the first time

METAMORPHOSIS

In growing larvae – whether caterpillars or the maggots of flies – hormones trigger growth spurts. When they reach maximum larval size, the juvenile-controlling hormone disappears, triggering a complete metamorphosis into adult form. During these final stages, adult body parts develop from clusters of cells in the larva, known as imaginal discs.

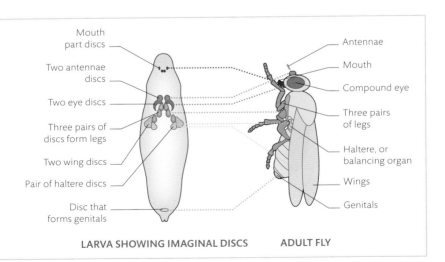

Mouth part discs

Two antennae discs

Two eye discs

Three pairs of discs form legs

Two wing discs

Pair of haltere discs

Disc that forms genitals

Antennae

Mouth

Compound eye

Three pairs of legs

Haltere, or balancing organ

Wings

Genitals

LARVA SHOWING IMAGINAL DISCS   **ADULT FLY**

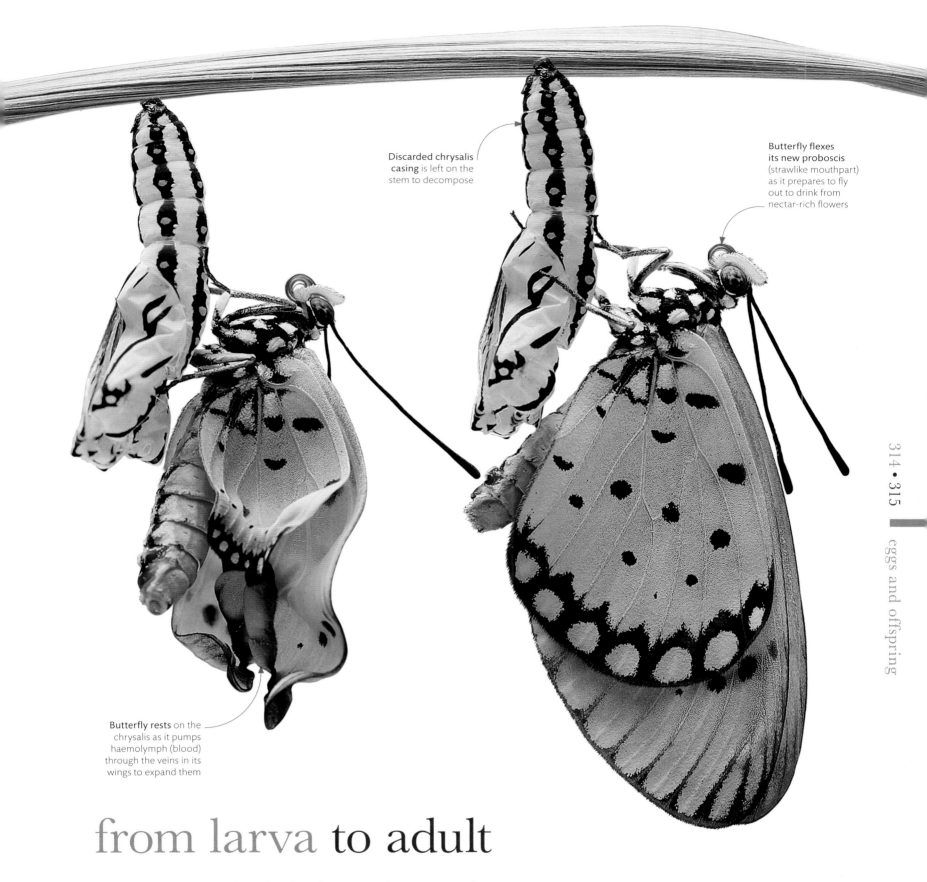

Discarded chrysalis casing is left on the stem to decompose

Butterfly flexes its new proboscis (strawlike mouthpart) as it prepares to fly out to drink from nectar-rich flowers

Butterfly rests on the chrysalis as it pumps haemolymph (blood) through the veins in its wings to expand them

from larva to adult

All animals change as they develop from juveniles into sexually mature adults, but in insects the transformation can be dramatic. For some, for example cockroaches and grasshoppers, the young are small, flightless versions of adults, but in others, such as the butterfly, metamorphosis from a caterpillar involves a complete overhaul of body structure.

amphibian
metamorphosis

When animals develop by metamorphosis, their juvenile and adult forms can use different habitats and resources in contrasting ways. In frogs this means that swimming, underwater larvae transform into adults that can walk on dry land. It is a process that involves replacing fins with legs and gills with lungs for breathing in air. Their entire behaviour – from how they move around to what they can eat – changes as much as their shape-shifting bodies.

Fully aquatic tadpole
The larva, or tadpole, of a common European frog (*Rana temporaria*) is adapted for life in water. More than half its length is a muscular tail, while gills – protected inside a swollen chamber – extract oxygen from the water. Horn-rimmed jaws are used first for grazing on algae, before later attacking animal prey.

Blocks of muscle, like those of a fish, contract to move tail from side to side

Broad tail fin pushes through water to provide propulsion

Gradual shape-shifting
Metamorphosis depends on the hormone thyroxine – the same one that speeds metabolism and controls growth in humans. In the common frog, this hormone triggers the genes that launch remodelling – making its limbs grow and tail shrink. The speed at which this transformation takes place depends on temperature, food, and oxygen – but by summertime most of the tadpoles that hatched in spring have become froglets.

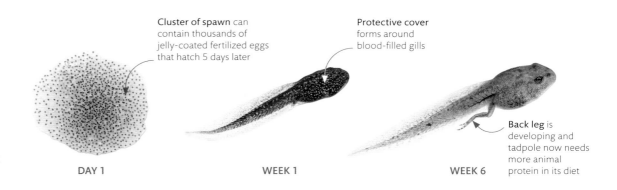

Cluster of spawn can contain thousands of jelly-coated fertilized eggs that hatch 5 days later

Protective cover forms around blood-filled gills

Back leg is developing and tadpole now needs more animal protein in its diet

DAY 1

WEEK 1

WEEK 6

PARENTAL CARE

Most amphibians leave the fate of eggs and larvae to chance, but the parents of some species take greater care of their brood. A few protect their young inside their vocal sacs, while others carry them on their backs. In Surinam toads (*Pipa* sp.), the newly fertilized eggs roll onto the mother's back during mating, where they then sink and become embedded in her skin, forming small pockets. Depending on species, they then hatch either as swimming tadpoles or, with some, as tiny toadlets. The mother then slowly sheds the thin layer of skin that was used to birth them.

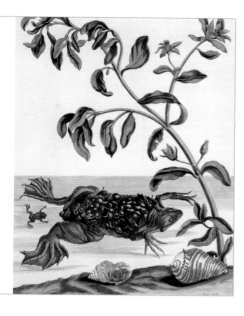

SURINAM TOADLETS HATCHING FROM A FEMALE'S BACK

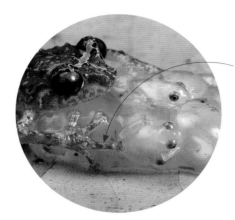

The male cross frog (*Oreophryne*) holds on to its eggs, guarding them until the froglets hatch

Metamorphosing within the egg
Many amphibian species complete their development inside eggs. This allows them to lay their eggs out of water in places that stay moist, such as the floor of tropical rainforests.

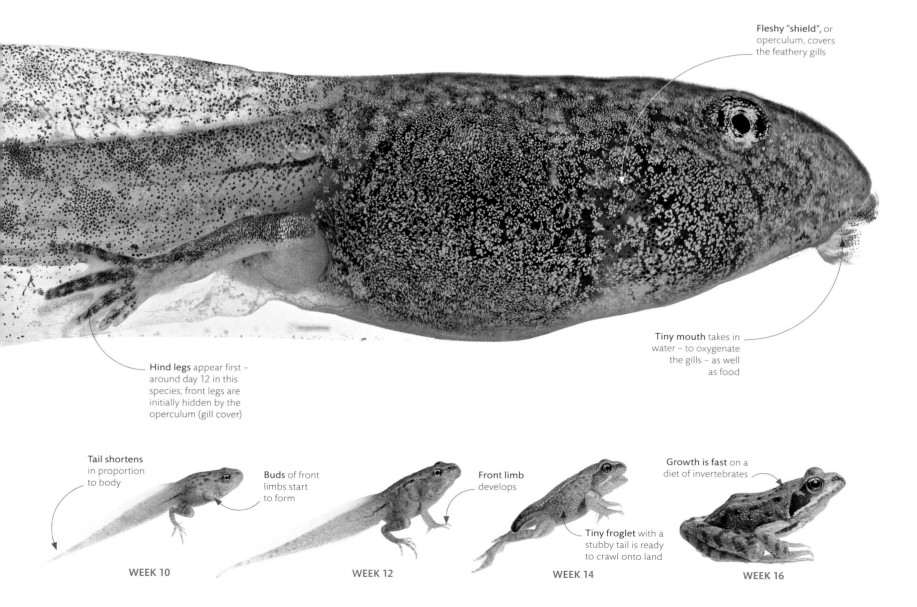

Fleshy "shield", or operculum, covers the feathery gills

Tiny mouth takes in water – to oxygenate the gills – as well as food

Hind legs appear first – around day 12 in this species; front legs are initially hidden by the operculum (gill cover)

Tail shortens in proportion to body

Buds of front limbs start to form

Front limb develops

Growth is fast on a diet of invertebrates

Tiny froglet with a stubby tail is ready to crawl onto land

WEEK 10

WEEK 12

WEEK 14

WEEK 16

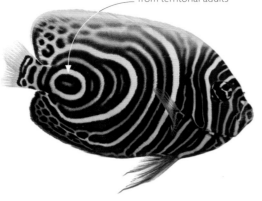

Dark blue colouring and white circles help to protect young emperor angelfish from territorial adults

JUVENILE EMPEROR ANGELFISH

reaching maturity

It takes time for animals to develop to a stage where they can reproduce for themselves. Their sex organs must mature before they can show others, by appearance or behaviour, that they can produce eggs or sperm and are ready for a mate. An individual's sex is genetic and pre-programmed in some animals, such as mammals and birds, and in others is determined by an environmental trigger, such as temperature. But a few animals, such as the emperor angelfish (*Pomacanthus imperator*), even switch sex at maturity.

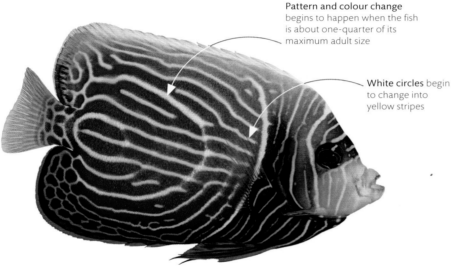

Pattern and colour change begins to happen when the fish is about one-quarter of its maximum adult size

White circles begin to change into yellow stripes

Patterns of growth
The difference between emperor angelfish of different ages is striking; the tiny, white-ringed juveniles could be mistaken for a different species to the large, yellow-striped adults.

SUBADULT EMPEROR ANGELFISH

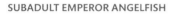

Yellow stripes signify that the emperor angelfish has reached sexual maturity

Dark, masklike band hides eyes, which may help to confuse predators

Circles to stripes
In many coral reef fish, reaching maturity involves undergoing a shift of colour and pattern. This may be so juveniles are not seen as competitors by adults, so are tolerated on a busy coral reef. Adult emperor angelfish can also change sex, from female to male. This may occur when the dominant male dies and another fish seizes the chance to replace him.

MATURE EMPEROR ANGELFISH

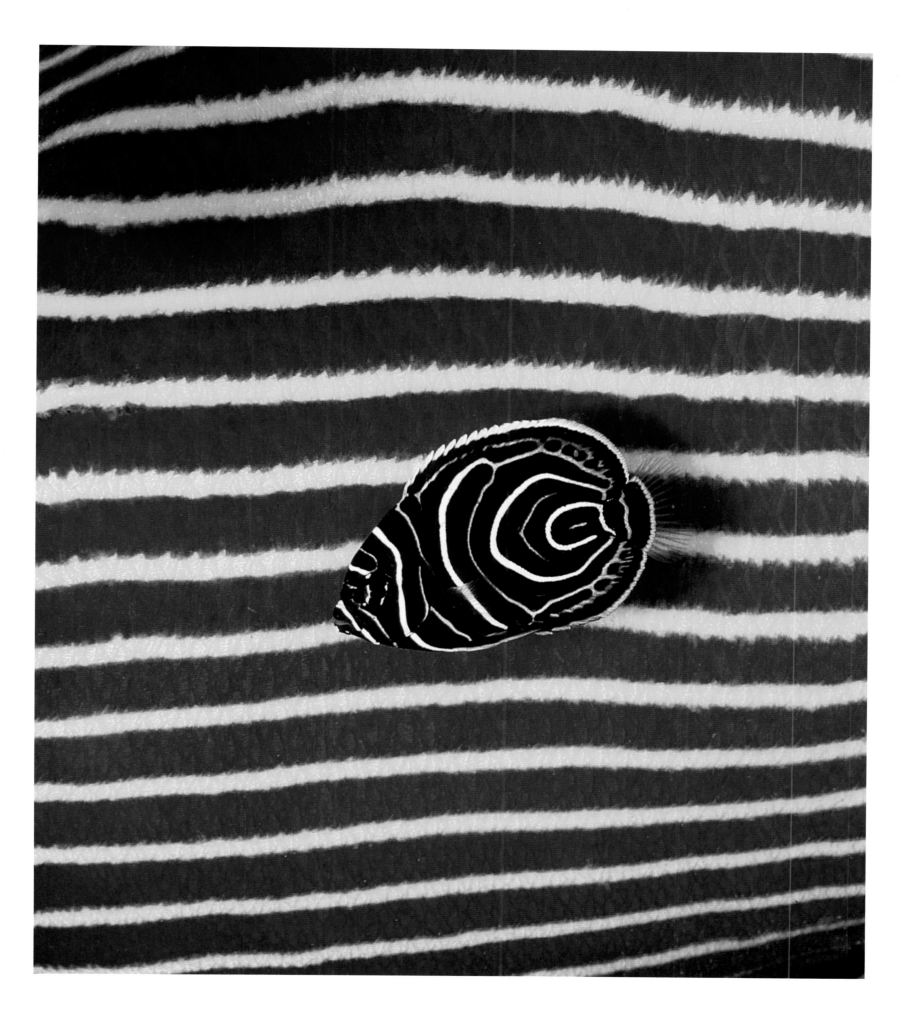

glossary

ABDOMEN The hind part of the body, lying below the ribcage in mammals, and behind the thorax in arthropods.

ABORAL The region of the body that is furthest from the mouth, especially in animals that do not have an obvious upper and lower side, such as in echinoderms.

ADIPOSE FIN A small fin lying between the dorsal and caudal (tail) fins of some fish.

ALULA A small, bony projection that forms the first digit, or "thumb" on a bird's wing.

AMPLEXUS A breeding position adopted by frogs and toads, with the male using its front legs to hold the female. Fertilization usually occurs outside the female's body.

AMPULLARY ORGANS Special sensing organs made up of jelly-filled canals containing electroreceptors, which help some fish – especially sharks, rays, and chimaeras – detect prey.

ANTENNA (pl. ANTENNAE) A sensory feeler on the head of an arthropod. Antennae always occur in pairs, and they can be sensitive to touch, sound, heat, and also taste. Their size and shape varies according to the way in which they are used.

ANTLER A bony growth on the head of deer. Unlike horns, antlers often branch, and in most cases they are grown and shed every year, in a cycle linked with the breeding season.

APOSEMATIC See *Warning coloration*.

ARBOREAL Living fully or partly in trees.

ARTHROPOD A major group of invertebrate animals with jointed legs and a hard outer skeleton. It includes crustaceans, insects, and spiders.

ARTICULATION A joint – for example, between adjacent bones.

ASEXUAL REPRODUCTION A form of reproduction that involves just one organism. Asexual reproduction is most common in invertebrates, and is used as a way of boosting numbers quickly in favourable conditions. See also *Parthenogenetic, Sexual reproduction*.

BEAK A set of narrow, protruding jaws, usually without teeth. Beaks have evolved separately in many groups of vertebrates, including tortoises and turtles, and some whales.

BILATERAL SYMMETRY A form of symmetry in which the body consists of two equal halves, on either side of a midline. Most animals show this kind of symmetry.

BILL An alternative name for a bird's beak. See also *Beak*.

BINOCULAR VISION Vision in which the two eyes face forwards, giving overlapping fields of view. This allows an animal to judge depth.

BIPEDAL Moving on two legs.

BIVALVES Molluscs, such as clams, mussels, and oysters, that have a shell made up of two hinged halves. Most bivalves move slowly or not at all, and are filter feeders. See also *Filter feeder, Mollusc*.

BLOWHOLE The nostrils of whales and their relatives, positioned on top of the head. Blowholes can be single or paired.

BRACHIATION The arm-swinging movement used by primates, such as gibbons, to move through trees.

BREEDING COLONY A large gathering of nesting birds.

BROOD PARASITE An animal – often a bird – that tricks other species into raising its young. In many cases, a young brood parasite kills all its nest-mates, so that it has sole access to all the food that its foster parents provide.

BROWSING Feeding on the leaves of trees and shrubs, rather than on grasses. See also *Grazing*.

CALCAREOUS Containing calcium. Calcareous structures – such as shells, exoskeletons, and bones – are formed by many animals, either for support or for protection.

CAMOUFLAGE Colours or patterns that enable an animal to merge with its background. Camouflage is very widespread in the animal kingdom – particularly among invertebrates – and is used both for protection against predators, and for concealment when approaching prey. See also *Mimicry, Cryptic coloration*.

CANINE TOOTH In mammals, a tooth with a single sharp point that is suited to piercing and gripping. Canine teeth are towards the front of the jaws, and are highly developed in carnivores.

CARAPACE A hard shield on the back of an animal's body.

CARNASSIAL TOOTH In mammalian carnivores, a bladelike cheek tooth that is suited to slice through flesh.

CARNIVORE Any animal that eats meat. The word carnivore can also be used in a more restricted sense, to mean mammals in the order Carnivora.

CARRION The remains of dead animals.

CARTILAGE A rubbery substance that forms part of vertebrate skeletons. In most vertebrates, cartilage lines the joints, but in cartilaginous fish – for example, sharks – it forms the whole of the skeleton.

CASQUE A bony growth on the head of an animal.

CAUDAL Relating to an animal's tail.

CELLULOSE A complex carbohydrate found in plants. Cellulose is used by plants as a building material, and it has a resilient chemical structure that animals find hard to break down. Plant-eating animals, such as ruminants, digest it with the help of microorganisms.

CEPHALOTHORAX In some arthropods, a part of the body that combines the head and the thorax. Animals that have a cephalothorax include crustaceans and arachnids.

CHEEK TOOTH See *Carnassial tooth, Molar tooth, Premolar tooth*.

CHELICERA (pl. CHELICERAE) In arachnids, each one of the first pair of appendages, at the front of the body. Chelicerae often end in pincers, and in spiders they are able to inject venom. In mites, they are sharply pointed and are used for piercing food.

CHELIPEDS Any limbs in crustaceans that bear pincers, or chela.

CHITIN A tough, fibrous substance that forms the exoskeleton of arthropods, including the shells of crabs; and the communal skeleton of some corals. See also *Exoskeleton*.

CHORDATE An animal belonging to the phylum Chordata, which includes all vertebrates. A key feature of chordates is their notochord, which runs the length of their bodies; it reinforces the body, yet allows it to move by bending.

CHRYSALIS A hard and often shiny case that protects an insect pupa. Chrysalises are often found attached to plants, or buried close to the surface of the soil.

CLASPERS A structure in some male invertebrates used to hold the female during mating; or a pair of modified pelvic fins in some male fish, such as sharks, used to direct sperm into the female's reproductive tract. See also *Pelvic fins*.

CLASS A level used in classification. In the sequence of classification levels, a class forms part of a phylum and is subdivided into one or more orders.

CLOACA An opening towards the rear of the body that is shared by several body systems. In some vertebrates – such as bony fish and amphibians – the gut, kidneys, and reproductive systems all use this single opening.

CLONE An asexually produced animal, genetically identical to its parent.

CLOVEN-HOOFED Having hooves that look as if they are split in two. Most cloven-hoofed mammals, such as deer and antelope, actually have two hooves, arranged on either side of a line that divides the foot in two.

COCOON A case made of open, woven silk. Many insects spin a cocoon before they begin pupation, and many spiders spin one to hold their eggs.

COLONY A group of animals, belonging to the same species, that spend their lives together, often dividing up the tasks involved in survival. In some colonial species, particularly aquatic invertebrates, the colony members are permanently fastened together. In others, such as ants, bees, and wasps, members forage independently, but live in the same nest together.

COMPOUND EYE An eye that is divided into separate compartments, each with its own set of lenses. Compound eyes are a common feature of arthropods. The number of compartments they contain varies from a few dozen to thousands.

COUNTERSHADING A type of camouflage pattern in which an animal is typically darker above and lighter beneath. It tends to counter the effect of shadows, making the animal more difficult to see.

COVERT A feather that covers the base of a bird's flight feather.

CRYPTIC COLORATION Coloration and markings that make an animal difficult to see against its background.

DACTYLUS In insects, one or more of the tarsal joints following the first, modified joint. In crustaceans, the moveable finger of a pincer, which swivels up to open and down to close the claw. See also *Pincer*.

DEWLAP A fold of loose skin hanging from an animal's throat.

DIASTEMA A wide gap separating rows of teeth. In plant-eating mammals, the diastema separates the biting teeth, at the front of the jaw, from the chewing teeth at the rear. In many rodents, the cheeks can be folded into the diastema, to shut off the back of the mouth while the animal is gnawing.

DIGIT A finger or toe.

DIGITIGRADE A gait in which only the fingers or toes touch the ground. See also *Plantigrade, Unguligrade*.

DIPLOSEGMENT A pair of fused segments in the body of some arthropods, such as millipedes.

DORSAL On or near an animal's back.

ECHINODERMS A major group of marine invertebrates that includes starfish, brittle stars, sea urchins, sea lilies, and sea cucumbers. Echinoderm bodies have radial symmetry. They have chalky protective plates under their skin and use hydraulic "tube feet" to move and to catch prey.

ECHOLOCATION A method of sensing nearby objects by using pulses of high-frequency sound. Echoes bounce back from obstacles and other animals, allowing the sender to build up a "picture" of its surroundings. Echolocation is used by several groups of animals, including mammals and a small number of cave-dwelling birds.

ELYTRON (pl. ELYTRA) A hardened forewing in beetles, earwigs, and some bugs. The two elytra usually fit together like a case, protecting the more delicate hindwings underneath.

EMBRYO A young animal or plant in a rudimentary stage of development.

ENDOPARASITE An animal that lives parasitically inside another (host) animal's body, either by feeding directly on its host's tissues or by stealing some of its food. Endoparasites frequently have complex life cycles, involving more than one host.

ENDOSKELETON An internal skeleton, typically made of bone. Unlike an exoskeleton, this kind of skeleton can grow in step with the rest of the body.

EPITHELIUM Covering or lining tissue that forms sheets and layers around and within many organs and other tissues in animals.

EVOLUTION Any change in the average genetic makeup of a population of living things between one generation and the next. The "theory of evolution" is based on the idea, supported by various lines of evidence, that such genetic change is not random but largely the result of natural selection, and that the operation of such processes over time can account for the huge variety of species found on Earth.

EXOSKELETON An external skeleton that supports and protects an animal's body. The most complex exoskeletons, formed by arthropods, consist of rigid plates that meet at flexible joints. This kind of skeleton cannot grow, and has to be shed and replaced (ecdysis) at periodic intervals. See also *Endoskeleton*.

FAMILY A level used in classification. In the sequence of classification levels, a family forms part of an order and is subdivided into one or more genera.

FEMUR (pl. FEMORA) The thigh bone in four-limbed vertebrates. In insects, the femur is the third segment of the leg, immediately above the tibia.

FERTILIZATION The union of an egg cell and sperm, which creates a cell capable of developing into a new animal. In external fertilization, the process occurs outside the body (usually in water), but in internal fertilization, it takes place in the female's reproductive system.

FIBULA (pl. FIBULAE) The outermost of the two lower leg or hind limb bones. See also *Tibia*.

FILTER FEEDER An animal that eats by sieving small food items from water. Many invertebrate filter feeders, such as bivalve molluscs and sea squirts, are stationary animals that collect food by pumping water through or across their bodies. Vertebrate filter feeders, such as baleen whales, collect their food by trapping it while they are on the move.

FLAGELLUM (pl. FLAGELLA) A long, hairlike projection from a cell. A flagellum can flick from side to side, moving the cell along. Sperm cells use flagella to swim.

FLIGHT FEATHERS A bird's wing and tail feathers, used in flight.

FLIPPER In aquatic mammals, a paddle-shaped limb. See also *Fluke*.

FLUKE A rubbery tail flipper in whales and their relatives. Unlike the tail fins of fish, flukes are horizontal, and they beat up and down instead of side to side.

FOOD CHAIN A food pathway that links two or more different species, in that each forms food for the next species higher up in the chain. In land-based food chains, the first link is usually a plant. In aquatic food chains, it is usually an alga or other form of single-celled life.

FORCIPULE In centipedes, the modified pincerlike first pair of legs, which are used to inject venom. See also *Pincer*.

FURCULA The forked, springlike organ attached to a springtail's abdomen.

GASTROPODS The group of molluscs that includes snails and slugs. See also *Molluscs*.

GENUS (pl. **GENERA**) A level used in classification. In the sequence of classification levels, a genus forms part of a family and is subdivided into one or more species.

GILL An organ used for extracting oxygen from water. Gills are usually positioned on or near the head, or – in aquatic insects – towards the end of the abdomen.

GRAZING Feeding on grass. See also *Browsing*.

GUARD HAIR A long hair in a mammal's coat, which projects beyond the underfur to protect it and help to keep the animal dry.

HERBIVORE An animal that feeds on plants or plantlike plankton.

HIBERNATION A period of dormancy in winter. During hibernation, an animal's body processes drop to a low level.

HORN In mammals, a pointed growth on the head. True horns are hollow sheaths of keratin covering a bony horn core.

HOST An animal on or in which a parasite feeds.

INCISOR TOOTH In mammals, a tooth at the front of the jaw, used in biting, slicing, or gnawing.

INCUBATION In birds, the period when a parent sits on the eggs and warms them, allowing them to develop. Incubation periods range from under 14 days to several months.

INTERNAL FERTILIZATION In reproduction, a form of fertilization that takes place inside the female's body. Internal fertilization is a characteristic of many land animals, including insects and vertebrates. See also *Sexual reproduction*.

IRIDOPHORE A type of specialized skin cell that contains light-reflecting guanine crystals. These are found in some species of crustaceans, cephalopods, fish, amphibians, and reptiles, such as chameleons.

JACOBSON'S ORGAN An organ in the roof of the mouth that is sensitive to airborne scents. Snakes often employ this organ to detect their prey, while some male mammals use it to find females that are ready to mate.

KEEL In birds, an enlargement of the breastbone that anchors the muscles used in flight.

KERATIN A tough structural protein found in hair, claws, and horns.

KINGDOM In classification, one of the six fundamental divisions of the natural world.

LATERAL LINE SYSTEM The bodily mechanism by which fish detect movement, vibration, and pressure under water. Canals running under the skin channel water to move sensory cells back and forth, which then triggers them to send nerve impulses to the brain.

LARVA (pl. **LARVAE**) An immature but independent animal that looks completely different from an adult. A larva develops the adult shape by metamorphosis; in many insects, the change takes place in a resting stage that is called a pupa. See also *Cocoon, Metamorphosis, Nymph*.

LEK A communal display area used by male animals – particularly birds – during courtship. The same location is often revisited for many years.

MANDIBLE The paired jaws of an arthropod, or a bone that makes up all or part of the lower jaw in vertebrates.

MANTLE In molluscs, an outer fold of skin covering the mantle cavity.

MELON A bulbous swelling in the heads of many toothed whales and dolphins. The melon is filled with a fatty fluid, and is believed to focus the sounds used in echolocation.

METABOLISM The complete array of chemical processes that take place inside an animal's body. Some of these processes release energy by breaking down food, while others use energy by making muscles contract.

METACARPAL In four-limbed animals, one of a set of bones in the front leg or arm, forming a joint with a digit at the end. In most primates, the metacarpals form the palm of the hand.

METAMORPHOSIS A change in body shape shown by many animals – particularly invertebrates – as they grow from juvenile to adult. In insects, metamorphosis can be complete or incomplete. Complete metamorphosis involves a total reorganization of the organism during a resting stage, called a pupa. Incomplete metamorphosis embraces a series of less drastic changes; these occur each time the young animal moults. See also *Chrysalis, Cocoon, Larva, Nymph*.

MIMICRY A form of camouflage in which an animal resembles another animal or an inanimate object, such as a twig or a leaf. Mimicry is very common in insects, with many harmless species imitating those that have dangerous bites or stings.

MOLAR TOOTH In mammals, a tooth at the rear of the jaw. Molar teeth may have a flattened or ridged surface used for chewing vegetation. The sharper molar teeth of meat-eaters can cut through hide and bone.

MOLLUSCS A major group of invertebrate animals that includes gastropods (slugs and snails), bivalves (clams and relatives), and cephalopods (squids, octopuses, cuttlefish, and nautiluses). Molluscs are soft-bodied and typically have hard shells, though some subgroups have lost the shell during their evolution.

MONOCULAR VISION A type of vision in which each eye is used independently, such as in chameleons. This gives a wide field of vision but limited depth perception. See also *Binocular vision, Stereoscopic vision*.

MOULT Shedding fur, feathers, or skin so that it can be replaced. Mammals and birds moult to keep their fur and feathers in good condition, to adjust their insulation, or so that they can be ready to breed. Arthropods – such as insects – moult in order to grow.

NEMATOCYST The coiled structure within the stinging cell of a jellyfish or other cnidarian that shoots out and injects toxin via a dartlike tip.

NEUROMAST Sensory cells that form part of the lateral line system in fish. They are stimulated by the motion of water to help fish detect movement. See also *Lateral line system*.

NOSE LEAF A facial structure in some bat species that focuses sound pulses emitted through the nostrils.

NICHE An animal's place and role in its habitat. Although two species may share the same habitat, they never share the same niche.

NOCTURNAL Active at night and sleeping during the day, as opposed to diurnal – active during the day.

NOTOCHORD A reinforcing rod that runs the length of the body. The notochord is a distinctive feature of chordates, although in some it is present only in early life. In vertebrates, the notochord becomes incorporated into the backbone during the development of the embryo.

NYMPH An immature insect that looks similar to its parents but that does not have functioning wings or reproductive organs. A nymph develops the adult shape by metamorphosis, changing slightly each time it moults.

OLFACTORY LOBE The area of the brain that receives and processes scent information from the olfactory nerves. In most vertebrates it is situated at the front of the brain.

OMMATIDIA The facets of light-receiving cells that make up the lenses of a compound eye, such as those common to many arthropods. See also *Compound eye, Photoreceptor.*

OMNIVORE An animal that eats both plant and animal food.

OPERCULUM A cover or lid. In some gastropod molluscs, an operculum is used to seal the shell when the animal has withdrawn inside. In bony fish, an operculum on each side of the body protects the chamber that contains the gills.

OPISTHOSOMA The abdomen, or back part of the body, behind the prosoma of arthropods including arachnids, such as spiders, and horseshoe crabs. See also *Prosoma.*

ORDER A level used in classification. In the sequence of classification levels, an order forms part of a class, and is subdivided into one or more families.

ORGAN A structure in the body, composed of several kinds of tissues, that carries out specific tasks.

OSSICLE A minute bone. The ear ossicles of mammals, which transmit sound from the eardrum to the inner ear, are the smallest bones in the body.

PALPS Pairs of long, sensory appendages that stem from near the mouths of arthropods. Similar to antennae, they have touch sensors and serve a variety of purposes, including touch and taste, and some are used for predation. See also *Pedipalps.*

PAPILLA (pl. **PAPILLAE**) A small, fleshy protruberance on an animal's body. Papillae often have a sensory function; for example, detecting chemicals that help to pinpoint food.

PARAPODIUM (pl. **PARAPODIA**) A leg- or paddlelike flap found in some worms. Parapodia are used for moving, or for pumping water past the body.

PARASITE An animal that lives on or inside another animal (its host), and that feeds either on its host or on food that its host has swallowed. The majority of parasites are much smaller than their hosts, and many have complex life cycles involving the production of huge numbers of eggs. Parasites often weaken their hosts, but generally do not kill them. See also *Endoparasite.*

PARATOID GLAND In amphibians, a gland behind the eyes that secretes poison onto the surface of the skin.

PARTHENOGENETIC Relating to reproduction from an unfertilized egg cell. Females of some invertebrates, such as aphids, produce young parthenogenetically only during summer months, when food is abundant. A few species always reproduce in this way, and form all-female populations. Unfertilized parthenogenetic eggs typically already have two copies of each chromosome. See also *Asexual reproduction.*

PATAGIUM In bats, the flap of double-sided skin that forms the wing. This term is also used for the parachutelike skin flaps of colugos and other gliding mammals.

PECTORAL FIN One of the two paired fins positioned towards the front of a fish's body, often just behind its head. Pectoral fins are usually highly mobile and are normally used for steering but sometimes for propulsion too.

PECTORAL GIRDLE In four-limbed vertebrates, the arrangement of bones that anchors the front limbs to the backbone. In most mammals, the pectoral girdle consists of two clavicles, or collar-bones, and two scapulae, or shoulder-blades.

PEDICEL The region of the skull from which antlers grow and from which they are shed at the end of the mating season. See also *Antler.*

PEDIPALPS In arachnids, the second pair of appendages, near the front of the body. Depending on the species, they are used for walking, sperm transfer, or attacking prey. See also *Claspers, Palps.*

PELVIC FINS One of the two paired fins in fish, which are normally close to the underside, sometimes near the head, but more often towards the tail. Pelvic fins are generally used as stabilizers. In some species, such as sharks, they are also used to transfer sperm. See also *Claspers.*

PELVIC GIRDLE In four-limbed vertebrates, the arrangement of bones that anchors the hind limbs to the backbone. The bones of the pelvic girdle are often fused, forming a weight-bearing ring called the pelvis.

PENTADACTYL The characteristic of having five digits – toes or fingers – common to many four-limbed vertebrates, or having evolved from such. See also *Digits, Tetrapods.*

PHEROMONE A chemical produced by one animal that has an effect on other members of its species. Pheromones are often volatile substances that spread through the air, triggering a response from animals some distance away.

PHOTORECEPTOR A type of specialized, light-sensitive cell that forms the retinal layer at the back of animals' eyes. In many animals photoreceptor cells contain different pigments, which allow for colour vision. See also *Ommatidia, Retina.*

PHOTOSYNTHESIS A series of chemical processes that enable plants to capture the energy in sunlight and convert it into chemical form.

PHYLUM (pl. **PHYLA**) A level used in classification. In the sequence of classification levels, a phylum forms part of a kingdom and is subdivided into one or more classes.

PINCER In arthropods, pincers are pointed, hinged organs used for feeding or defence, such as the mandibles of an insect, or the chelae of crustaceans. See also *Dactylus.*

PINNA (pl. **PINNAE**) The external ear flaps found in mammals.

PHARYNX The throat.

PLACENTA An organ developed by an embryonic mammal that allows it to absorb nutrients and oxygen from its mother's bloodstream, before it is born. Young that grow in this way are known as placental mammals.

PLACENTAL MAMMAL See *Placenta*.

PLANKTON Floating organisms – many of them microscopic – that drift in open water, particularly near the surface of the sea. Planktonic organisms can often move, but most are too small to make any headway against strong currents. Planktonic animals are collectively known as zooplankton.

PLANTIGRADE A gait in which the sole of the foot is in contact with the ground. See also *Digitigrade, Unguligrade*.

PLASTRON The lower part of the shell in tortoises and turtles.

POLYP In cnidarians, a body form that has a hollow cylindrical trunk, ending in a central mouth surrounded by a circle of tentacles. Polyps are frequently attached to solid objects by their base.

PREDATOR An animal that catches and kills others, known as its prey. Some predators catch their prey by lying in wait, but most actively pursue and attack other animals. See also *Prey*.

PREHENSILE Able to curl around objects and grip them.

PREMOLAR TOOTH In mammals, a tooth positioned midway along the jaw, between the canines and the molars. See also *Canine Tooth, Molar Tooth*.

PREY Any animal that is eaten by a predator. See also *Predator*.

PROBOSCIS An animal's nose, or a set of mouthparts with a noselike shape. In insects that feed on fluids, the proboscis is often long and slender, and can usually be stowed away when not in use.

PROPODUS The fixed part of a pincer that cannot be moved. It consists of a wide, muscular palm. See also *Dactylus, Pincers*.

PROSOMA The front part of the body before the opisthoma in arthropods such as arachnids and horseshoe crabs. See also *Cephalothorax, Opisthosoma*.

PTEROSTIGMA A coloured, weighted panel near the leading edge on some insects' wings, such as dragonflies.

PUPIL The hole in the centre of the eye that allows light to enter.

RADIAL SYMMETRY A form of symmetry in which the body is arranged like a wheel, often with the mouth at the centre.

RADULA A mouthpart that many molluscs use for rasping away at food. The radula is often shaped like a belt, and armed with many microscopic toothlike denticles.

RESPIRATION This refers to both the action of breathing itself and to cellular respiration, which is the biomechanical processes that take place within cells, which break down food molecules – usually by combining them with oxygen to provide energy for an organism.

RETINA A layer of light-sensitive cells lining the back of the eye that converts optical images into nerve impulses, which travel to the brain via the optic nerve. See also *Photoreceptor*.

RICTAL BRISTLES Modified feathers that project from the base of the bill in some birds such as nightjars and kiwis. The feathers have a stiff shaft that lacks barbs. They may work like whiskers and help the birds to detect prey when hunting.

RODENT A large and adaptable group of mostly small, four-limbed mammals with a long tail, clawed feet, and long whiskers and teeth – especially the large incisors. Their jaws are adapted for gnawing. They are found worldwide, expect for Antarctica and represent over 40 per cent of mammal species.

ROSTRUM In bugs and some other insects, a set of sucking mouthparts that looks similar to a beak.

RUMINANT A hoofed mammal that has a specialized digestive system, with several stomach chambers. One of these – the rumen – contains huge numbers of micro-organisms that help to break down the cellulose in plant cell walls. To speed this process, a ruminant usually regurgitates its food and rechews it, a process called "chewing the cud".

RUTTING SEASON In deer, a period during the breeding season when males clash with each other for the opportunity to mate.

SALIVA A watery fluid secreted by the salivary glands in the mouth to aid chewing, tasting, and digestion.

SALIVARY GLAND Sets of paired glands in the mouth that produce saliva. See also *Saliva*.

SCALES The thin horny or bony plates that cover and protect the skin of fish and reptiles. Scales are typically arranged so that they overlap.

SCAPE The first segment in an insect's antennae, nearest its head.

SCUTE A shieldlike plate or scale that forms a bony covering on some animals.

SEBACEOUS GLAND A skin gland in mammals that normally opens near the root of a hair. Sebaceous glands produce substances that keep skin and hair in good condition.

SEX CELLS The reproductive cells of all animals – the male sperm and female egg – also known as gametes. See also *Sexual reproduction*.

SEXUAL DIMORPHISM Showing physical differences between males and females. In animals that have separate sexes, males and females always differ, but in highly dimorphic species, such as elephant seals, the two sexes look different and are often unequal in size.

SEXUAL REPRODUCTION A form of reproduction that involves fertilization of a female cell or egg, by a male cell or sperm. This is the most common form of reproduction in animals. It usually involves two parents – one of either sex – but in some species individual animals are hermaphrodite. See also *Asexual reproduction*.

SHELL A hard, protective outer casing found in many molluscs and crustaceans, as well as some reptiles – the turtles and tortoises.

SILK A protein-based fibrous material produced by spiders and some insects. Silk is liquid when it is squeezed out of a spinneret but turns into elastic fibres when stretched and exposed to air. It has a wide variety of uses. Some animals use silk to protect themselves or their eggs, to catch prey, to glide on air currents, or to lower themselves through the air.

SIPHONOPHORE A group of cnidarians that live as floating colonies of connected individual polyps, which can form very long strings. Species include the Portuguese man o' war. See also *Polyp*.

SPAWN The release or deposit of eggs in aquatic animals such as crustaceans, molluscs, fish, and amphibians.

SPERM Male reproductive cells. See also *Sex cells, Sexual reproduction*.

SPECIES A group of similar organisms that are capable of interbreeding in the wild, and of producing fertile offspring that resemble themselves. Species are the fundamental units used in biological classification. Some species have distinct populations that vary from each other. Where the differences between the populations are significant, and they are biologically isolated, these forms are classified as separate subspecies.

SPICULE In sponges, a needlelike sliver of silica or calcium carbonate that forms part of the internal skeleton. Spicules have a wide variety of shapes.

SPIRACLE In rays and some other fish, an opening behind the eye that lets water flow into the gills. In insects, an opening on the thorax or abdomen that lets air into the tracheal system.

STEREOSCOPIC VISION The ability to use forward-facing eyes, as in humans or predators like tigers, to see in very similar but slightly differing ways with each eye. This allows accurate depth perception. See also *Binocular vision*.

STERNUM The breastbone in four-limbed vertebrates, or the thickened underside of a body segment in arthropods. In birds, the sternum has a narrow flap.

SUCKER A round, concave suction cup on the tentacles of squids and octopuses. Each sucker is highly flexible and has a ring of muscle that can squeeze tightly. It is also equipped with taste receptors. See also *Tentacle*.

SWIM BLADDER A gas-filled bladder that most bony fish use to regulate their buoyancy. By adjusting the gas pressure inside the bladder, a fish can become neutrally buoyant, meaning that it neither rises nor sinks in the water column.

TAGMA (pl. TAGMATA) A distinct region in the bodies of arthropods and other segmented animals that consists of several connected parts, such as the head, thorax, and abdomen of insects. See also *Cephalothorax*.

TARSUS (pl. TARSI) A part of the leg. In insects, the tarsus is the equivalent of the foot, while in vertebrates, it forms the lower part of the leg or the ankle.

TELSON In arthropods, the last part of the abdomen, or the last appendage to it, as in horseshoe crabs.

TENDON A strong band of tough collagen fibres that usually joins muscle to bone and transmits the pull caused by muscle contraction, allowing movement of the skeleton.

TENTACLE One of two longest flexible appendages of squid and cuttlefish or the stinging appendages of jellyfish

TERRESTRIAL Living wholly or mainly on the ground.

TERRITORY An area defended by an animal, or group of animals, against other members of the same species. Territories often include some useful resources that help the male to attract a mate.

TEST In echinoderms, a skeleton made of small calcareous plates.

TETRAPOD A member of the group of animals consisting of four-limbed vertebrates, or those that have evolved from them, such as snakes.

THORAX The middle region of an arthropod's body. The thorax contains powerful muscles, and bears the legs and wings, if the animal has any. In four-limbed vertebrates, the thorax is the chest. See also *Cephalothorax*, *Prosoma*.

TIBIA (pl. TIBIAE) The shin bone in four-limbed vertebrates. In insects, the tibia forms the part of the leg immediately above the tarsus, or foot. See also *Fibula*.

TINE A point that branches off from the main beam of an antler. See also *Antler*.

TISSUE A layer of cells in an animal's body. In a tissue, the cells are of the same type, and carry out the same work. See also *Organ*.

TORPOR A sleeplike state in which the body processes slow to a fraction of their normal rate. Animals usually become torpid to survive difficult conditions, such as cold or lack of food.

TRACHEA (pl. TRACHEAE) Part of the respiratory system, a breathing tube, known in vertebrates as the windpipe.

TUSK In mammals, a modified tooth that often projects outside the mouth. Tusks have a variety of uses, including defence and digging up food. In some species, only the males have them – in this case, their use is often for sexual display or competition.

TYMPANUM The external ear membrane of frogs and insects.

UNDERFUR The dense fur that makes up the innermost part of a mammal's coat. Underfur is usually soft, and is a good insulator. See also *Guard hair*.

UNGULIGRADE A gait in which only the hooves touch the ground. See also *Digitigrade*, *Plantigrade*.

UROPYGIAL GLAND Also known as the preening or oil gland, the uropygial gland is located at the base of the tail in most birds. It produces sebaceous oils that birds rub over their feathers with their beak to keep them waterproof. See also *Sebaceous gland*.

UTERUS In female mammals, the part of the body that houses developing young. In placental mammals, the young are connected to the wall of the uterus via a placenta.

VENTRAL On or near the underside.

VERTEBRA One of the bones that form the vertebral column (backbone or spine) in vertebrate animals.

VERTEBRATE An animal with a backbone. Vertebrates include fish, amphibians, reptiles, birds, and mammals.

VIBRISSA See *Whiskers*.

WARNING COLORATION A combination of contrasting colours that warns that an animal is dangerous. Bands of black and yellow are a typical form of warning coloration, found in stinging insects. Also known as aposematic coloration.

WHISKER One of the long, stiffened hairs growing on the face, and particularly around the snout, of many mammals. Whiskers allow animals to sense vibrations in the water or air and are used as organs of touch. Also known as a vibrissa. See also *Rictal bristles*.

YOLK The part of an egg that provides nutrients to the developing embryo.

ZOOID An individual animal in a colony of invertebrates. Zooids are often directly linked to each other, and may function like a single animal.

ZOOPLANKTON See *Plankton*.

index

acknowledgments

Dorling Kindersley would like to thank the directors and staff at the Natural History Museum, London, including Trudy Brannan and Colin Ziegler, for reading and correcting earlier versions of this book and providing help and support with photoshoots, particularly Senior Curator in Charge of Mammals, Roberto Portelo Miguez.

DK would also like to thank others who provided help and support with photoshoots –Barry Allday, Ping Low, and the staff of The Goldfish Bowl, Oxford; and Mark Amey and the staff of Ameyzoo, Bovington, Hertfordshire.

DK would also like to thank the following:

Senior Editor:
Hugo Wilkinson

Senior Art Editor:
Duncan Turner

Senior DTP Designer:
Harish Aggarwal

DTP Designers: Mohammad Rizwan, Anita Yadav

Senior Jacket Designer:
Suhita Dharamjit

Managing Jackets Editor:
Saloni Singh

Jackets Editorial Coordinator:
Priyanka Sharma

Image retoucher: Steve Crozier

Illustrator:
Phil Gamble

Additional illustrations:
Shahid Mahmood

Indexer: Elizabeth Wise

The publisher would like to thank the following for their kind permission to reproduce their photographs:

(Key: a-above; b-below/bottom; c-centre; f-far; l-left; r-right; t-top)

The publisher would like to thank the following for their kind permission to reproduce their photographs:

(Key: a-above; b-below/bottom; c-centre; f-far; l-left; r-right; t-top)

1 Getty Images: Tim Flach / Stone / Getty Images Plus. **2-3 Getty Images:** Tim Flach / Stone / Getty Images Plus. **4-5 Getty Images:** Barcroft Media. **6-7 Brad Wilson Photography. 8-9 Alamy Stock Photo:** Biosphoto / Alejandro Prieto. **10-11 Dreamstime.com:** Andrii Oliinyk. **12-13 Alexander Semenov. 12 Alamy Stock Photo:** Blickwinkel. **14-15 Getty Images:** Eric Van Den Brulle / Oxford Scientific / Getty Images Plus. **16 Getty Images:** David Liittschwager / National Geographic Image Collection Magazines. **17 Alamy Stock Photo:** Roberto Nistri (tr). **Getty Images:** David Liittschwager / National Geographic Image Collection (cl); Nature / Universal Images Group / Getty Images Plus (tl). **naturepl.com:** Piotr Naskrecki (cr). **18 Dorling Kindersley:** Maxim Koval (Turbosquid) (bc); Jerry Young (tr). **naturepl.com:** MYN / Javier Aznar (tl); Kim Taylor (c). **19 Philippe Bourdon / www. coleoptera-atlas.com:** (r). **Dorling Kindersley:** Natural History Museum, London (tl); Jerry Young (bl). **20 Alamy Stock Photo:** RGB Ventures / SuperStock. **21 naturepl.com:** MYN / Piotr Naskrecki. **22 Brad Wilson Photography. 23 Brad Wilson Photography. 24 Brad Wilson Photography. 25 Dreamstime.com:** Abeselom Zerit. **26 Alamy Stock Photo:** Heritage Image Partnership Ltd. **26-27 Bridgeman Images:** Rock painting of a bull and horses, c.17000 BC (cave painting), Prehistoric / Caves of Lascaux, Dordogne, France. **28-29 Alamy Stock Photo:** 19th era 2. **31 Alamy Stock Photo:** SeaTops (br). **Getty Images:** Auscape / Universal Images Group (bl). **NOAA:** NOAA Office of Ocean Exploration and Research, 2017 American Samoa. (bc). **33 Alamy Stock Photo:** Science History Images (br). **34 Alamy Stock Photo:** Science History Images (cra). **36 iStockphoto.com:** GlobalP / Getty Images Plus (bl, c, crb). **Kunstformen der Natur by Ernst Haeckel:** (cr). **36-37 iStockphoto.com:** GlobalP / Getty Images Plus. **37 iStockphoto. com:** GlobalP / Getty Images Plus (tc, tr, cra, bc). **38 Kunstformen der Natur by Ernst Haeckel. 39 Alamy Stock Photo:** Chronicle (tr); The Natural History Museum (clb). **40-41 Alexander Semenov. 41 Alexander Semenov. 42-43 Alexander Semenov. 42 naturepl.com:** Jurgen Freund (br). **44-45 Getty Images:** Gert Lavsen / 500Px Plus. **45 Carlsberg Foundation:** (crb). **Getty Images:** GP232 / E+ (cra). **46 Getty Images:** Heritage Images / Hulton Archive. **47 Alamy Stock Photo:** Heritage Image Partnership Ltd (tr). **Photo Scala, Florence:** (cl). **50-51 Dreamstime. com:** Dream69. **51 Science Photo Library:** Science Stock Photography. **52 Dorling Kindersley:** Jerry Young (cl). **Dreamstime.com:** Verastuchelova (tr). **iStockphoto.com:** Farinosa / Getty Images Plus (bl). **naturepl.com:** MYN / JP Lawrence (tl). **53 Getty Images:** Design Pics / Corey Hochachka (c). **naturepl.com:** MYN / Alfonso Lario (bl); Piotr Naskrecki (cl). **54 Alamy Stock Photo:** Biosphoto (cra, crb, clb, cla). **55 Alamy Stock Photo:** Biosphoto. **56 naturepl.com:** Daniel Heuclin (cla). **Science Photo Library:** Ted Kinsman (crb). **56-57 Brad Wilson Photography. 58-59 Alamy Stock Photo:** Granger Historical Picture Archive. **59 Bridgeman Images:** British Library, London, UK / © British Library Board (tr). **60-61 Alamy Stock Photo:** Denis-Huot / Nature Picture Library. **61 Getty Images:** Mik Peach / 500Px Plus (crb). **62-63 Getty Images:** Nastasic / DigitalVision Vectors. **65 Dreamstime.com:** Erin Donalson (br). **66-67 Alexander Semenov. 67 Alexander Semenov. 69 Alamy Stock Photo:** The Natural History Museum (bc). **70-71 Igor Siwanowicz. 72 Image courtesy of Derek Dunlop:** (bl). **74-75 Igor Siwanowicz. 76 Getty Images:** Education Images / Universal Images Group (l). **77 Alamy Stock Photo:** Age Fotostock (t). **Getty Images:** Werner Forman / Universal Images Group (bl). **78-79 All images © Iori Tomita / http://www.shinsekai-th.com/. 81 Alamy Stock Photo:** Blickwinkel (br); Florilegius (bc). **82-83 Science Photo Library:** Arie Van 'T Riet. **84-85 Thomas Vijayan. 86 Image from the Biodiversity Heritage Library:** The great and small game of India, Burma, & Tibet (tl). **88-89 Getty Images:** Jim Cumming / Moment Open. **90-91 Dreamstime.com:** Channarong Pherngjanda. **92-93 Getty Images:** Matthieu Berroneau / 500Px Plus. **92 National Geographic Creative:** Joel Sartore, National Geographic Photo Ark (tl). **94-95 National Geographic Creative:** David Liittschwager. **94 Getty Images:** David Doubilet / National Geographic Image Collection (br). **naturepl.com:** MYN / Sheri Mandel (crb). **96 FLPA:** Piotr Naskrecki / Minden Pictures (tr). **Image from the Biodiversity Heritage Library:** Proceedings of the Zoological Society of London. (cla). **98-99 Greg Lecoeur Underwater and Wildlife Photography. 98 Dreamstime.com:** Isselee (clb, cla). **Getty Images:** Life On White / Photodisc (cb). **99 Alamy Stock Photo:** WaterFrame (tr). **100-101 Getty Images:** David Liittschwager / National Geographic Image Collection. **100 Getty Images:** David Liittschwager / National Geographic Image Collection Magazines (bc, br). **102 Dorling Kindersley:** The Natural History Museum, London (fbr). **103 Science Photo Library:** Gilles Mermet (c). **104-105 Brad Wilson Photography. 105 Alamy Stock Photo:** Martin Harvey (crb); www.pqpictures.co.uk (tc); Ron Steiner (tr). **Dreamstime.com:** Narint Asawaphisith (c). **Getty Images:**

Norbert Probst (cr). **106 naturepl.com:** Chris Mattison (clb). **106-107 Brad Wilson Photography. 107 Brad Wilson Photography:** (cra). **108 Alamy Stock Photo:** Imagebroker (cla, ca). **110 Alamy Stock Photo:** The History Collection (tl); The Picture Art Collection (crb). **111 Alamy Stock Photo:** Granger Historical Picture Archive. **113 akg-images:** Florilegius (cla). **114-115 Igor Siwanowicz. 115 FLPA:** Piotr Naskrecki / Minden Pictures (cra). **118-119 David Weiller / www.davidweiller.com. 119 FLPA:** Thomas Marent (cb). **National Geographic Creative:** Frans Lanting (crb). **120-121 Paul Hollingworth Photography. 122-123 iStockphoto.com. 122 iStockphoto.com:** GlobalP / Getty Images Plus (crb). **124-125 Dreamstime.com:** Isselee. **124 Bridgeman Images:** Natural History Museum, London, UK (tl). **125 Alamy Stock Photo:** Minden Pictures (br). **126 Getty Images:** Arterra / Universal Images Group (ca). **126-127 FLPA:** Sergey Gorshkov / Minden Pictures. **128 123RF.com:** Graphiquez (tc). **Alamy Stock Photo:** Nature Picture Library / Radomir Jakubowski (cr). **Dreamstime.com:** Lukas Blazek (bc); Volodymyr Byrdyak (c). **129 123RF.com:** Theeraphan Satrakom (tc); Dennis van de Water (tr). **Dreamstime.com:** Isselee (c); Scheriton (tl); Klaphat (cl). **National Geographic Creative:** Joel Sartore, National Geographic Photo Ark (br). **naturepl.com:** Suzi Eszterhas (cr). **130-131 Alamy Stock Photo:** Heritage Image Partnership Ltd. **131 Alamy Stock Photo:** Niday Picture Library (tr). **132-133 Alamy Stock Photo:** Mauritius images GmbH. **132 Mogens Trolle:** (bc). **134-135 Brad Wilson Photography. 135 Getty Images:** Parameswaran Pillai Karunakaran / Corbis NX / Getty Images Plus (tr). **136-137 National Geographic Creative. 137 Image from the Biodiversity Heritage Library:** Compléments de Buffon, 1838 / Lesson, R. P. (René Primevère), 1794-1849 (tc). **138-139 123RF.com:** Patrick Guenette (c). **Dreamstime.com:** Jeneses Imre. **140-141 Leonardo Capradossi /**

500px.com/leonardohortu96. 141 Dreamstime.com: Photolandbest (br). **142-143 Sebastián Jimenez Lopez. 142 naturepl.com:** MYN / Gil Wizen (tl). **144 Alamy Stock Photo:** ART Collection (crb); Pictorial Press Ltd (tr). **145 Image from the Biodiversity Heritage Library:** Uniform: Metamorphosis insectorum Surinamensium.. **146 naturepl.com:** Tui De Roy (bc). **146-147 Dreamstime.com:** Anolis01. **148 Duncan Leitch, David Julius Lab, University of California - San Francisco. 149 Getty Images:** SeaTops (tr). **152-153 © Ty Foster Photography. 153 Dreamstime.com:** Isselee (b). **154-155 National Geographic Creative:** Joel Sartore, National Geographic Photo Ark. **155 Bridgeman Images:** © Purix Verlag Volker Christen (crb). **156 Getty Images:** David Liittschwager / National Geographic Image Collection (tl). **156-157 Javier Rupérez Bermejo. 158 Dreamstime.com:** Martin Eager / Runique (c); Gaschwald (tc); Joanna Zaleska (bc). **Getty Images:** David Liittschwager / National Geographic Image Collection (cr, br); Diana Robinson / 500px Prime (tl). **naturepl.com:** Roland Seitre (tr). **159 Dorling Kindersley:** Thomas Marent (cl). **Getty Images:** Fabrice Cahez / Nature Picture Library (bl); Daniel Parent / 500Px Plus (c); Anton Eine / EyeEm (cr); Joel Sartore / National Geographic Image Collection (bc); Jeff Rotman / Oxford Scientific / Getty Images Plus (br). **160-161 John Joslin. 161 John Hallmen. 164 iStockphoto.com:** Arnowssr / Getty Images Plus (ca). **165 Getty Images:** Joel Sartore, National Geographic Photo Ark / National Geographic Image Collection. **166-167 FLPA:** Paul Sawer. **168 naturepl.com:** John Abbott (tl). **168-169 naturepl.com:** John Abbott. **170 National Geographic Creative:** Joel Sartore, National Geographic Photo Ark. **171 Getty Images:** DEA / A. De Gregorio / De Agostini (cra). **National Geographic Creative:** Joel Sartore, National Geographic Photo Ark (cla). **172 Alamy Stock Photo:** The Picture Art

Collection. **173 Alamy Stock Photo:** Artokoloro Quint Lox Limited (r). **174-175 Alamy Stock Photo:** Biosphoto. **174 Getty Images:** Visuals Unlimited, Inc. / Ken Catania. **178 Dorling Kindersley:** Jerry Young (ca). **178-179 Brad Wilson Photography. 180 FLPA:** Piotr Naskrecki / Minden Pictures (bl). **Science Photo Library:** Merlin D. Tuttle (bc); Merlin Tuttle (br). **181 Kunstformen der Natur by Ernst Haeckel. 182-183 Getty Images:** Alexander Safonov / Moment Select / Getty Images Plus. **182 Image from the Biodiversity Heritage Library:** Transactions of the Zoological Society of London.. **184-185 123RF.com:** Andreyoleynik. **186-187 naturepl.com:** ZSSD. **187 Alamy Stock Photo:** WaterFrame (crb). **188-189 Dreamstime.com:** WetLizardPhotography. **190 Science Photo Library:** Scubazoo. **191 Casper Douma:** (ca). **192-193 Arno van Zon. 193 Getty Images:** Tim Flach / Stone / Getty Images Plus (bl). **194 Alamy Stock Photo:** Nature Photographers Ltd (tl). **196-197 Robert Rendaric. 197 Getty Images:** Erik Bevaart / EyeEm (tr). **198 123RF.com:** Eric Isselee / isselee (bc). **Dreamstime.com:** Assoonas (cb); Nikolai Sorokin (cr); Ozflash (clb); Gualberto Becerra (crb); Birdiegal717 (bl). **199 123RF.com:** Gidtiya Buasai (cr); Miroslav Liska / mircol (c); Jacoba Susanna Maria Swanepoel (br). **Dreamstime.com:** Sergio Boccardo (bl); Isselee (clb, cb, bc); Jixin Yu (crb). **200 National Audubon Society:** John J. Audubon's Birds of America. **201 Image from the Biodiversity Heritage Library:** The Birds of Australia, John Gould. (cr); The Zoology of the Voyage of H.M.S. Beagle (tl). **202-203 Brad Wilson Photography. 204-205 National Geographic Creative:** Ami Vitale. **205 Image from the Biodiversity Heritage Library:** Recherches Sur Les Mammifères / Henri Milne-Edwards, 1800-1885. **207 Alamy Stock Photo:** Danielle Davies (br). **208-209 Brad Wilson Photography. 210-211 123RF.com:** Channarong Pherngjanda. **213 Alamy Stock Photo:** WaterFrame

(br). **216 Science Photo Library:** (cb); D. Roberts (tc). **217 FLPA:** Photo Researchers. **218 National Geographic Creative:** Joel Sartore, National Geographic Photo Ark (ca). **218-219 National Geographic Creative:** Joel Sartore, National Geographic Photo Ark. **219 National Geographic Creative:** Christian Ziegler (tr). **220-221 Solent Picture Desk / Solent News & Photo Agency, Southampton:** Shivang Mehta. **223 Science Photo Library:** Power And Syred (tr). **225 © caronsteelephotography.com:** (tr). **Mary Evans Picture Library:** Natural History Museum (br). **227 FLPA:** Steve Gettle / Minden Pictures (l). **Tony Beck & Nina Stavlund / Always An Adventure Inc.:** (r). **228 Getty Images:** Joel Sartore, National Geographic Photo Ark / National Geographic Image Collection (ca). **228-229 Thomas Vijayan. 230-231 Alamy Stock Photo:** Andrea Battisti. **232 Dorling Kindersley:** Jerry Young (tl). **232-233 Getty Images:** Joel Sartore, National Geographic Photo Ark / National Geographic Image Collection. **234-235 Alamy Stock Photo:** Artokoloro Quint Lox Limited. **235 Rijksmuseum, Amsterdam:** Katsushika Hokusai (tc). **236-237 National Geographic Creative:** Joel Sartore, National Geographic Photo Ark. **237 National Geographic Creative:** Joel Sartore, National Geographic Photo Ark (cr). **238 Dreamstime.com:** Isselee. **238-239 Science Photo Library:** Tony Camacho. **240-241 National Geographic Creative:** Tim Laman. **240 Image from the Biodiversity Heritage Library:** The Cambridge Natural History / S. F. Harmer (bc). **242-243 National Geographic Creative:** © David Liittschwager 2015. **242 Hideki Abe Photo Office / Hideki Abe:** (tl). **244-245 Aaron Ansarov. 245 Aaron Ansarov. 246 Alamy Stock Photo:** Arco Images GmbH (tr). **248-249 David Yeo. 248 Alamy Stock Photo:** Saverio Gatto. **250-251 Dreamstime.com:** Channarong Pherngjanda. **252 Getty Images:** George Grall / National Geographic Image Collection Magazines (tc). **Kunstformen der Natur by Ernst Haeckel:** (bc). **252-253 Alexander Semenov. 254 National Geographic Creative:** David Liittschwager. **254-255 naturepl.com:** Tony Wu. **257 UvA, Bijzondere Collecties, Artis Bibliotheek. 258-259 Erik Almqvist Photography. 259 Alamy Stock Photo:** Sergey Uryadnikov (tr). **260 Dreamstime.com:** Isselee (c); Johannesk (bc). **261 Dorling Kindersley:** Professor Michael M. Mincarone (cr); Jerry Young (bl). **Dreamstime.com:** Deepcameo (clb); Isselee (cl, bc); Zweizug (c); Martinlisner (crb); Sneekerp (br). **264 Getty Images:** Leemage / Corbis Historical (bc). **264-265 SeaPics.com:** Blue Planet Archive. **266 Dreamstime.com:** Isselee (tc). **267 Alamy Stock Photo:** Hemis (tr). **268 Alamy Stock Photo:** Science History Images. **269 Digital image courtesy of the Getty's Open Content Program.:** Creative Commons Attribution 4.0 International License (cl). **Photo Scala, Florence:**

(tr). **272 naturepl.com:** Pascal Kobeh (cb). **272-273 Greg Lecoeur Underwater and Wildlife Photography. 274-275 Jorge Hauser. 274 Alamy Stock Photo:** The History Collection (bc). **276 Dreamstime.com:** Evgeny Turaev (bl, t). **276-277 Dreamstime.com:** Evgeny Turaev. **278-279 naturepl.com:** MYN / Dimitris Poursanidis. **279 Dreamstime.com:** Isselee. **280-281 Alamy Stock Photo:** Razvan Cornel Constantin. **281 Bridgeman Images:** © Florilegius (br). **Getty Images:** Ricardo Jimenez / 500px Prime (tr). **282 Solent Picture Desk / Solent News & Photo Agency, Southampton:** © Hendy MP. **283 Getty Images:** Florilegius / SSPL (bc). **Solent Picture Desk / Solent News & Photo Agency, Southampton:** © Hendy MP (t). **284-285 Alamy Stock Photo:** Avalon / Photoshot License. **286-287 naturepl.com:** Markus Varesvuo. **287 Dreamstime.com:** Mikelane45 (br). **288 Shutterstock:** Sanit Fuangnakhon (cl); Independent birds (cr). **Slater**

Museum of Natural History / University of Puget Sound: (tl, tr, clb, crb, bl, br). **289 123RF.com:** Pakhnyushchyy (cr). **Slater Museum of Natural History / University of Puget Sound:** (tl, tr, cl, clb, crb). **290-291 Getty Images:** Paul Nicklen / National Geographic Image Collection. **292-293 National Geographic Creative:** Anand Varma. **293 National Geographic Creative:** Anand Varma. **294 Bridgeman Images:** British Museum, London, UK. **295 akg-images:** François Guénet (cr). **Getty Images:** Heritage Images / Hulton Archive (t). **297 naturepl.com:** Piotr Naskrecki (tc). **298-299 123RF.com:** Patrick Guenette. **300-301 Alamy Stock Photo:** Blickwinkel. **302-303 Alamy Stock Photo:** Life on white. **303 Alamy Stock Photo:** Life on white. **304-305 John Hallmen. 305 Alamy Stock Photo:** Age Fotostock (br). **306-307 Sergey Dolya. Travelling photographer. 307 Getty Images:** Jenny E. Ross / Corbis Documentary / Getty Images Plus.

308-309 naturepl.com: Paul Marcellini. **310 Dorling Kindersley:** Natural History Museum (tl); Natural History Museum, London (tc, tr, cl, c, cr, br); Time Parmenter (bc). **311 Alamy Stock Photo:** Nature Photographers Ltd / Paul R. Sterry (tr). **Dorling Kindersley:** Natural History Museum, London (ftl, tl, tc, fcl, fbl, cl, c, cr, bl, bc, br). **312-313 Michael Schwab. 312 Alamy Stock Photo:** Gerry Pearce (bc). **314-315 Getty Images:** Rhonny Dayusasono / 500Px Plus. **316 naturepl.com:** MYN / Tim Hunt (bc, br). **317 Alamy Stock Photo:** The Natural History Museum (tc). **National Geographic Creative:** George Grall (tr). **naturepl.com:** MYN / Tim Hunt (fbl, bc, br, fbr). **319 Alamy Stock Photo:** Images & Stories.

Endpaper images: *Front and Back:* **Aaron Ansarov**

All other images © Dorling Kindersley
For further information see:
www.dkimages.com

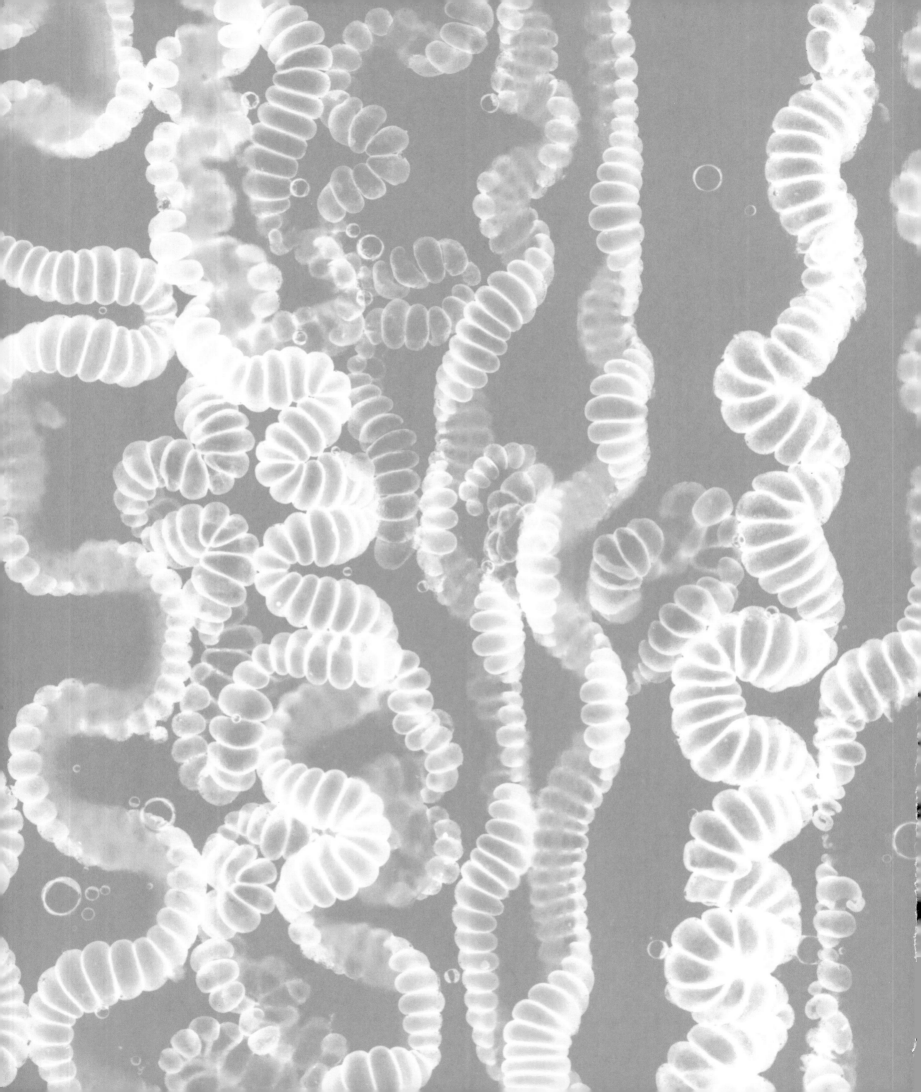